# MODELLING AND SIMULATION

*By*

**PUSHPA SINGH**
Deptt. of Computer Application
Accurate Institute of Management & Technologyy
Greater Noida

**NARENDRA SINGH**
Deptt. of Computer Application
Sri Ram Institute of Management & Technology
Greater Noida

S.K. KATARIA & SONS®
**Publisher of Engineering Books**
4885/109, Prakash Mahal, Dr. Subhash Bhargav Lane,
Opposite Delhi Medical Association, Daryaganj, New Delhi-110002 (India)
Phone: +91-11-43551243, Mobile: +91-9871775858, 7678425209
e-mail: katariabook@gmail.com, katariabooks@yahoo.com
Website: www.skkatariaandsons.com
***(e-books, subscriptions and bulk buying also available on our website)***

**Published By :**

**S.K. KATARIA & SONS®**

4885/109, Prakash Mahal, Dr. Subhash Bhargav Lane,
Opposite Delhi Medical Association, Daryaganj, New Delhi-110002
Phone: +91-11-43551243, Mobile: +91-9871775858, 7678425209
e-mail: katariabook@gmail.com, katariabooks@yahoo.com





First Edition, 2008; Fourth Edition, 2014; Reprint: 2024, Fifth Edition: 2026

*Laser Typesetting at:* Goswami Associates, Delhi. *Printed at:* N.K. Printers

# PREFACE TO THE FOURTH EDITION

Simulation is a powerful technique for solving the wide variety of problems. Recently, modelling and simulation has been slated to become the computing paradigm of the future. As a paradigm, it is a way of representing problems and thinking about them, as much as a solution method. The problems span the analysis and design of complex dynamical systems. Simulation is generally referred to as computer simulation which becomes a useful part of modeling many natural systems in physics, chemistry and biology, human systems in economics and social science and in the process of new technology in the field of engineering, to gain insight into the operation of those systems. This fourth edition of modeling and simulation reflects the changes. Changes of this edition are related to the new syllabus of **B.Tech** and **MCA** of Uttar Pradesh Technical University (UPTU), Lucknow and other Universities.

The brief content of the book is as follows:

**Chapter 1 & 2:** Cover basics of system and model.

**Chapter 3:** Covers different types of simulation, cobweb and distributed lag model.

**Chapter 4:** Introduces the concept of Input Data Model.

**Chapter 5:** Gives concept of validation and verification.

**Chapter 6:** Introduces the basic of probability.

**Chapter 7:** Covers Continuous and Discrete system simulation while chapter-8 gives knowledge about system dynamics.

**Chapter 9:** Examines random number generation and testing of random number.

**Chapter 10:** Covers queuing theory and chapter 11, gives idea about arrival process and simulation programming task.

**Chapter 12:** Presents an overview of continuous, discrete and object oriented simulation language.

**Chapter 13:** Devoted to GPSS and basics of SIMSCRIPT-II.

**Chapter 14:** Introduces the estimation and output analysis.

**Chapter 15:** Deals simulation of PERT network. The last chapter discusses the various real world application of simulation.

We have tried our best to organize the contents of this book with sufficient examples, diagrams and each chapter concludes with number of exercises.

We would like to extend our thanks to Dr. S.K. Dubey, Director & Dr. Anurag Jain, H.O.D (I.T), Accurate Institute of Management & Technology, Gr. Noida and Prof. G.K. Tripathi, Executive Director & Prof. Soumitra Chakraborty, Dean, SRIRAM Institute of Management & Technology, Gr. Noida, for providing us such a nice working environment for extending the work of this revised book.

We shall greatly appreciate the comments, any error and all suggestions for improving the future editions of the book from the reader at narendra.naman09@gmail.com or pushpa.gla@gmail.com.

**—AUTHORS**

# PREFACE TO THE FIRST EDITION

The objective of the first edition is to provide a basic treatment of all the important aspects of modeling and simulation. The book is written to address undergraduate and post-graduate students with a quick introduction to the major issues and topics in modeling and simulation. The prime purpose is to provide material for system simulation, discrete system simulation and system software.

Some familiarity with programming concepts is assumed but no knowledge of any particular programming language is needed as a prerequisite. In addition, familiarity with the concepts of statistics and probability theory will be helpful. Many areas of application are discussed, but the technical details are kept simple so that the book will be useful to students of different backgrounds.

The first chapter discusses the system definition, components and type of models that are basic to any simulation.

The second chapter considers the topic of corporate model and organizing a system study.

The third chapter deals with why and when simulation is required, different types of simulation, simulation of a pure pursuit, queuing system, on inventory system. This chapter also deals with distributed lag model, cobweb model and Monte-Carlo simulation with various examples.

The fourth chapter explains about simulation of continuous system alongwith simulation of water reservoir system, servo-system and autopilot system.

The fifth chapter examines simulation of discrete system, generation and testing of random numbers, generation of non-uniformly distributed random number and stochastic simulation.

The sixth chapter discusses about the system dynamics studies-alongwith various growth model and feedback in socio-economic system.

The seven chapter deals critical path methods, analysis of activity networks, uncertainties in activity duration, PERT and resource allocation.

The eight chapter presents an overview of various special purpose simulation languages, continuous and discrete simulation language, SIMSCRIPT II.5, MODSIM III and CSMPIII.

The chapter nine deals with process of model and simulation, verification and validation of simulation model.

Appendix A discusses the use of database and AI in different areas of simulation.

There are some solved question-answers and exercises provided at the end of each chapter are an integral part of the text and must not be ignored.

Great care has been taken while editing this book, still I would whole heartedly accept suggestions for improvement offered by the reader. I hope this book will meet all the requirements of the readers and come upto their expectations.

**—AUTHORS**

# ACKNOWLEDGEMENT

We express our deep sense of love and affection to our parents for their strength and support, failing which this wouldn't have been complete.

We are grateful to Dr. B.N. Roy, Executive Director, IEC College of Engineering & Technology, Greater Noida, for their constant encouragement and motivation. We are also grateful to Mr. K.L. Luthra, Chairman, and Dr. P.V. Kapoor, Director, Sri Ram Institute of Management & Technology, Greater Noida, for their nice corporation.

We would like to express our sincere gratitude to our colleagues and friends who contributed in many ways towards completion of this book.

We extend our gratitude to our publisher of this book, for publishing this work in present form.

Finally, we are thankful to Almighty God, without whose grace nothing is possible for any one.

**Pushpa Singh**
**Narendra Singh**

*Dedicated*
*to*
*Our Parents*

# CONTENTS

## 13. GPSS 247–268



## 14. ESTIMATION AND OUTPUT ANALYSIS OF SIMULATION 269–286



## 15. SIMULATION OF PERT NETWORK 287–316



## 16. REAL WORLD APPLICATION OF SIMULATION 317–330

1

# Introduction of System and Models

**INSIDE THIS CHAPTER**

**1.1** System : Definition; **1.2** Components of System; **1.3** State of the System; **1.4** System Environment; **1.5** Types of Activities; **1.6** Types of System; **1.7** Purpose of Model; **1.8** System Modeling; **1.9** Model Definition; **1.10** Types of Models; **1.11** Principles used in Modeling; **1.12** Concept of Simulation; **1.13** Simplified Representation of Complex and Large Systems; **1.14** Advantages and Disadvantages of Simulation; **1.15** Pitfalls of Simulation; **1.16** Selection Criteria of Simulation and Modeling Languages.

## 1.1 SYSTEM: DEFINITION

A system is defined as an aggregation of objects joined in some regular interaction or interdependence for achievement of a common goal. We can also define a system as organized set of interrelated ideas or principles. All system have input, output and feedback, and maintain a basic level of equilibrium.

**Example:** Consider a factory system that makes and assembles parts into a product. Two major components of the factory system are:

- Fabrication department: making the parts
- Assembly department: producing the products.

However, some other departments like:

- Purchasing department: It maintains a supply of raw material
- Shipping department: It dispatches the finished products
- Production control department: It receives orders and assigns task to other department.

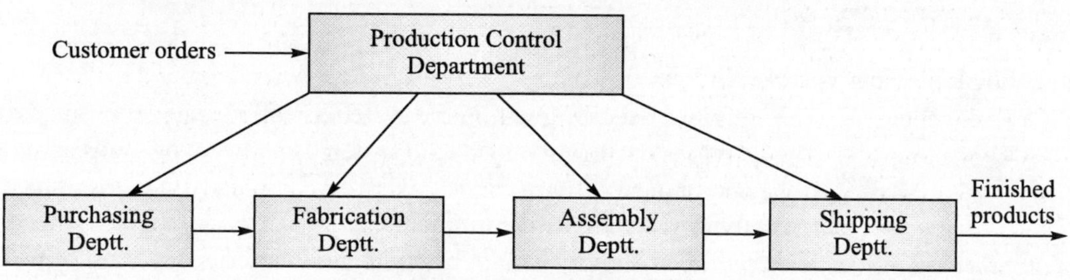


**Fig. 1.1** A factory system

We observe that these different departments are like distinct objects, each of which possesses properties of interest. There are also certain interactions occurring in the system that cause changes in the system.

## 1.2 COMPONENTS OF SYSTEM [*UPTU 2005-06*]

Number of terms are needed to understand and analyze a system. These terms can be ask as components of the system, these are:

(*i*) **Entity:** It is used to denote an object of interest in system.

(*ii*) **Attribute:** Attribute is used to denote a property of an entity.

(*iii*) **Activity:** Process that causes changes in the system is called an activity.

## 1.3 STATE OF THE SYSTEM [*UPTU 2005–06*]

Collection of all the entities, attributes, and activities at one point in time is called *state of the system.*

For example in banking system, the entities of the system are the customers. Their attributes are such factors as balance, account numbers and credit status. The activities are deposit or withdrawal of money, etc.

Table 1.1 lists examples of what might be considered entities, attributes, and activities for a number of other systems.

**Table 1.1.** Examples of system

| S.No. | System | Entities | Attributes | Activities |
|---|---|---|---|---|
| 1. | Supermarket | Customers | Shopping list | Checking-out |
| 2. | Communication | Messages | Length, priority | Transmitting |
| 3. | Rapid rail | Passengers | Origination, destination | Travelling |
| 4. | Traffic | Cars | Speed, distance | Driving |

The above table do not have a complete list of all entities, attributes, and activities for the systems. We cannot be made complete list without knowing the purpose of the system description. Depending upon that purpose, various aspects of the system will be of interest and will determine what needs to be identified.

## 1.4 SYSTEM ENVIRONMENT [*UPTU 2005–06*]

A system is generally affected by changes occurring outside the system. Such changes occurring outside the system are said to occur in the **system environment.**

## 1.5 TYPES OF ACTIVITIES

We can define following type of activity:

(*i*) Endogenous : The activities occurring within the system is called endogenous activities.

(*ii*) Exogenous : The activities occurring outside the system is called exogenous activities.

(*iii*) Deterministic : If the output of an activity can be described completely in terms of its input, the activity is said to be deterministic.

(*iv*) Stochastic : If output of an activity cannot be completely described in term of its input or its vary randomly over various possible outcomes, the activity is said to be stochastic.

The randomness of a stochastic activity would seem to imply that the activity is part of the system environment because the exact output at any time is not known. If an activity is **truly stochastic** then there is no known explanation for its randomness.

## 1.6 TYPES OF SYSTEM

We can define following types of system:

- **Open system:** A system which does have exogenous activities is called as open system.
- **Closed system:** A system which don't have any exogenous activities is called as closed system.
- **Continuous system:** A system, in which the changes are predominantly smooth, are called as continuous system. *Example:* Aircraft system.
- **Discrete system:** A system, in which the changes are predominantly discontinuous, are called as discrete system. *Example:* Factory system.
- **Sampled-data system:** There are also some system that are intrinsically continuous but information about them is only available at discrete points in time. These system are called as sampled-data system.

There is no specific rules can be given as to how a particular system is to represented and what particular model we have to use to represent the system. The description of a system, rather than the nature of the system itself, determines what type of model will be used.

## 1.7 PURPOSE OF MODEL

The purpose of the model, coupled with the general principle that a model should not be more complicated than is needed. It must determine the level of detail and the accuracy with which a model needs to be developed. On the basis of experience of knowledgeable people we will decide the type of model that is needed.

## 1.8 SYSTEM MODELING

Sometime it is possible to experiment with the system itself, and sometime it is not possible to experiment with the system while it is in hypothetical form. Sometime we have to construct a number of prototypes and test them, but this can be very expensive and time-consuming. So we need model of the system, model is not only a substitute for a system but it is also a simplification of the system.

## 1.9 MODEL DEFINITION

Model can be define as the body of information about a system gathered for the purpose of studying the system. There is no unique model of a system. Different models of the same system will be produced by different analysts.

- The task of deriving a model of a system may be divided broadly into two subtasks:
  (*i*) Establishing the model structure
  (*ii*) Supplying the data
- Establishing the structure is possible by—
  (*a*) Determining the system boundary
  (*b*) Identifying the entities, attributes, and activities of the system.
- Supplying the data means—
  (*a*) It provides the values of the attributes
  (*b*) It defines the relationships involved in the activities.

Establishing the structure and supplying the data are not two separate tasks, infact they are defined as parts of one task, because they are usually so intimately related that one task can not be performed by other task.

## 1.10 TYPES OF MODELS

Models can be divided in two parts: that is Physical model and Mathematical model.

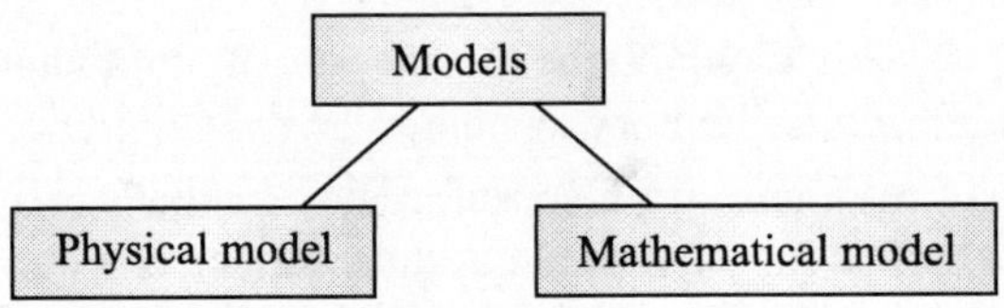


**Fig. 1.2**

Both the models can be static and dynamic. In mathematical model, static and dynamic can be further divided in two parts—Numerical and Analytical. Finally, mathematical, dynamic, numerical model can further represent as **system simulation.** This whole can be represented diagrammatically as follows.

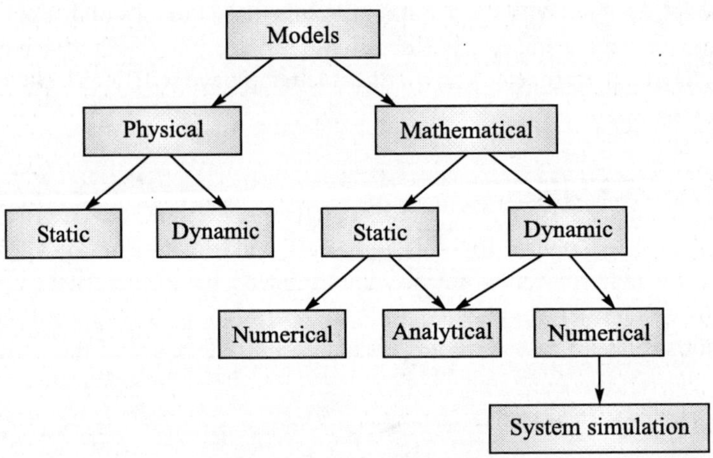


**Fig. 1.3.** Types of models

Numerical and analytical methods are techniques by which mathematical model can be solved. Numerical technique is further identified as simulation.

### Physical Models

Physical models are based on some analogy between such systems as mechanical and electrical or electrical and hydraulic.

System attribute are represented by voltage or position of a shaft. System activities are reflected in physical laws that derive the model. Example: The rate of which the shaft of a direct current motor turns depends upon the voltage applied to the motor.

### Mathematical Models

It use symbolic notation and mathematical equation to represent a system. System attributes are represented by variables. System activities are reflected by mathematical functions that interrelate the variables.

### Static Models

Static models can only represented the values that system attributes take when the system is in balance. This model is a representation of a system at a particular time.

It may be used to represent a system in which time simply plays no role.

### Dynamic Models

It follows the changes over time that result from the system activities.

In this type of model, time plays a critical role.

### Analytical Technique

Analytical techniques, use **deductive reasoning** of mathematical theory to solve a problem in a model. It is a matter of finding the model that can be solved and best fits the system being studied.

It requires a considerable amount of computation.

### Numerical Technique

This techniques involve to apply computational procedures to solve equations. Any assignment of numerical values that uses mathematical tables involves numerical methods, since tables are derived numerically, so it produce accurate solution.

However, we can also classified our models as stochastic and deterministic.

### What is System Simulation?

System simulation is considered to be a numerical computation technique used in conjunction with dynamic mathematical models.

## 1.10.1 Static Physical Models

Example of physical model are scale models. In construction of college building, making a scale model provides a simple way of determining exact measurements of the classrooms, library, conference room, playground, etc., rather than having to produce drawings of complicated, 3*D* shapes.

- Scientists have used models in which spheres represent atoms and rods (sheet of metals) represent atomic bonds.
- Scale models are used to represent:
  - DNA molecule structure.
  - Wind tunnels in designing of aircraft.
  - Water tanks in designing of ships.

Although air is blown over the model, or the model is pulled through water, these are static physical models because the measurements that are taken to represent attributes of the system being studied under one set of balance conditions.

Sometimes, a static physical model is used as a means of solving equations with particular boundary conditions.

There are many physics where the same equations apply to different physical phenomena.

### 1.10.2 Dynamic Physical Models

Dynamic physical model represent a analogy between what the system we have studied with some other system that is different in nature.

**Example:** Consider the two system shown in Fig. 1.4, *i.e.,* one is mechanical system (*a*) and other is electrical system (*b*).

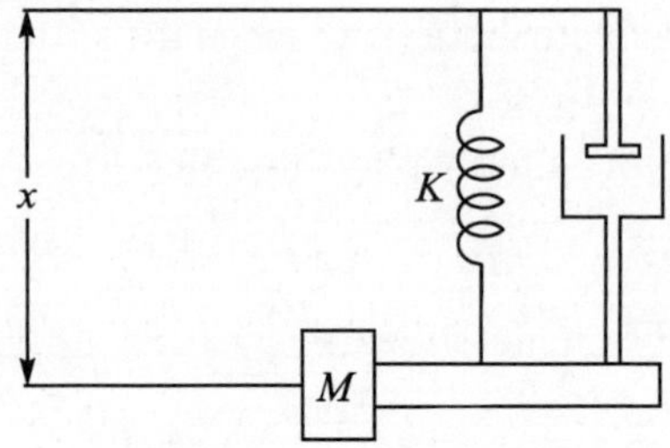

(*a*) Mechanical system

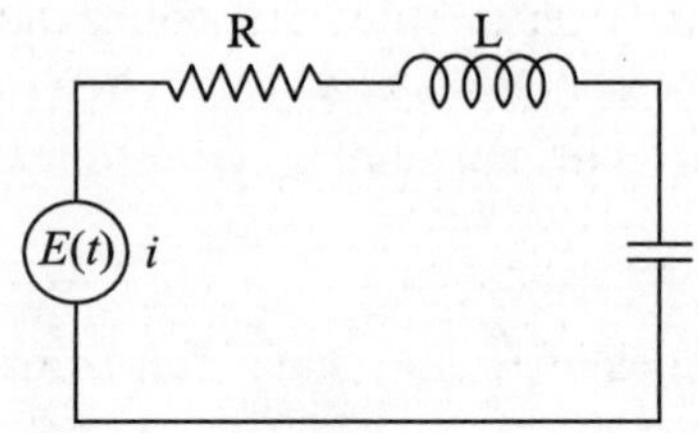

(*b*) Electrical system

**Fig. 1.4**

#### Mechanical System

- It represents a mass (*M*) that is subject to an applied force changing with time, *i.e.,* $F_1 = M\ddot{x}$.
- A spring whose force is proportional to its extension or contraction,

  $$F_2 \propto x$$
  $$F_2 = Kx$$

  where *K* is stiffness constant of the spring.
- A shock absorber that exerts a damping force proportional to the velocity of mass.

  $$F_3 \propto \dot{x} \quad \text{or} \quad F_3 = D\dot{x}$$

  *D* is damping factor of the shock absorber.

  Let *F*(*t*) be total system force, varying with time.

  The whole system is described by the following differential equation

  $$KF(t) = F_1 + F_2 + F_3$$
  $$KF(t) = M\ddot{x} + Kx + D\dot{x}$$

  or $$KF(t) = M\ddot{x} + D\dot{x} + Kx \qquad ...(1)$$

#### Electrical System

Figure (*b*), represents an electrical circuit with an inductance *L*, a resistance *R*, and a capacitance *C*, connected in series with a voltage source that varies in time according to the function *E*(*t*).

If *q* is the charge on the capacitance then it can be shown that the behaviour of the circuit is governed by the following differential equation

$$L\ddot{q} + R\dot{q} + q/C = \frac{E(t)}{C}\begin{pmatrix}\dot{q} = dq/dt \\ \ddot{q} = d^2q/dt^2\end{pmatrix} \qquad ...(2)$$

Equations (1) and (2) shows that they are in same form. Following equivalence occur between the quantities in the two system.

| Displacement | $x$ | Charge | $q$ |
|---|---|---|---|
| Velocity | $\dot{x}$ | Current | $\dot{q}$ or I ($I = \dot{q}$) |
| Force | $F$ | Voltage | $E$ |
| Mass | $M$ | Inductance | $L$ |
| Damping factor | $D$ | Resistance | $R$ |
| Spring stiffness | $K$ | 1/capacitance | 1/C |

### Conclusions

- The mechanical system and the electrical system are analogs of each other.
- It is simpler to modify the electrical system than to change the mechanical system.
- It is more comfortable to study electrical system, so whenever we need to built mechanical system, we can study electrical system.

### 1.10.3 Static Mathematical Models *[UPTU 2006-07]*

This model gives the relationships between the system attributes when the system is in equilibrium. If, due to some reason this equilibrium point is change then model have to be derived new values for all the attributes. But it does not represent the way in which they changed to their new values.

**Example:** In marketing commodity there is a equilibrium point between demand (*D*), and supply (*S*). Both demand and supply depends upon price (*P*). A simple market model will represent that at what price, equilibrium occurs.

**Consider following point:**

- Demand for the commodity will be low when the price is high, and demand will increase if price is low.
- Supply can be increase as price increases, because the suppliers see an opportunity for more revenue.
- **In Diagram,** Supply (*S*) and demand (*D*) is plotted against price (*P*). The relationship is denoted by two straight-line that cross at a point to each other as in Fig. 1.5.

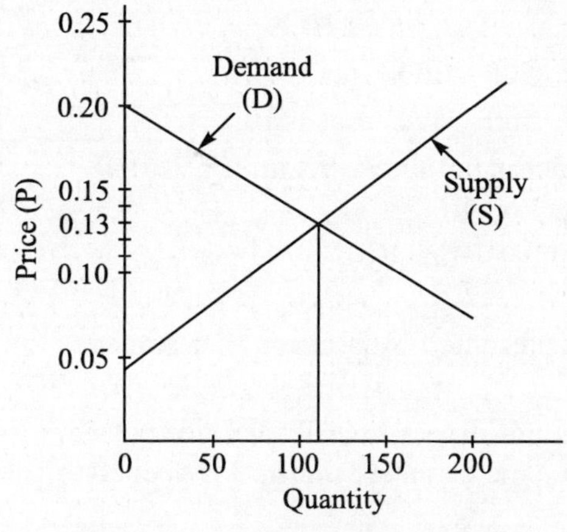


**Fig. 1.5.** Linear market model

If condition remain stable, the price will settle to this point. This point represent that supply equals to the demand. If relationship have been assumed linear, the complete market model can be written mathematically as

$$D = a - b\,P$$
$$S = c + d\,P$$
$$S = D$$

**Note:** *b* and *d* must be positive number because if price increases, demand decreases and supply increases.

Figure 1.4 has been plotted for the following values of the coefficients:

$$a = 500 \qquad b = 3{,}000$$
$$c = -150 \qquad d = 2{,}000$$

The equilibrium market price, in fact is given by

$$S = D$$
$$c + d\,P = a - b\,P$$
$$d\,P + b\,P = a - c$$
$$P(b + d) = a - c$$
$$P = (a - c)/(b + d)$$

Put the value

$$P = \frac{500 - (-150)}{3000 + 2000} = \frac{650}{5000}$$

$$P = 0.13$$

which is correspond to supply 110.

**Note:** Demand and supply both can represent as no. of quantity.

**How?**

$$S = c + d$$
$$S = -150 + 2000 \times P.13 \;\Rightarrow\; -150 + 260$$
$$\Rightarrow\; 110$$
$$D = a - b\,P \;\Rightarrow\; 500 - 3000 \times .13$$
$$\Rightarrow\; 500 - 390 \;\Rightarrow\; 110$$

But generally market model is not linear (as in Fig. 1.6), it will be represented by followings Fig. 1.6.

- Demand (*D*), will be represented by a curve that slopes downwards.
- Supply (*S*), will be represented by a curve that slopes upwards.

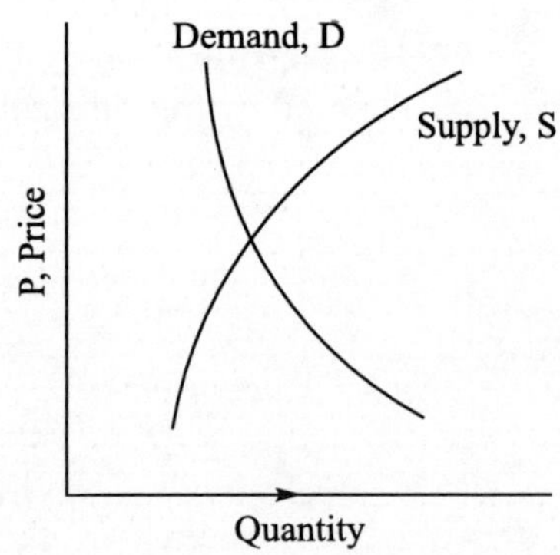

**Fig. 1.6** Non-linear market model.

*OR* we can say market is non-linear then it may not possible to express the relationships by equations that can be solved. So some **numeric method** is needed to solve these equations.

## 1.10.4 Dynamic Mathematical Models **[*UPTU 2006-07*]**

This model allows the changes of system attributes to be derived as a function of time. The derivation may be made with an analytical solution or with a numerical computation, depending upon the complexity of the model.

**Example:** The equation that was derived to describe the behaviour of a car wheel, is an example of a dynamic mathematical model.

**Equation:** As previously we have described, motion of the system by the following differential equation

$$Md^2x/dt^2 + D\,dx/dt + Kx = KF(t)$$

or

$$M\ddot{x} + D\dot{x} + Kx = KF(t)$$

Divide by $M$ on both side of equation and get

$$\ddot{x} + D/M\,\dot{x} + K/Mx = \frac{K}{M}F(t)$$

where

$$\frac{D}{M} = 2\zeta\omega \quad \text{and} \quad \frac{K}{M} = \omega^2$$

the equation become

$$\ddot{x} + 2\zeta\omega\,\dot{x} + \omega^2 x = \omega^2 F(t) \qquad ...(1)$$

The factor $\zeta$ is called the damping ratio.

We can write equation (1) as follows, showing how $x$ varies in response to a steady force applied at time $t = 0$

$$\ddot{x} + 2\zeta\omega\,\dot{x} + \omega^2 x = 0$$

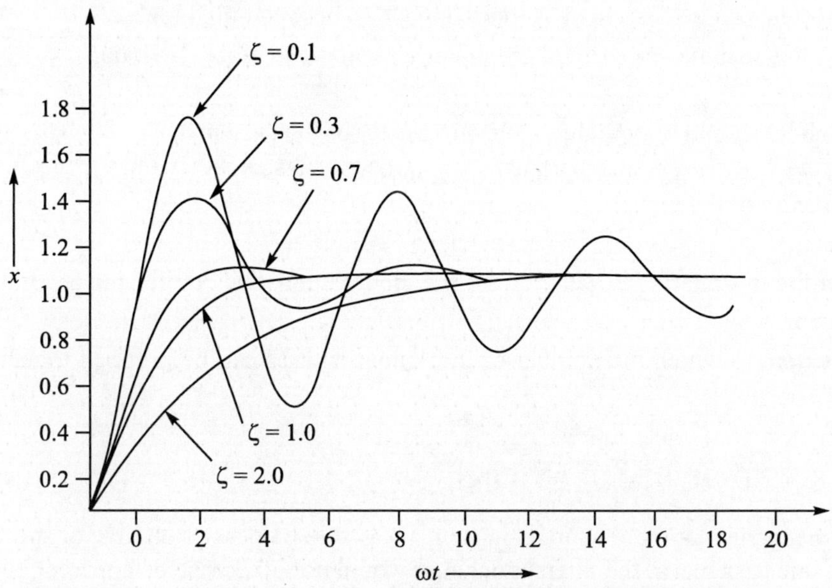


**Fig. 1.7.** Solution of second order equation.

Solution can be given in terms of the variable $\omega t$. This equation can be solved with solution of homogeneous linear differential equation.

$$\boxed{x(t) = Ce^{-\omega t}}$$

where $C$ and $\omega$ are both complex constants.

Solution are shown for several values of $\zeta$, and it can be seen that when it is less than 1, the motion is oscillatory (See in Fig. 1.7).

When the motion is oscillatory, the frequency of oscillation is determined from the formula

$$\boxed{\omega = 2\pi f}$$

where $f$ is the number of cycles per second.

## 1.11 PRINCIPLES USED IN MODELING

There is no exact rules by which we can construct the model. However no. of guidelines are given, they describe different perspective to analyze the information in the model. These are:

- Block building
- Relevance
- Accuracy
- Aggregation

1. **Block Building**
   - We can describe a system in series of blocks.
   - Each block describe a part of the system.
   - It depends upon few input variables and result in a few output variables.
   - The system can be described as interconnection between the blocks.
   - The system can be represented graphically as a simple block diagram.
2. **Relevance**
   - The model should only include relevant aspect of the system.
   - However, irrelevant information in the model may not do any harm but it increases the complexity of the model.
3. **Accuracy**

   Whatever the information is gathered for the model it must be correct and accurate.
4. **Aggregation**

   It is the extent to which the number of individual entities can be grouped together into larger entities.

## 1.12 CONCEPT OF SIMULATION

A simulation is the manipulation of a model in such a way that it operates on time or space to compress it, thus enabling one to perceive the interactions that would not otherwise be apparent because of their separation in time or space.

Simulation and modeling is a concept for developing a level of understanding of the interaction of the parts of a system, and of the system as a whole. A system is understood to be an entity which maintains its existence through the interaction of its parts. A model is a simplified representation of the actual system intended to promote understanding. Whether a model is a good model or not depends on the extent to which it promotes understanding.

A simulation generally refers to a computerized version of the model which is run over time to study the implications of the defined interactions. Simulations are generally iterative in there development.

One develops a model, simulates it, learns from the simulation, revises the model, and continues the iterations until an adequate level of understanding is developed.

Modeling and Simulation follows as much as the reality. But it depends upon the system study because model is constructed on the basis of system study. If model is correct and represent actual system then simulation is much more according the reality.

Key issues in simulation include acquisition of valid source information about the relevant selection of key characteristics and behaviours, the use of simplifying approximations and assumptions within the simulation, and fidelity and validity of the simulation outcomes.

## 1.13 SIMPLIFIED REPRESENTATION OF COMPLEX AND LARGE SYSTEMS

Models used to study large-scale systems tend to be very complex, and writing computer programs to execute them can be very complicated task. But this task can be simplified by the development of excellent software products that automatically provide many features needed to 'program' a simulation model.

A problem with simulation of complex system is that a large amount of computer time is required. However, this problem has becomes less severe as a computers become faster and cheaper. Consequently, many simulation studies have been composed of heuristic model building, programming, and a single run of the program to obtain the answer.

## 1.14 ADVANTAGES AND DISADVANTAGES OF SIMULATION

**Advantages**

Application areas for simulation are numerous and diverse. Following is a list of some problems for which simulation has been found to be a useful and powerful tool:

1. Analyzing military weapons systems.
2. Analyzing manufacturing systems.
3. Simulation play more vital role in biology, sociology, economics medicine, psychology, etc.
4. Designing and analyzing inventory control system.
5. Reengineering of business process.
6. It may be less expensive in several real-world situations.
7. Using simulation, performance of the existing system can be estimated for a range of operating conditions.
8. Alternatives to existing system could be tested with relative ease.
9. Control in simulation experiment is better than that in experiment with the system itself.
10. Mathematical models which can give analytical solutions, fail to describe many real-life systems and complex system accurately.

**Disadvantages**

1. Models used to study large scale systems tend to be very complex, and writing computer programs to execute them can be a tedious task.
2. Simulation models are usually expensive and time consuming to develop.
3. Level of model detail is inadequate.
4. Failure to collect good system data.

5. Simulation software is not appropriate.
6. Generally people don't have knowledge about simulation methodology.
7. Large volume of numerical data and animation (with the use of latest software) create—at times, false—impression of accuracy and hence, confidence.
8. Large number of variables makes simulation unwieldy and difficult.

## 1.15 PITFALLS OF SIMULATION

If we want to decide whether simulation study is appropriate according in given-situation, then it is suggested that remember advantages and disadvantages of simulation and all other relevant facets of one's particular situation. Both simulation and analytic model might be useful in some situations. Simulation can be used to check the validity of assumption needed in an analytical model there may be following pitfalls to the successful completion of a simulation study.

- Failure to have clearly defined objective at the beginning of the study.
- Inappropriate level of modeling details.
- Failure to communicate with the management on regular basis.
- Treating it as if it is a complex exercise in computer programming.
- Failure to have people with a knowledge of simulation methodology and statistics background in modeling teams.
- Using commercial simulation software without knowing its limitations.
- Misuse of animation.
- Failure to collect good system data.

## 1.16 SELECTION CRITERIA OF SIMULATION AND MODELING LANGUAGES

In conducting a simulation modeling study one or more software tool is needed. There are many (87+) packages available for Discrete Event Simulation (DES) alone. Selecting the appropriate language or package for your study can be a study of its own. Tools for this selection are generally scarce. Nance (1993) and (1995) identified the six characteristics of a discrete event simulation language. He proposed that discrete event simulation programming languages must meet a minimum of six requirements.

1. Generation of random numbers to represent uncertainty.
2. Process transformers, to permit other than uniform random varieties to be used.
3. List processing capability, so that objects can be created, manimulated, and deleted.
4. Statistical generation, to provide the descriptive summary of model behaviour.
5. Report generation to provide the presentation of potentially large reams of data in an effective way for decision-making.
6. A timing executive or time flow mechanism.

Many software tools can and do meet these requirements. Finding them and evaluating them can be tedious.

## SUMMARY

- System is defined as an collection of objects joined in some regular interaction or interdependence.

- The state of the system is collection of all the entities, attributes, and activities at any point of time.
- Changes occurring outside the system are said to occur in the system environment.
- If the output of an activity can be described completely in terms of its input, the activity is called to be deterministic.
- If the output of an activity cannot be described completely in terms of its input, is called stochastic activity.
- System, in which changes are predominantly smooth, are called continuous otherwise discrete system.
- The task of deriving a model of a system is divided into two subtasks:
  - ➢ establishing the model structure
  - ➢ supplying the data
- Physical models are based on some analogy between two system such as mechanical and electrical, etc.
- Mathematical models, use symbolic notation and mathematical equations to represent a system.
- Static models show the values that system attributes take when the system is in balance.
- Dynamic models, follow the changes over time that result from the system activities.
- In analytical technique, we use deductive reasoning of mathematical theory to solve a model.
- In numerical technique, we use computational procedures to solve the equations.
- There are four principles used in modeling:
  - ➢ Block-building
  - ➢ Relevance
  - ➢ Accuracy
  - ➢ Aggregation
- Simulation and modeling is a concept for developing a level of understanding of the interaction of the parts of system, and of the system as a whole.
- Complex and large system can be simplified by the development of excellent software product.

## SOLVED QUESTIONS

**Q. 1. Name three or four of the principles entities, attributes, and activities to be considered if you were to simulate the operation of:**

**(*a*) A factory system**

**(*b*) A aircraft system**

**(*c*) A grocery shop**

**Ans.** **(*a*) In factory system:**

Entities — Departments, orders, parts and products

Attributes — Quantities for each order, type of part, no. of machines

Activities — Manufacturing processes of departments

**(*b*) In aircraft system:**

Entities — Airframe, control surfaces, gyroscope

Attributes — Speed, control surface angle, and gyroscope setting

Activities — Driving of the control surfaces and the response of the airframe to the control surface movements

**(*c*) In grocery shop:**

Entities — Shopper, customer, counter

Attributes — Name of item, no. of items, availability

Activity — Arrival, purchasing, check-out.

**Q. 2. In supermarket, identify following things:**

**(*i*) Exogenous activity**

**(*ii*) Endogenous activity**

**Ans.** To identify these activity, first we have to define system boundary, so that it can distinguish between system and its environment. It also depends upon our objective of system study.

**Generally we can say**

(*i*) Exogenous activity — Arrival of customer in supermarket

(*ii*) Endogenous activity — Any activity within the supermarket. For example, purchasing, check-out.

But, if our purpose of study is effects of car parking facilities on supermarket business, the boundary of the system would need to include the parking place then—

Exogenous activity — Arrival of cars

Endogenous activity — Arrival of customer in supermarket.

**Q. 3. "The complete aircraft system might even be regarded as a discrete system", proof this statement.**

**Ans.** We know that aircraft system are example of continuous system because the changes are predominantly smooth. If our purpose of studying the aircraft were to follow its progress along its scheduled route, perhaps to studying air traffic problems, there would be no point in following precisely how the aircraft turns. It would be sufficiently accurate to treat changes of heading at scheduled turning points as being made instantaneously and so regard the system as being discrete.

**Q. 4. What point we have to keep in our mind, while creating the model?**

**Ans.** Following points are there:

(*i*) A model should not be more complicated.

(*ii*) It will determine the level of detail and the accuracy with which a model needs to be developed.

(*iii*) Model needs, the experience of knowledgeable people.

**Q. 5. How models are classified?**

**Ans.** There are three ways on which models are classified:

(*i*) Models will first be separated into **physical models** or **mathematical models.**

(*ii*) *A* second difference will be between **static models** and **dynamic models.**

(*iii*) A third difference is on the basis of technique by which the model is "Solved". A difference is made between **analytical** and **numerical methods.**

**Q. 6. What is stochastic activities?**

**Ans.** If the outcome of an activity cannot be described completely in terms of its input or vary randomly over various possible outcomes, the activity is called stochastic activity.

**Example:** In factory system, the time taken for a machining operation may need to be described by a probability distribution.

**Q. 7. What is damping ratio?**

**Ans.** **Damping Ratio:** Damping Ratio is a parameter $\zeta$ that characterize the frequency response of a second order ordinary differential equation. It is particularly important in the study of control theory and harmonic oscillator.

**Q. 8. Name three principal entities, attributes and activities to be considered for the simulation of the following system:**

**(*a*) Railway Reservation System,** **[*UPTU 2006-07*]**

**(*b*) Banking System**

**Ans.** (*a*) *In Railway Reservation System—*

Entity — Passenger, Ticket, Clerk

Attribute — PNR. No., source, destination, no. of person, birth no., distance

Activity — Reservation, cancellation, check the status

(*b*) *In Banking System—*

Entity — Customer, Account

Attribute — Name, Account_No., Account_type, amount

Activity — Deposit, withdrawal, check the status.

**Q. 9. Discuss entitilies, attributes, activities, events and state variables of following system:**

**(*a*) Fast Food Resturant,**

**(*b*) Hospital Emergency Room**

**Ans.** (*a*) *System: Fast Food Resturant:*

Entity — Customers

Attribute — Size of order desired

Activity — (*i*) Placing the order
(*ii*) Paying for the order

Events — (*i*) Arrival of the customer
(*ii*) Completion of purchase

*State Veriable*— (*i*) Number of customers waiting
(*ii*) Number of positions operating

(*b*) *Hospital Emergecy Room:*

Entity — Patients

Attribute — Attention level required

Activity — Providing service required

Events — (*i*) Arrival of the patient
(*ii*) Departure of the patient

Statevariable — (*i*) No. of patients waiting
(*ii*) No. of physicians working.

## EXERCISES

**Q. 1.** Name three of the principal entities attributes and activities to be considered of the operation of following is to be carried out. [*UPTU 2007-08*]

(*i*) Computer centre (*ii*) Bank
(*iii*) Reservation counter (*iv*) Hospital emergency room

**Q. 2.** Define following terms:

(*a*) System environment (*b*) Activity of system
(*c*) System modeling

**Q. 3.** Differentiate between endogenous system and exogenous system.

**Q. 4.** Differentiate between continuous and discrete systems with suitable examples. *[UPTU 2006-07]*

**Q. 5.** Differentiate between static mathematical and dynamic mathematical model. Take suitable examples to illustrate the use of these model. *[UPTU 2006-07]*

**Q. 6.** Name three or four of the principal entities, attributes, and activities to be considered if you were to simulate the operation of:

(*a*) University registration system

(*b*) Examination system

(*c*) A cafeteria

(*d*) *A* gasoline filling station

**Q. 7.** In the aircraft system, suppose the control surface angle *y* is made to be A times the error signal. The response of the aircraft to the control surface is found to be $I\theta_0 + D\theta_0 = ky$. Find the conditions under which the aircraft motion is oscillatory.

**Q. 8.** Discuss various types of models.

**Q. 9.** What do you mean by deterministic and stochastic activities.

**Q. 10.** What is system modeling? Why it is useful?

**Q. 11.** What is the difference between static physical models and dynamic physical models? Explain with suitable example.

**Q. 12.** Discuss the pitfalls of simulation.

**Q. 13.** Discuss the various advantages of simulation.

# 2

# System Study

**INSIDE THIS CHAPTER**

**2.1** Corporate Model; **2.2** Types of System Study; **2.3** Full Corporate Model.

## 2.1 CORPORATE MODEL [*UPTU 2007-08*]

Corporate model is used by various corporations. It provide helps in:

(*a*) Planning of their operations.

(*b*) Understand what conditions will prevail under different set of assumption.

Corporate model has three main parts:

- Environment
- Management
- Plant or physical plant

By looking at the corporation itself, it divides into two major part or segment.

- *Physical plant or company:* It provides the means for production.
- *Management:* It provides decision-making and policy-making aspects of the corporation.

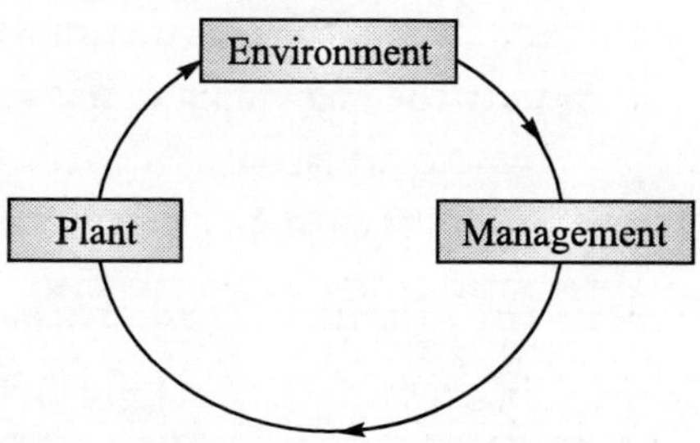


**Fig. 2.1.** Major part of corporate model.

The corporation, as a whole, operates within an environment, these parts can be viewed as subsystem also.

### 2.1.1 Environment Segment

The environment segment can be further broken down. The major part is the market model, the set of relationships that determine what the demand for the product will be.

**Input:** Input of this segment will be:

- Price
- Supply of product

**Output:** Output of this segment will be **demand.**

### Factors to Influence the Market Environment

- Population
- Competition
- National economy

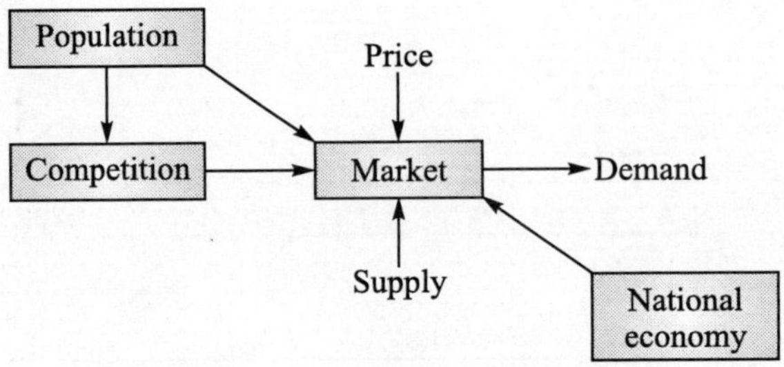


**Fig. 2.2.** The environment segment.

**Population** — Influence of population on the market is through knowing how many people are likely to buy the product.

**Competition** — Competition from other producers or from similar products.

**National Economy** — It affects the market by influencing the amount of money available to target customers.

## 2.1.2 Production Segment

The breakdown of the production subsystem is shown in Fig. 2.3.

**Input:** The main input of this segment is

(*a*) No. of labour (*b*) No. of machinery

**Output:** The main output of the segment is **supply of product.**

### Factors to Influence Production Segment

There are two main factors to influence.

1. **National Economy:** Wages affect production through labour supply, and they are an important item in any national economy model.
2. **World Economy:** The world economy can also be important, if the cost of imported raw materials has to be considered.

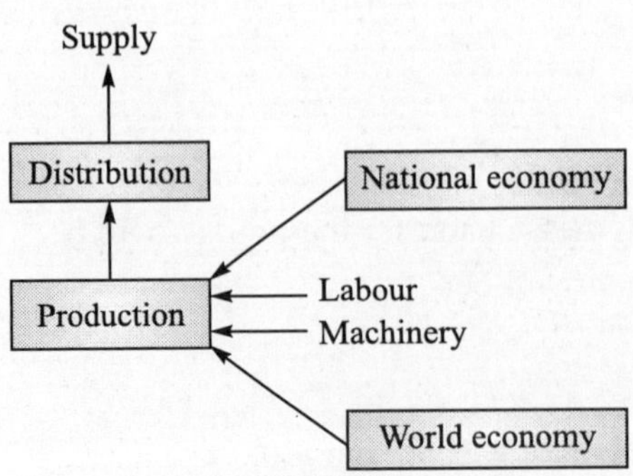


**Fig. 2.3.** The production segment.

### 2.1.3 Management Segment

The management segment is further broken down as in Fig. 2.4.

**Input:** The main input of this segment are:

- Demand for product
- Capital investment

**Output:** The main output, of this segment are:

- Price
- Expected profit

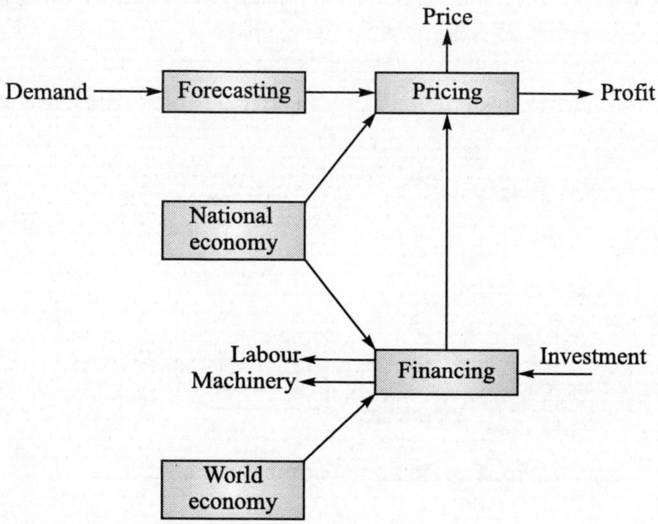


**Fig. 2.4.** The management segment.

- A pricing model sets the **price.**
- A financial model decides how the investment capital is to divided between **labour** and machinery.
- A forecasting model is needed to reflect the way, to make the ***prediction.***
- Another activity of management segment is to predict the **future demand** for the product.

**Factor to Influence Management Segment**

- National Economy
- World Economy.

## 2.2 TYPES OF SYSTEM STUDY [*UPTU 2005-06*]

There are three types of system study:

- System analysis
- System design
- System postulation.

1. **System analysis:** It analysis the system to understand how an existing system is working and what a proposed system will do? Analysts understand the system behaviour by making experiment

with the system itself. For this purpose first of all, analyst construct a model of system and find out or study the behaviour of the model. This result is interpreted in terms of system performance.

2. **System design:** On the basis of analysis result, we design a system that meets some specifications. System designer decides—what components and parameters are needed for the system. The proposed system is modeled and its performance predicted from knowledge of the model's behaviour. If predicted performance is same as analysts want or according to specification, the design is accepted otherwise rejected and again repeat the whole process.
3. **System postulation:** System postulation, described how the models are implemented in social, economic, political and medical studies, where the behaviour of the system is known but the how this behaviour is produce (*i.e.,* process) is not known. Number of hypothesis are made that can explain the behaviour. If a hypothesis can closely explain the behaviour of system or resemblance to the actual system, we postulates the system structure.

   We may need to postulate the system that is, we need to be able to present a reasonable explanation of system structure.

### 2.2.1 Example of System Analysis

To explain the system analysis study let take some segment of corporate model.

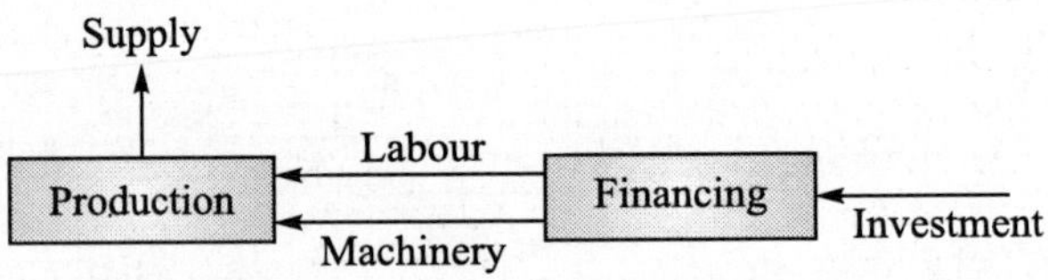


**Fig. 2.5.** Portion of corporate model.

$C$ — Capital investment

$L$ — Labour

$M$ — Machinery

$S$ — Supply

Relationship between investment in labour and machinery is given by:

$$S = g\, L^{x1}\, M^{x2} \qquad \text{where } g \text{ is a constant}$$

Models of this form are known as **Cobb-Douglas models.** Exponents $x_1$ and $x_2$, are both equal to one then supply can be:

$$S = gLM$$

The financial model assumes that investment is divided into two subparts— number of labour and machinery. So we can assume a linear relationship as

$$C = eL + M$$

- $e$ is a constant, describes that one unit of investment in labour is equivalent to investing $e$ units in machinery.
- Both equations depend upon $L$ and $M$.
- Supply $S$, depends upon the amount invested in $L\&M$.
- $e = 0.75$ (assumed) and $g = 0.1$ (assumed)
- One line is for $C = 50$ and another line is for $C = 100$.
- There is three curve, for $S = 50$, 83.3 and 200.
- The maximum supply occurs when $L = 33.3$ and $M = 25$.

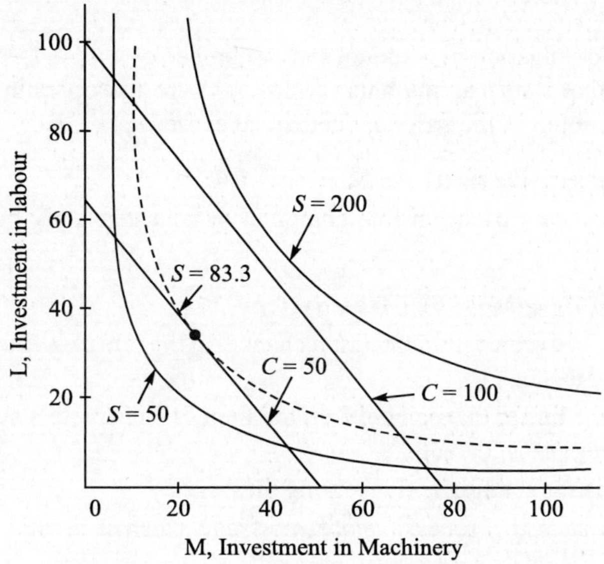


**Fig. 2.6.** Finding optimum allocation of capital.

## ILLUSTRATIONS

**For Supply $S = 50$**

Supply $S = 50$ is possible in case of $C = 50$ means there are some combinations of values of $L$ and $M$ which produce a supply of 50.

(*i*) If $L = 55, \quad M = 9$ and $e = 0.75$ (given)

$$C = 55 * .75 + 9 = 41.25 + 9$$
$$= 50.25 \text{ (Approx.)}$$

If labour is 55 and machinery is 9 then supply $S = 50$ ($S = gLM$, *i.e.*, $0.1 * 55 * 9 = 49.5 \approx 50$) is possible within capital investment $C$ or budget.

(*ii*) If $L = 12.12, \quad M = 41$ and $e = 0.75$ (given)

$$C = 12.12. * 75 + 41$$
$$C = 0.09 + 41 = 50.09 \approx 50$$

So if labour is 12.12 and machinery is 41 then supply $S = 50$ ($S = gLM$, *i.e.*, $0.1 * 12.12 * 31 = 49.69 \approx 50$) is also possible within budget.

**For $S = 200$**

The curve for $S = 200$ does not touches the line for $C = 50$ and $C = 100$. Means it is impossible to produce a supply of 200, within the above budget, *i.e.*, $C = 50$ and 100.

**For $S = 83.3$**

The most that can be produced with $C = 50$ is given by the curve which just touches or is tangent to, the line for $C = 50$. It happens to be curve for 83.3. The maximum supply occurs when

$$L = 33.3 \quad M = 25$$
$$C = 0.75 * 33.3 + 25 = 49.975 \equiv 50$$
$$S = 0.1 * 33.3 * 25$$
$$S = 83.3$$

**Suboptimization**

It is a way to divide the optimization of a model into a number of simple steps. Generally when the system as a whole is operating at its maximum its various parts are also operating at their maximum but this is not satisfactory maximum in the strict mathematical sense.

### 2.2.2 Example of System Design

Here we will design a small part of an on line computer system that reply immediately to received messages.

**Step by Step Procedure of Designing the Given System**

1. Messages are received over a communication channel at the rate of $M$ messages a second. There are $n$ character in a message.
2. They are received in a buffer that can hold a maximum of $b$ characters. A part $p$ of the messages need to replies average of $r$ character.

   **Note:** Buffer is same for sending and receiving messages.
3. Now processing is started. Processing of a message when it received takes 2,000 approx. instructions.
4. Next step is for preparation of reply that requires a program of about 10,000 instructions.
5. As buffer is of finite size so messages and replies must be broken into group of not more than $b$ characters.
6. Servicing of this group needed some interruption for processing, so it require the execution of 1,000 instruction for interruption.

**Suppose there are 3 Category of Computer**

1. Slow : Execution rate is 25,000 instruction per sec.
2. Medium : Execution rate is 50,000 instruction per sec.
3. Fast : Execution rate is 100,000 instruction per sec.

**Four Category of Buffer Sizes**

(*a*) One character buffer (*b*) Two character buffer

(*c*) Five character buffer (*d*) Ten character buffer.

So total 12 (4 buffer size × 3 category of computer) combinations of computer speed, and buffer size wills be capable of carrying out the processing.

**Cheapest Design can be Determined by Following Method**

We have to calculate total no. of characters transferred in and out of the computer. We have to also calculate no. of instruction executed per second.

M messages are coming in computer and pM replies are going out every second.

It means Mn + pMr character are require to pass through the buffer per second.

As buffer can holds maximum $b$ characters then

$M(n + pr)/b$ interruption needed per second

Let total instructions that are require to process messages is given by $N$.

$N$ = Instruction require to [receive messages + process messages + interruptions]

$$N = 2{,}000\ M + 10{,}000\ Mp + \frac{1000\ M\ (n + pr)}{b}$$

Let $S$ is a no. of instructions per second, can execute the computer then the require condition for the computer to keep up with the data flow is that

$$N \leq S$$

For the simplification let take following values

$$M = 5 \qquad n = 15$$
$$p = 0.1 \qquad r = 50$$

Now buffer size $b$ and the computer speed $S$, we have to select. Putting these values in formula.

$$N = 2000 * 5 + 10{,}000 \times 5 * 0.1 + \frac{1000 * 5(15 + 0.1 * 50)}{b}$$

$$N = 10000 + 5000 + \frac{5000(20)}{b}$$

$$N = 5{,}000\left(2 + 1 + \frac{20}{b}\right) \Rightarrow 5{,}000\left(3 + \frac{20}{b}\right)$$

$$N = 5{,}000\left(3 + \frac{20}{b}\right)$$

$$N \leq S$$

$$5{,}000\left(3 + \frac{20}{b}\right) \leq S$$

$$3 + \frac{20}{b} \leq \frac{S}{5{,}000}$$

$$\boxed{\frac{20}{b} \leq \frac{S}{5{,}000} - 3}$$

Plots figure on log-log paper between $20/b$ and $b$ as in Fig. 2.7.

Value of $20/b$ = Plotted vertical line on graph

Value of $b$ = Plotted horizontal line on graph

Result = Plotted by diagonal line, sloping downward to the right.

Dotted vertical line ⇒ Plotted at the point 1, 2, 5 and 10 (buffer sizes)

Dotted horizontal line ⇒ Plotted at the three values of computer speed by ($S$/5000 – 3).

- 12 points of intersection of the vertical and horizontal lines represent the possible combinations of system components.
- It can be seen that three combinations are satisfactory:

The fast speed computer can operate with: 2 character buffer

The medium speed computer can operate with: 5 character buffer

The low speed computer can operate with: 10 character buffer like for vertical line

Low speed computer $S = 25{,}000$

$$\frac{20}{b} \Rightarrow \frac{S}{5{,}000} - 3$$

$$\frac{20}{b} \Rightarrow \frac{25{,}000}{5{,}000} - 3 \Rightarrow 2$$

Medium speed computer $S = 50{,}000$

$$\frac{20}{b} \Rightarrow \frac{50{,}000}{5{,}000} - 3 \Rightarrow 10 - 3 \Rightarrow 7$$

Fast speed computer $S = 100{,}000$

$$\frac{20}{b} \Rightarrow \frac{100{,}000}{5{,}000} - 3 \Rightarrow 17$$

**12 COMBINATION:**

**Low Speed Computer: $S = 25{,}000$**

$$20/b \leq S/5{,}000 - 3, \text{ (i.e., At } b = 1 \; \frac{20}{1} \leq 25000/5{,}000 - 3)$$

| | | | | | |
|---|---|---|---|---|---|
| At $b = 1$ | $20/1 \leq 5 - 3$ | $\Rightarrow$ | $20 \leq 2$ | × | (not possible) |
| At $b = 2$ | $20/2 \leq 2$ | $\Rightarrow$ | $10 \leq 2$ | × | (not possible) |
| At $b = 5$ | $20/5 \leq 2$ | $\Rightarrow$ | $4 \leq 2$ | × | (not possible) |
| At $b = 10$ | $20/10 \leq 2$ | $\Rightarrow$ | $2 = 2$ | ✓ | (possible)* |

**Medium Speed Computer: $S = 50{,}000$**

| | | | | | |
|---|---|---|---|---|---|
| At $b = 1$ | $\frac{20}{1} \leq \frac{50{,}000}{5{,}000} - 3$ | $\Rightarrow$ | $20 \leq 10 - 3$ | × | (not possible) |
| At $b = 2$ | $\frac{20}{2} \leq 10 - 3$ | $\Rightarrow$ | $10 \leq 7$ | × | (not possible) |
| At $b = 5$ | $\frac{20}{2} \leq 7$ | $\Rightarrow$ | $4 \leq 7$ | ✓ | (possible)* |
| At $b = 10$ | $\frac{20}{10} \leq 7$ | $\Rightarrow$ | $2 \leq 7$ | ✓ | (possible) |

**High Speed Computer: $S = 100{,}000$**

| | | | | | |
|---|---|---|---|---|---|
| At $b = 1$ | $\frac{20}{1} \leq \frac{100{,}000}{5{,}000} - 3$ | $\Rightarrow$ | $20 \leq 17$ | × | (not possible) |
| At $b = 2$ | $\frac{20}{2} \leq 17$ | $\Rightarrow$ | $10 \leq 17$ | ✓ | (possible)* |
| At $b = 5$ | $\frac{20}{5} \leq 17$ | $\Rightarrow$ | $4 \leq 17$ | ✓ | (possible) |
| At $b = 10$ | $\frac{20}{10} \leq 17$ | $\Rightarrow$ | $2 \leq 17$ | ✓ | (possible) |

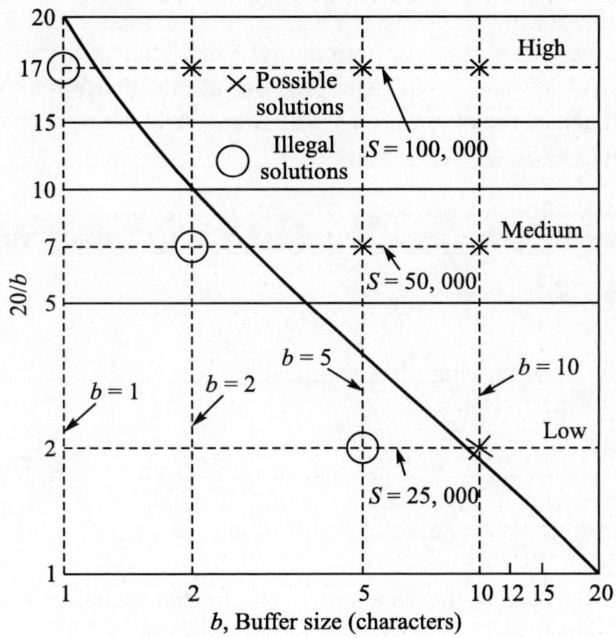


**Fig. 2.7.** Solutions of the input-output model.

## 2.3 FULL CORPORATE MODEL [*UPTU 2005-06*]

By combining together the expansion of previous three segments gives the full corporate model of Fig. 2.8. We have now gone from three subsystems to ten, and it is possible to expand to much larger number.

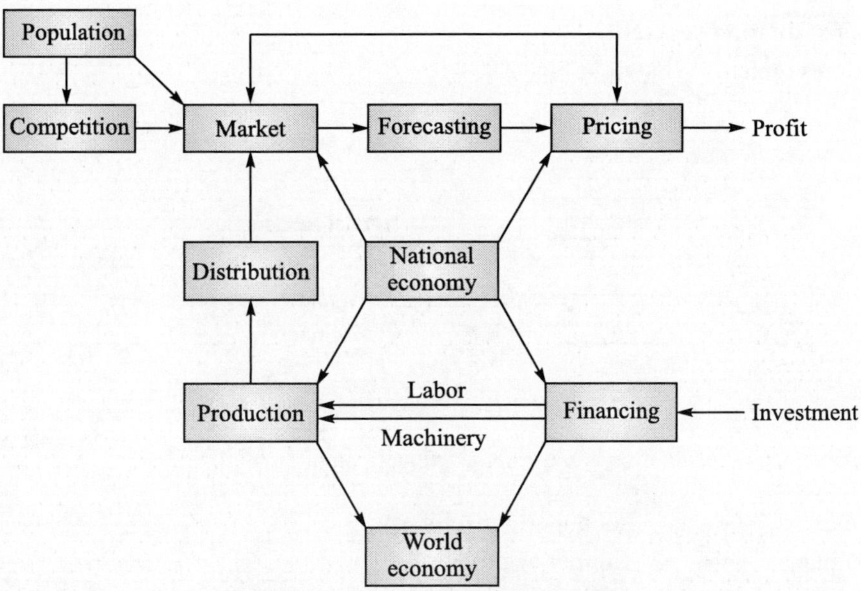


**Fig. 2.8.** The full corporate model.

If we don't know what is our purpose of study, then it is not possible to decide where the process of refinement should stop. Data analysis will help to make this kind of decision. Examining data and plotting one variable against another or using more formal statistical analysis techniques such as correlation or regression analysis, may uncover a significant relationship, or show whether a suggested relationship is strong enough to be worth considering.

## SUMMARY

- Corporate models are used by many corporations to help in various aspects of planning their operations.
- Corporate model has following three segment to:
  - ➢ Environment
  - ➢ Production
  - ➢ Management
- The input for environment segment is the price and supply of product, and output is demand of the product.
- The main input for production segment is labour and machinery, and the main output is supply of the product.
- The main input for management system is demand and capital investment while the main output is price and profit.
- Combining the expansions of three segments together is called as full corporate model.
- Three types of system study

  ⇒ System analysis, ⇒ System design ⇒ System postulation

## SOLVED QUESTIONS

**Q. 1. Discuss production segment.**

**Ans.** Production segment is shown in Fig. 2.9.

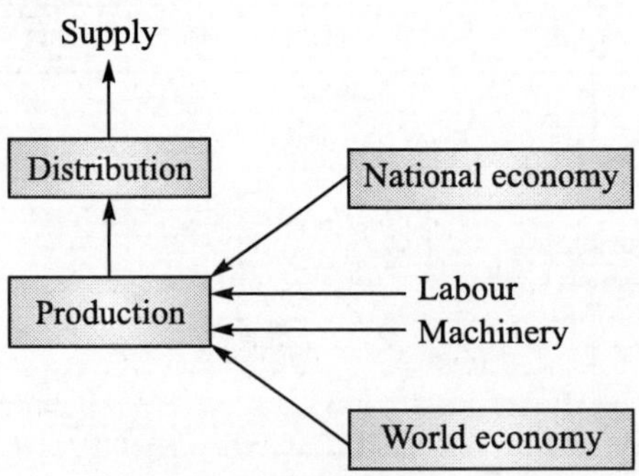


**Fig. 2.9.** The production segment.

| | | |
|---|---|---|
| Main input | — | Labour and machinary |
| Main output | — | Supply of product |
| Economic Factor | — | • National economy |
| | | • World economy |

**Q. 2. What is the objective of corporate model?**

**Ans.** It is used by many corporations to help in various aspects of planning their operations. It is use to understand what conditions will prevail under different sets of assumptions.

**Q. 3. What is Cobb-Douglas models?**

**Ans.** Supply is often related to the investment in labour and machinery. This relation is represented by following equation

$$S = g\ L^{x_1}\ M^{x_2}$$

where $g$ is a constant. Models of this form are known as Cobb-Douglas models.

**Q. 4. Where we can use compartment model?**

**Ans.** Compartment model is used extensively in medical, biological, and ecological studies. The study of compartment models is greatly enhanced by using simulation methods.

**Q. 5. Discuss system postulation.**

**Ans.** System postulation is characteristic of the way models are employed in social, economic, political and medical studies, where the behaviour of the system is known but the processes that produce the behaviour are not known. Hypothesis are made on a likely set of entities and activities that can explain the behaviour models is result are analysis and compare with hyothesis. If it is close to the actual system, then system structure is postulated.

## EXERCISES

1. Discuss full corporate model?
2. What is system analysis? Explain with suitable example.
3. Discuss the various types of system study.
4. What is system Design? Explain with example.
5. Write main input and output of following segments:
   (*i*) Environment segment
   (*ii*) Production segment
   (*iii*) Management segment.

# 3

# System Simulation

**INSIDE THIS CHAPTER**



## 3.1 SYSTEM SIMULATION [*UPTU 2005-06*]

Simulation is a powerful technique for solving a wide variety of problems. Simulation in general is to pretend that one deals with a real thing while really working with an imitation. Simulation done by hand or on a computer.

**Example:** A flight simulator on a PC is also a computer model of some aspects of the flight: it shows on the screen the **controls** and what the "pilot" **is** supposed to **see** from the "cockpit" (his arm chair).

Term "Simulation" is generally used to describe any procedure of establishing a model and deriving a solution numerically.

**Definitions of Simulation**

Following are a few definitions of simulation:

(*a*) Simulation is a representation of reality through the use of a model or other device which will react in the same manner as reality under a given set of conditions.

(*b*) A simulation is a numerical technique for conducting experiments on a digital computer, which involves certain types of mathematical and logical relationships necessary to describe the behaviour and structure of a complex real world system over extended periods of time. **(By Naylor *et. al.* (1966)**

(*c*) Simulation may be defined as one which depicts the working of a large-scale system of men, machines, materials and information operating over a period of time in a simulated environment of the actual real world conditions. **(By Donald G. Malcalm)**

### 3.1.1 Why to Use Models

To fly a simulator is safer and cheaper than real airplane. For precisely this reason, models are used in industry commerce and military. It is very costly, dangerous and often impossible to make experiments with real systems. By using models, we can map reality and check the validity of the real systems.

Simulation is basically an experimental technique. It is a fast and relatively inexpensive method of doing an 'experiment' on the computer for this experiment we need some models. By experimenting with models, we can save *money* and *time.*

## 3.2 WHY TO USE SIMULATION

For solving various types of managerial decision-making problems, following methods are used:

- Scientific method
- Analytic method
- Iterative method

But each method has its own drawbacks and limitations.

***(i)*** **Limitations of Scientific Method:** There are following drawbacks and limitations of scientific method.

- It may be either impossible or too expensive to observe certain processes in real-life situations.
- The observed system may be too complex that cannot be described in term of mathematical equation.
- If mathematical model is formulated then it is very typical to find out a straight forward analytical solution.
- It is too costly to validate mathematical models describing the system.

Due to above drawbacks, the scientific method cannot be used to solve complex managerial decision-making processes.

***(ii)*** **Limitation of Analytic Method:** Analytical techniques used in queueing theory, network model and dynamic programming and hence are not sufficient to tackle problems requiring data analysis due to some limitation. For example, dynamic programming models can only be used to tackle very simple situations, involving very few variables. If the number of static variables becomes larger, the computation becomes complex and difficult.

***(iii)*** **Limitation of Iterative Method:** Generally iterative method is used in linear programming models. Linear programming assumes that the data should be known with certainty. In many real-life situations, the uncertainties about the data are such that they cannot be ignored. In case the uncertainty relates to only few variables, sensitivity analysis can be used to determine its effect on the decision. But in situations where uncertainty pervades the entire model; the sensitivity analysis may become too cumbersome and computationally difficult to determine the impact of uncertainty on the recommended plan.

Due to above discussed limitations, we can say that whenever characteristics like **uncertainty, complexity, dynamic interaction** between the decision and subsequent event, our model becomes too complex to be solved by any of the techniques of mathematical programming and probabilistic models. Such complex models must be solved by some other kind of quantitative technique which may give accurate and reliable results.

Many new techniques are there but among all, **simulation** is the best. The simulation technique is a dependable tool in situations where mathematical analysis is either too complex or too expensive.

## 3.3 WHEN TO USE SIMULATION

System that change with time, such as a gas station where cars come and go (Dynamic Systems) and involve randomness. No one can guess, when will next car come at the gas station or traffic jam, are good candidates for simulation.

These problems are mathematical in nature, and complex. The performance of such system can be difficult to predict by some mathematical analysis is due to:

- System itself is complex.
- No sufficient theory is available.
- Randomness.
- Dynamic interactions between decisions and subsequent events.
- Complex interdependencies among variables to the system.

How to solve these problems (Before Computer Simulation).

- An intuitive decision was made (if problem rely on intuition).
- Laboratory experiments had to be conducted.

Which is generally expensive and time consuming.

Simulation provides an alternative which is cheap and fast. It fills the gap between exact analysis and physical intuition. Simulation is also used even when an exact analytic solution is possible, but it is too costly in terms of computation time.

## 3.4 BASIC NATURE OF SIMULATION [*UPTU 2007-2008*]

The basic nature of simulation varies as types of system such as discrete or continuous, deterministic or stochastic and static or dynamic in nature. These systems are different in nature from each other but there are some common features, and these are the features essential to simulation.

Actually we started from a mathematical model of the system under study some initial conditions (state at time zero) were assumed and the change of state occured in accordance with some equations or laws. Using these rules numerically we continued changing the state of the system as time moved forward. This was done for as long a period as needed. At the end of this period we collected the desired information about the system. These calculations were programmed for a digital computer. It should be noted that we can also performed simulation manually with pencil and paper. In theory any system that can be simulated on a digital computer can also be simulated by hand. In practice, however, simulation as an analytic tool is useful only when done on a computer; because the practical problems that require simulation are complex and need a very large number of simple, repetitive calculations. We can say, simulation is one of the most important uses of the digital computer.

Simulation is basically an experimental technique. It is a fast and relatively inexpensive method of doing an 'experiment' on the computer. We conducted the experiment on the computer and obtained the results within a few minutes, at a very small cost. This is why computer simulation is often referred to as performing a simulation experiment.

However, there is no unifying theory of computer simulation. There are no underlying principles guiding the formulation of simulation models.

## 3.5 TECHNIQUE OF SIMULATION [*UPTU 2007-08*]

Simulation is a technique to use to describe any procedure of establishing a model and deriving a solution numerically.

Simulation is the process of solving the equations of model, step by step, with increasing values of time.

We generally define simulation as "System Simulation". We therefore define system simulation as the technique of solving problems by the observation of the performance, over time, of a dynamic model of the system. In some problems the time element is not significant, but a step-by-step calculation to represent the development process in a system.

**Example:** Result of a sequential decision process, in which a number of successive choices must be made with given probabilities assigned to each choice step-by-step to find the probability of reaching each particular final result.

Problems of this nature are often solved by simulation programs. In such simulations, time is not explicitly involved.

## 3.6 COMPARISON OF SIMULATION AND ANALYTICAL METHOD

[*UPTU 2005-06*]

In order to study a system we have to make a set of assumptions, about how it works. These assumptions, which usually take the form of mathematical or logical relationships, constitute a model.

If the relationship that constitute the model are simple and easy to understand then we can use **mathematical methods** (such as algebra, calculus or probability theory) to get exact information or answer of our desire question. This is called an analytic solution or **analytical method.**

Most real-world system are too complex to allow realistic models to be evaluated analytically. We have to use a computer to evaluate a model **numerically,** data are gathered in order to estimate the desired true properties. This is called **simulation.** Simulation is an extension of mathematical solution.

**Table 3.1.** Comparison between analytical methods and simulation methods

| Sl. No. | Analytical methods | Simulation methods |
|---|---|---|
| 1. | Analytical method, gives general solution. | 1. Simulation method, gives specific solutions. |
| 2. | It consider all the conditions to solve a problem. | 2. Simulation tells only about particular conditions. |
| 3. | There is limited problem that can be solved mathematically. | 3. Simulation is an extension of mathematical solution. |
| 4. | Analytical results occur in the form of complex series or integrals that still require extensive evaluation. | 4. Simulation will provide a quicker or more convenient way of deriving results. |
| 5. | Analytical methods are expensive and time consuming. | 5. Simulation give results in few minutes at a very low cost. |
| 6. | There are many simple limitations such as physical stops, finite time delays, or nonlinear forces. | 6. Simulation easily removes these limitation. |
| 7. | Analytical methods employ the deductive reasoning of mathematics to solve the model. | 7. Simulation model employ numerical methods, to solve the methods. |

## 3.7 TYPES OF SYSTEM SIMULATION [*UPTU 2006-07*]

System simulation can be classified on the basis of model selected. Since there is not a unique correspondence between types of systems and models. However, there are following types of system simulation.

### 3.7.1 Static and Dynamic Simulation Models

A static simulation model is used to represent a system in which time simply plays no role. It represents a system at a particular time.

**Example:** Monte Carlo models.

A dynamic simulation model is used to represent a system in which time plays critical role. It represents a system that changes over time.

**Example:** Conveyor system in a factory.

### 3.7.2 Deterministic and Stochastic Simulation Models

If activity of simulation model can be described completely in terms of its input. There is no randomness or probability. It is called deterministic.

In deterministic models, the output is "determined" when we have specified set of input and relationship in the model.

**Example:** Chemical reaction.

**Stochastic**

If activity of simulation model is random or probabilistic it is called stochastic. In stochastic simulation models, we can have at least some random input components.

These models produce output that is itself random, and hence can only estimate the true characteristics of model. This is main disadvantage of stochastic simulation.

**Example:** Most queueing and inventory systems are modeled stochastically.

### 3.7.3 Continuous and Discrete Simulation Models

Systems or model in which the changes are predominantly smooth are called continuous simulation models, *e.g.* Aircraft System. While in discrete simulation models, changes are predominantly discontinuous.

Discrete simulation or discrete event simulation model is represent in which the state variables changes instantaneously at separate points of time or we can say that system can change at only accountable number of points in time. These points in time are the ones at which an event occurs, where an event is defined as an instantaneous occurrence that may change the state of the system.

Arrivals and departures of customer occurred at distinct points of time this is what we refer to as events. When the number of events are finite, we call the simulation **"Discrete Event"**.

**Example:** Factory system.

In some system the state changes all the time, not just at the time of some discrete events.

**Example:** The water level in a reservoir with given in and outflows may change all the time. In such cases "continuous simulation" is more appropriate, although discrete event simulation can serve an approximation.

Few systems are wholly continuous or discrete, *e.g.,* aircraft system can be regarded as a discrete system. In aircraft we may take discrete adjustments to trim as altitude changes while in factory system machinery is continuous activity and start and finish of a job are as discrete changes.

But in most of system one type of change continuous or discrete predominates, so that we can categorized our system as being continuous or discrete.

**Sampled Data System**

There are also some system where information about them is only available at discrete points in time but system are intrinsically continuous.

# SOME OTHER TYPE OF SIMULATIONS

## 3.8 HYBRID SIMULATION [*UPTU 2007-08*]

When an analog and digital computer are combined to provide a simulation. This simulation is called **hybrid simulation.**

One computer may be simulating the system being studied, while the other is providing a simulation of the environment in which the system is to operate. System can be simulated as interconnection of continuous and discrete subsystems which can be best modeled by an analog and digital computer being linked together.

**Following Technological Advancement is Needed for Hybrid Simulation**

- High-speed converters are needed to transform signals from one form of representation to the other.
- Availability of minicomputer can made hybrid simulation easier by lowering the costs.
- It is generally reserved for the case in which functionally distinct analog and digital computers are linked together for the purpose of simulation.
- Hybrid simulation may be used to simulate systems that are mainly continuous, but do, in fact, have some digital elements.

**Example:** An aircraft satellite for which both the continuous equation of motion, and the digital control signals must be simulated.

**Note:** Don't get confused with the use of digital elements added to the operational amplifiers of an analog computer (in addition of points).

## 3.9 REAL TIME SIMULATION [*UPTU 2007-08*]

The general process by which a mathematical model of an engineering system is constructed, it involves.

- Understanding of physical or chemical laws.
- Implies, number of experiment to derive the coefficient of the model.

This can be particularly time consuming of the model is not being simplified by assuming linearity. But this experiment or measurement and initial work could be avoided if an actual device can be used, rather than constructing a model. This approach is called **real time simulation.**

In real time simulation:

1. Actual devices are used in conjunction with either a digital or hybrid computer to provide real-time simulation.
2. It involve interaction with human being, so avoid the need to design and validate the model of human behaviour.

3. Real time simulation requires computers that can operate in real time.
4. They must be able to respond immediately to signals sent from the physical devices, and send out signals at specific points in time.

**Example:** Aerospace industry makes extensive use of real time simulation. There are some devices, called **simulators.** The main function of simulator is to provide human beings with a substitute for some environment or situation.

## 3.10 WEB BASED SIMULATION

Web based simulation is quickly emerging as an area of significant interest for both simulation researchers and simulation practitioners. This interest in web based simulation is a natural outgrowth of the WWW and its attendant technologies, *e.g.,* HTML, HTTP, CGI, etc.

The appearance of the network friendly programming language Java, and of distributed object technologies like the Common Object Request Broker Architecture (CORBA) and the Object Linking Embedding/Component Object Model (OLE/COM) have had particularly acute effects on the state of simulation practice.

Researchers in the field of web-based simulation are interested in dealing with topics such as methodologies for web-based model development, collaborative model development over the Internet, Java based modeling and simulation, distributed modeling and simulation using web technologies, and new applications.

## 3.11 RELATIONSHIP BETWEEN SELECTED MODELS AND SIMULATION

If the relationships that compose the model are simple enough, then you may use mathematical methods (such as calculus, algebra and probability) to obtain exact information, this is called an analytic solution. However, most real world systems are too complex to allow realistic models to be evaluated analytically and these models must be studied by means of simulation. In classical thinking there are three types of simulation:

- Discrete event
- Continuous
- Monte Carlo.

**Discrete event simulation** utilizes a mathematical/logical model of a physical system that portrays state changes at precise points in simulated time. Both the nature of the state change and the time at which the change occurs mandate precise description. Customers waiting for service, the management of parts inventory or military combat are typical domain of discrete event simulation.

**Continuous simulation** uses equational models, often of physical systems, which do not portray precise time and state relationships that result in discontinuities. The objectives of studies using such models do not require the explicit representation of state and time relationships. Examples of such systems are found in ecological modeling, ballistic reentry, or large scale economic models.

**Monte Carlo simulation,** the name given by John van Neumann and Stanislaw M.Ulam to reflect its gambling similarity, utilizes models of uncertainty where representation of time is unnecessary. The term originally attributed to "a situation in which a difficult non-probabilistic problem is solved through the invention of a stochastic process that satisfies the relations of the deterministic problem". A more recent characterization is that Monte Carlo is "the method of repetitive trials".

## 3.12 SIMULATION OF PURE-PURSUIT PROBLEM

Let consider what is pure-pursuit problem. In pure-pursuit problem a fighter aircraft wants to destroyed an enemy bomber (the target) and flies directly towards it. For this purpose, fighter has to change its direction to keep pointed toward the enemy. The fighter is continuously flying along a specified curve. We have to determine how long it would take to the catch of the bomber.

- If the target flies along a straight line, we can solve this problem directly with analytic techniques.
- If the target flies along a curved, then problem becomes more difficult and normally cannot be solved directly. We will use simulation to solve this problem with following condition.
  - ➢ The target or bomber and the fighter or pursuer are flying in the same horizontal plane. When the fighter first sights the bomber, and both stay in that plane. This make 2 Dimensional pursuit model.
  - ➢ The fighter's speed $V_{fs}$ is constant (22 kms/min).
  - ➢ The target's path is specified.
  - ➢ After a fixed time interval $\Delta t$, the fighter changes its direction in order to point itself toward the bomber.

Consider a rectangular coordinate system in which the two aircraft are flying. Distances are given in kilometers and time in minutes. Let we measure the time when the fighter first sees the bomber.

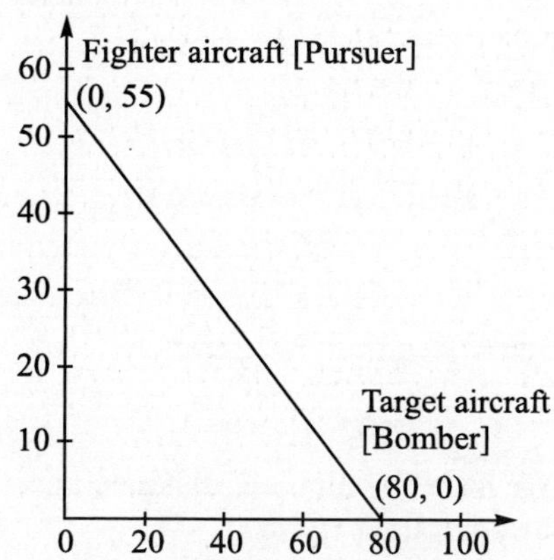


**Fig. 3.1**

We are representing the path of the bomber of two arrays $XB(t)$ and $YB(t)$ respectively. Similarly, the path of the fighter by two arrays $XF(t)$ and $YF(t)$ respectively.

The value of $XB(t)$ and $YB(t)$ will be given $t = 0, 1, .....$ 12. So we have to calculate $XF(t)$ and $YF(t)$ for $t = 0, 1, 2, ......$ 12 or until the fighter catches the bomber.

Initially, we are given: $YF(0) = 55$ kms, $XF(0) = 0$ kms.

**Now Following Assumption we have to Consider**

- Assuming that if the fighter is within 10 kms of the target, the fighter shoots down its target by firing a missile and the pursuit is over.

- If target is not caught up within 12 minutes, the pursuit is abandoned and the target is considered escaped.
- From the time $t = 0$ till the target is shot down, the fighter uses this strategy; it looks at the target at instant $t$, aligns its velocity vector with the line of sights.

It continues to fly in that direction for one minute till instant $(t + 1)$. At time $(t + 1)$ it looks at the target again and realigns itself.

The distance $D(t)$ at a given time $t$ between target and the fighter is given by

$$D(t) = \sqrt{(YB(t) - YF(t))^2 + (XB(t) - XF(t))^2} \quad ...(1)$$

The angle θ of the line from the fighter to the target at a given time $t$ is given by

$$\sin\theta = \frac{YB(t) - YF(t)}{D(t)}$$

$$\cos\theta = \frac{XB(t) - XF(t)}{D(t)}$$

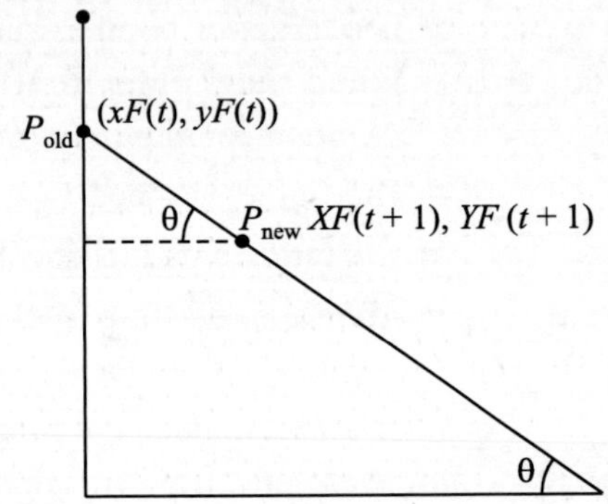

Fig. 3.2

Position after $(t + 1)$ time is $P_{new}$

$$\text{Coordinate of } P_{old} = XF(t), YF(t)$$

$$\text{Coordinate of } P_{new} = XF(t + 1), YF(t + 1)$$

$$\cos\theta = \frac{XF(t+1) - XF(t)}{(D'(t) \text{ is distance})}$$

$$D'(t)\cos\theta = XF(t + 1) - XF(t)$$

$$D'(t)\cos\theta + XF(t) = XF(t + 1)$$

or $$XF(t + 1) = XF(t) + D'(t)\cos\theta \quad ...(2)$$

and $$YF(t + 1) = YF(t) + D'(t)\sin\theta$$

Speed in 1 min

$$\text{Speed} = \frac{D'(t)}{\text{Time}} = \frac{D'(t)}{1}$$

$$V_{fs} = D(t) \quad ...(3)$$

Constant velocity is given. We have to calculate distance after one minute but it is equal to $V_{fs}$. So our equation (2) can be written as equation (4).

Put this value in equation (2)

$$XF(t + 1) = XF(t) + V_{fs}\cos\theta \quad ...(4)$$

Similarly $$YF(t + 1) = YF(t) + V_{fs}\sin\theta \quad ...(5)$$

With this new coordinates of the fighter, its distance from the target is again computed using $D(t)$. If this distance is 10 kms or less the pursuit is over, otherwise θ is recomputed, and the process continues.

Algorithm of pure pursuit problem:

- Read the bomber's path $XB(t)$, $YB(t)$ for $t = 0, 1, ... 12$. Initial position of fighter $XF(0)$, $YF(0)$ and fighter's velocity $V_{fs}$.
- Initialize $t = 0$.

- While ($t \leq 12$).
- Compute the current distance between the two and the next position at $(t + 1)$ of fighter (by equations (1) and (4), (5).
- If ($D(t) \leq 10$).
- Print $t$ and distance.
- Else if ($t > 11$)
  print 'target escaped'.
- Else
  $t = t + 1$.
- Stop.

The fighter redirecting himself toward the target at fixed intervals of time while the target at fixed intervals of time while the target goes on its predetermined path without making any effort to evade the fighter, is called **pure pursuit.**

## 3.13 SIMULATION OF SINGLE-SERVER QUEUEING SYSTEM

Queue represents a certain number of customers waiting for service. People form queues to be served in banks, post offices, ships, and railway ticket counters.

**Problem of Queuing System**

A queueing problem is essentially a problem of balancing the cost of waiting against the cost of idle time for the services facilities in the system. This balancing requires an analysis of the queueing system, like idle time, average waiting time, queue length, etc. The problem arises due to the stochastic nature of the times between the arrivals of customers as well as the time it takes to serve each customer. We can solve this problem by the help of simulation.

**Elements of Queuing Systems**

There are following three elements of queueing systems:

- Arrivals of customers
- Service mechanism
- Queueing Discipline

Figure. 3.3 shows the elements of single server queueing system.

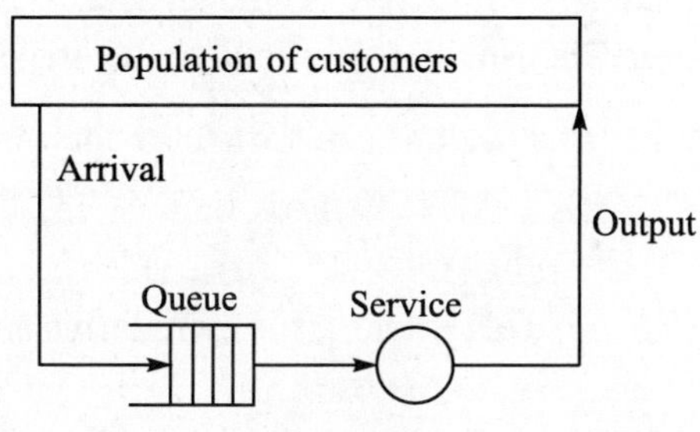


**Fig. 3.3**

***(i)* Arrivals of Customers:** It defines the way customers enter into the system. Mostly the arrivals are random with random intervals between two adjacent arrivals. Typically the arrivals is described by a random distribution of intervals also called *Arrival Pattern.*

***(ii)* Service Mechanism:** A customer who arrives and finds the server idle enters service immediately, and the service time $s_1$, $s_2$ .... of the successive customers are independent and identically distributed (IID) random variables that are independent of inter arrival times. A customer who arrives and finds the server busy joins the end of a single queue.

***(iii)* Queueing Discipline:** It represents the way the queue is organised or it is the order in which service is provided such as:

- FIFO – (First in First out) are also called FCFS.
- LIFO – (Last in First out) are also called LCFS.
- Priority Queue – It may be viewed as a no. of queues for various priorities.
- SIRO – Serve in Random order.

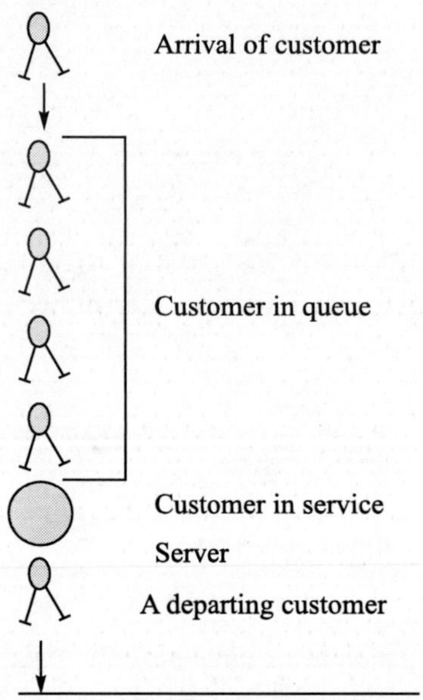


**Fig. 3.4.** A single server queueing system.

**Simulation**

A queueing system consists of one or more servers that provide service of some kind of arriving customers. Customers who arrive to find all servers busy (generally) then all the customer must be in a line in front of servers. Now let we simulate the arrival and servicing of N customers by a single server.

**Points that we have to Consider for Simulation**

- Let the interarrival times $AT_K$ time between the arrivals of the $(K-1)^{th}$ and $K^{th}$ customers into the system.
- $CAT_K$ denote the cumulative arrival time of the $K^{th}$ customer. If there is no queue, that means server is free.
- First customer arrives, at time zero, he goes directly to the service counter.
- After the service, this customer departs from the service counter (from the system) at time: $ST_1$.
- The second customer will arrive at

$$CAT_2 = AT_2$$

- If $ST_1 > CAT_2$, the second customer has to wait for a period

$$WT_2 = ST_1 - CAT_2$$

**Example:**

Server

6.02 min = $ST_1$ (Service time of customer 1st)

| Custom | III | II | I | |
|---|---|---|---|---|
| Time | 6.03 | 6.01 | 6.00 | Service time = 2 sec of customer 1st |

$ST_1 > CAT_2$ (that means second customer has to wait in queue) how much a second customer has to wait, *i.e.*, $WT_2$:

$$WT_2 = ST_1 - CAT_2 \quad \text{or} \quad 6{:}02 - 6{:}01 = 01 \text{ sec}$$

($WT_K$ denotes the waiting time of the $K^{th}$ customer in the queue).

- If $CAT_2 > ST_1$ then the departure of the first customer takes place before the arrival of the second. So the service counter remains idle awaiting the arrival of the second customer for a period.

$$IDT_2 = CAT_2 - ST_1$$

- $IDT_K$ = idle time of the server awaiting the arrival of the $K^{th}$ customer

Suppose, so far $(p-1)^{th}$ customers have arrived into the system and $(q-1)$ customers have left. The next customer due to arrive is $p^{th}$ and the next customers due to departs is $q^{th}$ (Clearly $1 \le q \le P \le N$). The length of queue is given by $p - q - 1$, if $p > q$ the next arrival time is $NAT = CAT_p$

The next departure time, *i.e.*,

$$NDT = CDT_q \text{ (cumulative arrival time of } q + \text{waiting time of } q + \text{service time of } q\text{)}$$

*i.e.,* $$= CAT_q + WT_q + ST_q$$

Now we have to check that whether $p$ would arrive first or $q$ would depart first. This is decided by comparing $NAT$ with $NDT$.

$$diff = NAT - NDT$$

**Consider following cases:**

**Case 1:** If difference diff, is positive and queue length $(p - q - 1)$ is also positive then:

- Departure would take place first
- Reduce the length of queue by 1 as departure take place first.

**Case 2:** If difference diff. is positive and queue length $(p - q - 1)$ is equal to zero then:

- The service counter will lie idle waiting for the $p^{th}$ customer for the duration $IDT_p$ = diff.

**Case 3:** If difference diff is zero then:

- Next arrival and departure will take simultaneously
- No change in the length of queue.

**Case 4:** If difference diff is negative then

- arrival takes place first
- length of the queue is increased by 1.

With this following output, one can calculate many different statistics. Some of these are:

1. Cumulative arrival times are easily calculated, using the relationship

$$CAT_K = CAT_{K-1} + AT_K$$
$$CAT_1 = AT_1 = 0$$

2. The average waiting time (AWT) of customers is calculated as follows:

$$AWT_{(\text{min})} = \frac{\text{Total time costomer, wait in queue (min)}}{\text{Total no. of customers}}$$

3. The probability that a customer has to wait in a queue is calculated as:

$$\text{Probability (wait)} = \frac{\text{No. of customers who wait}}{\text{Total no. of customers}}$$

4. The probability that a server is idle can be calculated as:

$$\text{Probability of idle server} = \frac{\text{Total idle time of server (min)}}{\text{Total run time of simulation (min)}}$$

5. The probability of server being busy is calculated as:

   Probability of busy server = 1 – probability of idle server

6. The average service time (AST) is calculated as

$$\text{Average service time (min)} = \frac{\text{Total service time (min)}}{\text{Total no. of customer}}$$

7. The average time between arrivals is given by:

$$\text{Average time between arrivals (min)} = \frac{\text{Sum of all times between arrivals (min)}}{\text{No. of arrivals} - 1}$$

**Note:** One is subtracted from no. of arrivals because the first arrival is assumed to occur at time 0.

8. The average waiting time of those who has to wait is given by:

   Average waiting time of those who wait (minutes)

$$= \frac{\text{Total time customers wait in queue (min)}}{\text{Total no. of customers who wait}}$$

9. The average time a customer spends in system is given by:

   Average time customer spends in the system (minutes)

$$= \frac{\text{Total time customers spend in the system (minutes)}}{\text{Total no. of customers}}$$

   or Average time customer spends in system

$$= \begin{matrix}\text{Average time customer}\\ \text{spends waiting in}\\ \text{the queue (min)}\end{matrix} + \begin{matrix}\text{Average time}\\ \text{customer spends}\\ \text{in service (min)}\end{matrix}$$

**Example 1:** Calculate different parameter for the single server queueing system when inter-arrival time and service times are given by the Tables 3.2 and 3.3.

**Table 3.2.** Inter-arrival and clock times

| Customer | Inter-arrival time | Arrival time on clock |
|---|---|---|
| 1 | – | 0 |
| 2 | 4 | 4 |
| 3 | 1 | 5 |
| 4 | 4 | 9 |
| 5 | 2 | 11 |
| 6 | 4 | 15 |

**Table 3.3.** Service Time of Each Customer

| Customer | Service time |
|---|---|
| 1 | 2 |
| 2 | 3 |
| 3 | 1 |
| 4 | 4 |
| 5 | 2 |
| 6 | 3 |

**Solution: Formula used in simulation Table 3.4**

1. Time since last arrival for $(i + 1)$ customer = Arrival time of $i^{th}$ customer – Arrival time of $(i – 1)^{th}$ customer
2. Time customer wait in queue for $i^{th}$ customer = Arrival time of $i^{th}$ customer – Time service begin of $i^{th}$ customer
3. Time service ends of $i^{th}$ customer = Time service begin of $i^{th}$ customer (min) + Service Time of $i^{th}$ customer
4. Time customer spends in system = Time service ends of $i^{th}$ customer – Arrival time of $i^{th}$ customer
5. Idle time of server for $i^{th}$ customer = Time service begins of $i^{th}$ customer – Time service ends of $(i – 1)^{th}$ customer
6. Time service Begins: For first customer it is zero (always) for rest of customer it is a maximum between arrival of $i^{th}$ customer and time service ends of $(i – 1)^{th}$ customer, *i.e.* Max. (Arrival of ($i^{th}$) customer, Time service ends of $(i – 1)^{th}$ customer)

**Example:** For second customer. Arrival time of customer 2 is 4 while service ends of first customer is 2 then

$$\text{Max}(4, 2)$$

Time service begin of $2^{nd}$ customer = 4

Now calculate the following parameters for short time simulation for this example:

1. Average waiting time $= \dfrac{\text{Total customer wait in queue}}{\text{Total no. of customer}}$

**Table 3.4:** Simulation for above queueing problem

| Customer | Time service last arrival | Arrival time | Service time | Time service begin | Waiting in queue | Time service ends | Time customer spends in system | Idle time of queue |
|---|---|---|---|---|---|---|---|---|
| 1 | – | 0 | 2 | 0 | (0 – 0) = 0 | (0 + 2) = 2 | (2 – 0) = 2 | 0 – 0 = 0 |
| 2 | 4 | 4 | 3 | 4 | (4 – 4) = 0 | (4 + 3) = 7 | (7 – 4) = 3 | 4 – 2 = 2 |
| 3 | 1 | 5 | 1 | 7 | (7 – 5) = 2 | (7 + 1) = 8 | (8 – 7) + 2 = 3 or 8 – 5 = 3 | 7 – 7 = 0 |
| 4 | 4 | 9 | 4 | 9 | (9 – 9) = 0 | (9 + 4) = 13 | (13 – 9) = 4 | 9 – 8 = 1 |
| 5 | 2 | 11 | 2 | 13 | (13 – 11) = 2 | (13 + 2) = 15 | (15 – 13) + 2 = 4 or 15 – 11 = 4 | 13 – 13 = 0 |
| 6 | 4 | 15 | 3 | 15 | (15 – 15) = 0 | (15 + 3) = 18 | (18 – 15) = 3 | 15 – 15 = 0 |
| | 15 | | 15 | | 4 | | 19 | 3 |

$$= \frac{4}{6} = 0.667 \text{ min}$$

2. Probability that a customer has to wait in the queue is

$$\text{Probability (wait)} = \frac{\text{No. of customer who has to wait}}{\text{Total no. of customer}}$$

$$= \frac{2}{6} = 0.33$$

3. Probability of idle server $= \dfrac{\text{Total run time of idle server (min)}}{\text{Total run time of simulation}}$

$$= \frac{3}{19} = 0.157$$

Probability of busy server is $= 1 - 0.157$
$= 0.84$

4. Average service time $= \dfrac{\text{Total service time (min)}}{\text{Total no. of customer}}$

$$= \frac{15}{6} = 2.5$$

5. Average time between arrivals (min)

$$= \frac{\text{Sum of all time between arrival}}{\text{No. of arrivals} - 1}$$

$$= \frac{15}{6-1} = 3$$

6. Average waiting time of those who wait

$$= \frac{\text{Sum of total waiting time in queue}}{\text{No. of customer who has to wait}}$$

$$= \frac{4}{2} = 2 \text{ min}$$

7. Average time customers spends in the system (min)

$$= \frac{\text{Total time customer spends in system}}{\text{Total no. of customer}}$$

$$= \frac{19}{6} = 3.16 \text{ minutes}$$

or by second alternative formula

= Average time customer spends waiting in the queue (min)
+ Average time customer spends in service (min) Average time customer

$= 0.667 + 2.5 = 3.167$

We can make following decision:

- Less customer has to wait with 2 minute average waiting time.
- Server is not often free, it is busy appropriate 84% of time.

**Note:** Simulation requiring variation of the arrival and service distribution as well as implementation in a spreadsheet for analysis.

**Example 2.** Calculate different parameter for a single server queueing system when arrival time and service times are given by the Table 3.5 and analysis the system.

**Table 3.5** For Arrival time and service time

| Customer | Arrival time | Service time |
|---|---|---|
| 1 | 0 | 4 |
| 2 | 8 | 1 |
| 3 | 14 | 4 |
| 4 | 15 | 3 |
| 5 | 23 | 2 |
| 6 | 26 | 4 |
| 7 | 34 | 5 |
| 8 | 41 | 4 |
| 9 | 43 | 5 |
| 10 | 46 | 3 |

**Solution:** Consider following parameter for single server queueing system.

1. The average waiting time for a customer, is determined in the following manner:

$$\text{Average waiting time} = \frac{\text{Total time customers wait in queue}}{\text{Total no. of customer}}$$

$$= \frac{9}{10} = 0.9 \text{ minutes}$$

2. The probability that a customer has to wait in the queue, is determined in the following manner

$$\text{Probability (wait)} = \frac{\text{No. of customers who wait}}{\text{Total no. of customers}}$$

$$= \frac{3}{10} = 0.3 \text{ minutes}$$

3. The fraction of idle-time of the server is determined in the following manner:

$$\text{Probability of idle} = \frac{\text{Total idle time of server (minutes)}}{\text{Total run time of simulation (minutes)}}$$

$$= \frac{18}{44} = .409 \approx .41$$

The probability of server being busy is

$$= 1 - .41$$

$$= .59$$

4. The average service time is determined as follows:

$$\text{Average service time (minutes)} = \frac{\text{Total service time (minutes)}}{\text{Total number of customers}}$$

$$= \frac{35}{10} = 3.5 \text{ minutes}$$

5. The average time between arrivals is determined in following manner:

Average time between arrivals (minutes)

$$= \frac{\text{Sum of all times between arrivals (min)}}{\text{Number of arrivals} - 1}$$

$$= \frac{46}{10-1} = \frac{46}{9} = 5.1 \text{ minutes}$$

6. The average waiting time of those who wait is determined in the following manner:

Average waiting time of those who wait (minutes)

$$= \frac{\text{Total time customers wait is queue (minutes)}}{\text{Total number of customers who wait}}$$

$$= \frac{9}{3} = 3 \text{ minutes}$$

7. The average time a customer spends in the system is determined in two ways

(*i*) Average time customer spends in system

$$= \frac{\text{Total time customers spend in the system (minutes)}}{\text{Total number of customers}}$$

$$= \frac{44}{10} = 4.4 \text{ minutes}$$

(*ii*) Average time customer spends in the system (minutes)

$$= \begin{array}{l}\text{Avg. time customer}\\ \text{spends waiting}\\ \text{in the queue (min)}\end{array} + \begin{array}{l}\text{Average time}\\ \text{customer spends}\\ \text{in service (min)}\end{array}$$

$$= 0.9 + 3.5$$

$$= 4.4$$

Now on the basis of above parameters. We can analysis following points:

- less customer has to wait
- server is quite idle, almost 41% it is idle
- average waiting time is not excessive.

**Table 3.6:** Simulation for Queueing Problem

| Customer | Time since last arrival (minutes) | Arrival time | Service time (minutes) | Time service begins | Time customer wait in queue | Time service ends | Time customer spends in system (minutes) | Idle time of server (minutes) |
|---|---|---|---|---|---|---|---|---|
| 1 | – | 0 | 4 | 0 | 0 | 4 | 4 | 0 |
| 2 | 8 | 8 | 1 | 8 | 0 | 9 | 1 | 4 |
| 3 | 6 | 14 | 4 | 14 | 0 | 18 | 4 | 5 |
| 4 | 1 | 15 | 3 | 18 | 3 | 21 | 6 | 0 |
| 5 | 8 | 23 | 2 | 23 | 0 | 25 | 2 | 2 |
| 6 | 3 | 26 | 4 | 26 | 0 | 30 | 4 | 1 |
| 7 | 8 | 34 | 5 | 34 | 0 | 39 | 5 | 4 |
| 8 | 7 | 41 | 4 | 41 | 0 | 45 | 4 | 2 |
| 9 | 2 | 43 | 5 | 45 | 2 | 50 | 7 | 0 |
| 10 | 3 | 46 | 3 | 50 | 4 | 53 | 7 | 0 |
|  |  |  | 35 |  | 9 |  | 44 | 18 |

So far, the queueing systems we studied were only single server systems: that is there was one service facility. But actual real world application or computer systems and communication networks are **multiple server systems.** To study multiple server queueing system you may refer chapter 10 (M/M/C and M/M/C/C queueing model).

## 3.14 SIMULATION OF AN INVENTORY PROBLEM [*UPTU 2005-06*]

Let take a example of a company that sells a single product, would like to decide what total number of item is required in inventory for each next months. The company reviews the inventory level at the beginning of each month and then decide what total number of item are ordered from its supplier.

So it is necessary to fill up again the inventory with certain item in the store by ordering it from wholesaler. So we have a simple policy. When inventory stock goes down to $X$ items, reorder point, they will order $Y$ more item from the wholesaler.

If demand is 125 and item in inventory is 100 then this excess (125 – 100 = 25) represents lost sale and also lost of goodwill. But we cannot store so many item or overstocking due to carrying cost.

| Carrying cost = cost of storage + insurance of item + interest + security. |
|---|

If we are making our order too frequently, it will result in excessive reorder cost. Then how we can keep our stock upto date, let consider following points for simulation.

(*i*) There is a 3 day lag between the order and arrival. The merchandise is ordered at the evening of the day and is received at the beginning of the fourth day. It means, merchandise ordered on the evening of the $i^{th}$ day is received on the morning of $(i + 3)^{th}$ day.

(*ii*) The carrying cost for each night for each unit of inventory is ₹ 0.95, means carrying cost = ₹ 0.95.

(*iii*) Each unit out of stock when ordered results into a loss of goodwill worth ₹ 2.00 per unit plus loss of ₹ 18.00 net income, the selling price of item, so total loss is of ₹ 20.00 per unit.

| Total loss = Loss of goodwill (₹) + Selling Price of an item (₹) |
|---|

= 2 + 18 = ₹ 20.

Lost sell is always lost.

(*iv*) Placement of order costs ₹ 80.00 regardless of the how many units are ordered.

(*v*) In any day, demand of units can be consider random, from 0 to 99, with equal probability.

(*vi*) Let initially we have 125 units on hand and no reorder is outstanding $S_k = 125$.

(*vii*) Total cost = Reorder cost + Carrying cost + Lost sales cost

with these condition, we have following six policies and we have to compare which policy is best, means at what reorder point, how much we have to order so that minimum total cost will occur.

| | **X (Reorder point)** | **Y (Reorder quantity)** | **(Cost in ₹)** |
|---|---|---|---|
| Policy I | 135 | 175 | 38361.76 |
| Policy II | 135 | 275 | 43983.69 |
| Policy III | 170 | 275 | 36283.91 **Best policy** |
| Policy IV | 200 | 275 | 41965.59 |
| Policy V | 200 | 350 | 45882.042 |
| Policy VI | 200 | 400 | 50404.859 |

Policy III is Best Policy because minimum cost occur in this policy. A Simulation Model of this inventory system can be easily constructed. We can start from day = 1 and increase time by fixed interval of time, *i.e.,* one day and continue this upto 200 day. Consider following way of simulation:

- First we have to check if today is due date for arrival of item. If yes, then existing stock $S_k$ is increased by $Y$ quantity. The quantity of item that was ordered.

  *i.e.,* $$S_k = S_k + Y$$
- Generate random demand $D$, for today. If $D \leq S_k$ then new stock at the end of today will be
$$S_k = S_k - D$$
- If $D > S_k$, then our new stock will be zero and add shortage cost or (lost sell) to the total cost
$$S_k = 0$$
- Compute total cost resulting from today's transactions, and add it to the total cost $C_t$ incurred till yesterday and determine. If the inventory on hand plus units on order is greater than $X$, the reorder point.
- If not, place an order (to be delivered three days hence), by stating the amount ordered and the day it is due to be received.
- Repeat this procedure for 200 days.
- Initially we set that units due ($UD$), and the day they are due ($DD_t$) is equal to zero, because there is no outstanding order.

**About Demand**

Actually demand is not fixed. In this system we have assume any integral value from 0 to 99. For this we have use a special function of $C$ language.

$$D = \text{random } (100)$$

This function **random (*int*)**; will generate random no. from 0 to 99 and each number is equiprobable. We will use this no. and produced is the daily demand.

**Procedure:** How to make inventory simulation model

**Step 1:** Read $X$, $Y$; initialize stock ($S_k$), Due Date ($DD_t$), Unit Due ($U_D$), starting day ($i$), total cost ($C_t$) as

$$DD_t = 0; \quad U_D = 0; \quad i = 1; \quad C_t = 0.0; \quad S_k = 125;$$

**Step 2:** If today is the due date ?

Yes — then (*i*) $S_k = S_k + Y$;

$U_D = 0$;

(*ii*) and goto step–4

**Step 3:** No—Then, generate random demand for today:

By using demand $D = \text{random } (100);$

**Step 4:** If demand ($D$) $\leq$ stock ($S_k$) ?

Yes — (*i*) Compute new stock by using

$$S_k = S_k - D$$

(*ii*) Add carrying cost by using

$$C_t = C_t + S_k * .95$$

(*iii*) Goto step–6

No — (*i*) Add shortage cost to the total cost by using

$$C_t = C_t + (D - S_k) * 20$$

(*ii*) Reset stock ($S_k$) = 0;

**Step 5:** If (stock + unit due) ≤ $X$ {Note: $ES_k$ (equivalent stock) = stock + unit due}.

Yes— (*i*) unit due ($U_D$) = $Y$

(*ii*) Add reorder cost [$C_t = C_t + 80$]

(*iii*) Due Date ($D_t$) = today + 3

(*iv*) Goto step–7

**Step 6:** No—Then increase the day by 1

*i.e.*, $i = i + 1$

**Step 7:** If 200 days passed ?

No — Goto step–2

Yes—Then goto step-8

**Step 8:** Print $X$, $Y$ and Total Cost ($C_t$).

### 3.14.1 C Program of Inventory Simulation

```
        # include <stdio. h>
        # include <conio. h>
        # include <stdlib. h>  void main ()
{
            float, D, X. Y, Sk, ESk, UD, DDt, Ct;
            int i;
        Ct = 0;
        Sk = 125;
        i = 1;
        UD = 0.0; // because there is no outstanding order
        DDt = 0; // due date is zero, because UD = 0
            clrscr ();
printf ("Enter the value of X & Y");
scanf ("% f %f", & X, & Y);
while (i <= 200)
        {if (DDt = = i) {
            Sk = Sk + Y;
            UD = 0;}
            D = random (100);
        If (D <= Sk)
            {
                Sk = Sk – D;
                C = C + Sk*0.95;
            }
                else
            }
                    Ct = Ct + (D – Sk) * 20
                    Sk = 0;
            }
```

```
        ESk = Sk + UD; // ESk is equivalent stock
    If (ESk ≤ X)
        {
            UD = Y;
            DDt = i + 3;
            Ct = Ct ÷ 80;
        }
            i = i + 1;}
  print f ("% f % f %f", X, Y, Ct);
        getch ();
  }
```

**Output**

| | *X* | *Y* | $C_t$ (in ₹) | |
|---|---|---|---|---|
| 1. | 135 | 175 | 38361.76 | |
| 2. | 135 | 275 | 43983.69 | |
| 3. | 170 | 275 | 36283.91 | ✓ Best Policy |
| 4. | 200 | 275 | 41965.59 | |
| 5. | 200 | 350 | 45882.042 | |
| 6. | 200 | 400 | 50404.859 | |

Minimum cost occur in 3rd policy *i.e.,* $X = 170$, $Y = 275$ so we have to select this policy. This is the result of simulation.

## 3.15 MONTE-CARLO SIMULATION [*UPTU 2007-08*]

The name "Monte-Carlo" simulation or method originated during World War-II, when this approach was applied to problems related to the development of the atomic bomb.

Monte-Carlo simulation is a method for iteratively evaluating a deterministic model using set of random no. as inputs. This method is often used when model is—complex, non-linear or involves uncertain parameters.

We can define Monte-Carlo simulation as a scheme of employing random number, which is used for solving certain stochastic or deterministic problems where the passage of time plays no substantive role.

**Six Steps of Monte-Carlo Simulation**

**Step 1:** Clearly define the problem.

(*a*) Identify the objectives of the problem.

(*b*) Identify the main factors which have the greatest effect on the objectives of the problem.

**Step 2:** Construct a suitable model.

(*a*) Specify the variables and parameters of the model.

(*b*) Formulate the decision rules, that is, state the conditions under which the experiment is to be performed.

(*c*) Identify the type of distribution that will be used—models use either theoretical distributions or empirical distributions.

(*d*) Specify in what way time will change.

(*e*) Define the relationship between the parameters and variables.

**Step 3:** Prepare the model for experimentation

(*a*) Define the starting conditions for the simulation.

(*b*) Specify the number of runs of simulation to be made.

**Step 4:** Using steps 1 to 3, experiment with the model

(*a*) Define a coding system that will correlate the factors defined in step 1 with some random number for the simulation.

(*b*) Choose a random number generator and create the random numbers to be used in the simulation.

(*c*) Associated the generated random numbers with the factors identified in step 1 and coded in step 4(*a*).

**Step 5:** Examine and summarise the results obtained in step 4.

**Step 6:** Evaluate the results of the simulation.

**Example 3:** Find the value of $\pi$, by using Monte-Carlo method of simulation.

**Solution:** Taking the origin *O* as the centre, draw an arc *PQ* of unit random cutting the coordinates axes *OX, OY* at *P* and *Q* and complete the square *OPRQ* as shown in Fig. 3.5.

Equation of arc *PQ* is $r_1^2 + r_2^2 = 1, 0 \leq r_1 \leq 1, 0 \leq r_2 \leq 1$ and $0 \leq r_1 \leq 1, 0 \leq r_2 \leq 1$ is nothing but all the points are inside and on the square *OPRQ*.

Using the random number table, select any two random numbers, say 0.7532 and 0.8362 and let $r_1 = 0.7532$ and $r_2 = 0.8362$. The point (0.7532, 0.8362) will lie inside or on the arc *PQ* if $r_1 + r_2 \leq 1$ or will be outside the arc and with in the square if $r_1 + r_2 > 1$.

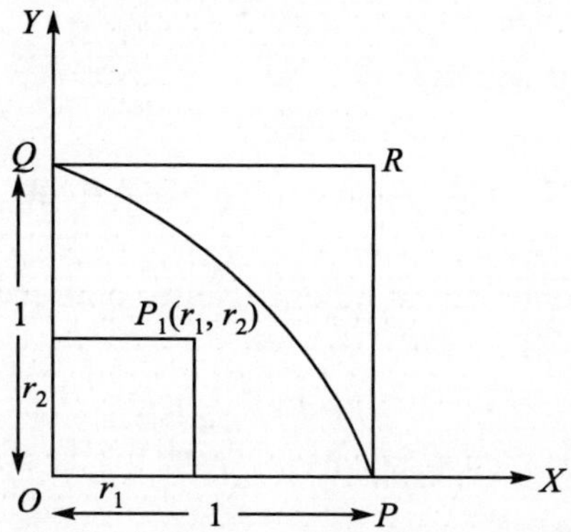


**Fig. 3.5.** Monte-Carlo evaluation of $\pi$.

By this method, a large number of pairs of random numbers are selected and whether the points representing the pairs lie in, on the arc or beyond the arc but inside the square is determined.

Let *m* be the total number of points considered (*m* pairs of random numbers are selected) and let *n* be the points which lie inside or in the arc.

Then, obviously,

$$\frac{n}{m} = \frac{\text{Area enclosed by the arc}}{\text{Area of the square}} \qquad ...(1)$$

$$\text{Area enclosed by the arc} = \frac{\pi}{4} * 1^2$$

$$\text{Area of the square} = OQ * OP = 1 * 1 = 1$$

$$\frac{n}{m} = \frac{\frac{\pi}{4} * 1^2}{1} \Rightarrow \frac{\pi}{4}$$

$$\frac{n}{m} = \frac{\pi}{4}$$

$$4n = m\pi$$

$$\pi = \frac{4n}{m}$$

**Note:** Monte-Carlo has been also applied to estimate the critical values for Kolmogorav Smirnov test for normality.

## 3.16 DISTRIBUTED LAG MODELS [*UPTU 2006-07, 2007-08*]

If model becomes larger or more complicated then numeric computation technique can be assumed that record keeping quickly becomes complicated. So addition of a computer and programming language is important. However, we have some simple technique in simulation that can be applied without difficulty, even for large models. If the events all occur synchronously, at fixed intervals of time, the computation remain simple.

### Properties of Distributed Lag Model

1. Model changes only at fixed intervals of times.
2. Model is based on current values of variables on other current values and values that occurred in previous intervals.
3. These are used mainly in econometric studies because some economic data are collected at fixed interval of time, *e.g.*, month or a year.
4. These model consist the rule of linear algebraic equation.
5. They represent a continuous system, but one in which the data is only available at fixed points in time.

### What is Lagged Variable?

If current values of variable can be represent in terms of values of one or more previous intervals. This variable is called as **lagged variable.** Its value in a previous interval is denoted by attaching the suffix–$n$, into the variable.

$$n = \text{interval.}$$

If $n = 1$, it means previous interval

If $n = 2$, it means One prior to above, and so on.

Consider following equations

$$C = 20 + 0.7\,(Y_R - T_X) \quad \ldots(i)$$

$$I = 2 + 0.1\, Y_R \qquad ...(ii)$$
$$T_X = 0.2\, Y_R \qquad ...(iii)$$
$$Y_R = C + I + G \qquad ...(iv)$$

where $C$ is consumption, $I$ is investment, $T_X$ is tax, $G$ is government expenditure, and $Y_R$ is national income. All quantities are expressed in billions of dollars.

Let simplify above equation by putting equation (*iii*) in equation (*i*)

$$C = 20 + 0.7\,(Y_R - 0.2\, Y_R)$$
$$C = 20 + 0.7\,(0.8\, Y_R)$$
$$C = 20 + 0.56\, Y_R \qquad ...(v)$$

Put this equation (*v*) in equation (*iv*) and get

$$Y_R = 20 + 0.56\, Y_R + I + G$$
$$Y_R - 0.56\, Y_R = 20 + I + G$$
$$0.44\, Y_R = 20 + I + G$$
$$Y_R = 45.45 + 2.27\,(I + G)$$

The set of equations can then be written in the following form:

$$I = 2 + 0.1\, Y_{R-1} \qquad ...(i)$$
$$Y_R = 45.45 + 2.27\,(I + G) \qquad ...(ii)$$
$$T_X = 0.2\, Y_R \qquad ...(iii)$$
$$C = 20 + 0.7\,(Y_R - T_X) \qquad ...(iv)$$

**Important**

1. There is only one variable $Y_R$ is taken as lagged variable.
2. To solve this equation we must have initial value of $Y_{R-1}$.

**Example 4.** For the above equation values of G are supplied for all year (time-intervals).

| Year | G |
|---|---|
| 1991 | 15 |
| 1992 | 20 |
| 1993 | 25 |
| 1994 | 30 |
| 1995 | 35 |

Initial value of $Y_{R-1}$ is 75. Then calculate consumption of Ist year.

**Solution:** Let take $Y_R$ as lagged variable in equation (*i*), then

$$I = 2 + 0.1\, Y_{R-1} \qquad ...(i)$$
$$Y_R = 45.45 + 2.27\,(I + G) \qquad ...(ii)$$
$$T_X = 0.2\, Y_R \qquad ...(iii)$$
$$C = 20 + 0.7\,(Y_R - T_X) \qquad ...(iv)$$

Given $Y_{R-1} = 75$ put this in equation (*i*)

$$I = 2 + 0.1 * 75$$
$$I = 2 + 7.5 = 9.5$$
$$Y_R = 45.45 + 2.27\,(9.5 + 15)$$

$$Y_R = 45.45 + 2.27 \times 24.5$$
$$Y_R = 45.45 + 55.615 = 101.065$$
$$T_X = 0.2 * 101.065$$
$$T_X = 20.213$$
$$C = 20 + 0.7\ (101.065 - 20.213)$$
$$C = 20 + 0.7 * 80.852$$
$$C = 20 + 56.5964$$
$$C = 76.596$$

**Note:** $C$, $I$, $T_X$, G and $Y_R$ are expressed in billions of dollors.

This is the consumption for the 1st year.

**Note:** Taking the current value of $Y_R$ as the new value of the lagged variable $Y_{R-1}$, and repeat same calculation for the next interval of year or time.

## 3.17 COBWEB MODEL

The Cobweb model was developed by the Hugorian economist, Nicholas Koldoar. It is an economic model of cyclical supply and demand in which there is a legs in producers response to change of price.

Consider following three points:

- If demand is less than the supply, the fluctuations would increase in magnitude per cycle, so diagram for each cycle would look like an outward spiral. This is called as an unstable Cobweb model.
- If the demand is more than the supply, fluctuations decrease in magnitude per cycle, so diagram for each cycle would look like an inward spiral. This is called as a stable cobweb model.
- Fluctuations may also remain of constant magnitude, so a diagram would produce a simple quadrangle. Such a result would be given from unit price of both supply and demand.

Points (*a*) and (*b*) is the combination of spiral and the supply and demand curves often looks like a cobweb, so we call this theory as cobweb model.

The most frequent "use" for first order linear differential equations supply *S* and demand *D*, to market price, *P*. Generally supply should be dependent upon the price from the previous marketing period. Again assuming market is cleared, the market model in distributed lag form is as follows:

Consider following equation:

$$D = a - b\,P \quad \text{...(1)}$$
$$S = c + d\,P_{-1} \quad \text{...(2)}$$
$$D = S \quad \text{...(3)}$$

Initial value of $P_0$ is given. This becomes value of $P_{-1}$ used to calculate the value of next interval and again this new value becomes $P_1$ and again calculate the value of next interval and so on.

Consider following initial values:

$$P_0 = 1.0$$
$$a = 12.4$$
$$b = 1.2$$
$$c = 1.0$$

$$d = 0.9$$

Equation can be modified as

$$D = a - b\,P_1$$
$$S = C + d\,P_0$$
$$D = S$$ We will calculate new value of $P_1$ with the help of $P_0$

$$a - b\,P_1 = c + d\,P_0$$
$$a - c = d\,P_0 + b\,P_1$$
$$a - c = 0.9 * 1 + 1.2 * P_1$$
$$12.4 - 1.0 = 0.9 + 1.2\,P_1$$
$$11.4 = 0.9 + 1.2\,P_1$$
$$11.4 - 0.9 = 1.2\,P_1 = 10.5/1.2 = 8.75$$

Now again by this $P_1$, we will calculate value for next interval and let new value is $P_2$.

$$11.4 = d\,P_1 + bP_2$$
$$11.4 = 0.9 \times 8.75 + 1.2 \times P_2$$
$$11.4 - 7.875 = 1.2\,P_2 \text{ or } P_2 = 2.9375$$

Similarly calculate following value of $P$.

| | | |
|---|---|---|
| $P_3 = 7.3$ | $P_9 = 5.75$ | $\mathbf{P}_{13} = 5.48$ |
| $P_4 = 4.027$ | $P_{10} = 5.18$ | |
| $P_5 = 6.5$ | $P_{11} = 5.61$ | |
| $P_6 = 4.640$ | $P_{12} = 5.29$ | $\mathbf{P}_{16}\ldots$ |
| $P_7 = 6.01$ | $P_{13} = 5.53$ | |
| $P_8 = 4.99$ | $P_{14} = 5.35$ | $\mathbf{P}_{17}\ldots$ |

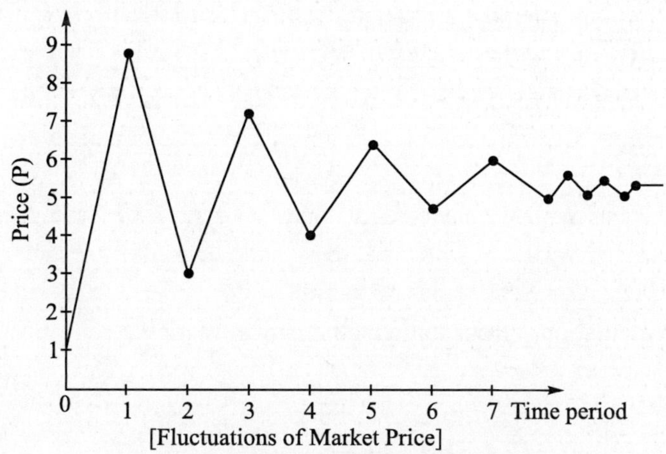


**Fig. 3.6.** Fluctuations of market price.

This is stable market because amplitude is continuously decreasing and after some intervals it becomes straight line. Cobweb graph is illustrated in Fig. 3.7 for the stable case just considered. Two straight lines plot the linear relationship representing supply and demand as we have plotted before in static mathematical model.

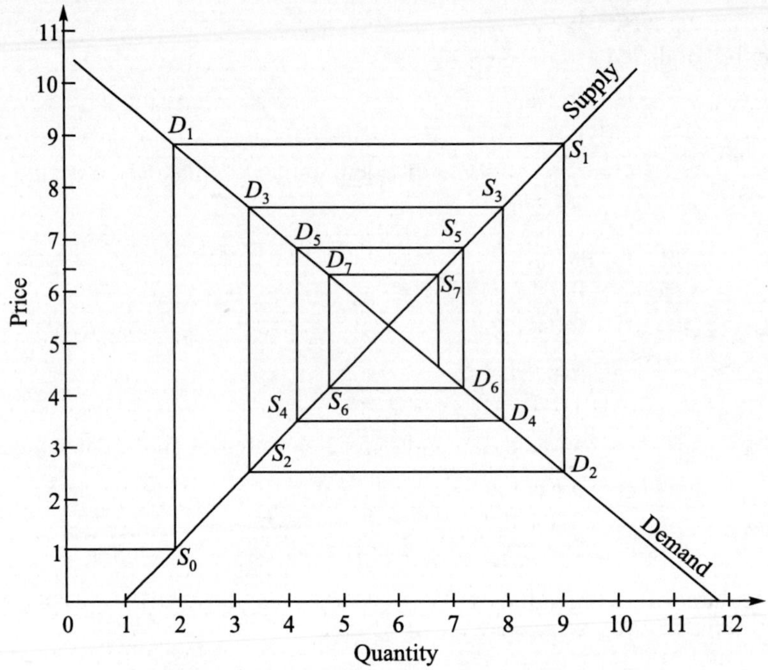


**Fig. 3.7.** Cobweb model of a market economy.

| | | |
|---|---|---|
| $P_0 = 0.1$ | $S_0 = 1.0 + 0.9 * 1 = 1.9$ | $D_0 = 12.4 - 1.2 * 1.0 = 11.2$ |
| $P_1 = 8.75$ | $S_1 = 1.0 + 8.75 * 1 = 8.87$ | $D_1 = 12.4 - 1.2 * 875 = 1.9$ |
| $P_2 = 2.94$ | $S_2 = 1.0 + 2.646 = 3.646$ | $D_2 = 12.4 - 1.2 * 2.94 = 8.87$ |
| $P_3 = 7.3$ | $S_2 = 1.0 + 0.9 * 7.3 = 7.57$ | $D_3 = 12.4 - 1.2 * 7.3 = 3.646$ |
| $P_4 = 4.0$ | $S_4 = 1.0 + 0.9 * 4.0 = 4.6$ | $D_2 = 12.4 - 1.2 * 4.0 = 7.6$ |
| $P_5 = 6.5$ | $S_5 = 1.0 + 5.85 = 6.85$ | $D_5 = 12.4 - 1.2 * 6.5 = 4.6$ |
| $P_6 = 4.64$ | $S_6 = 1.0 + 4.176 = 5.176$ | $D_6 = 12.4 - 1.2 * 4.64 = 6.83$ |
| and $P_7 = 6.01$ | $S_7 = 5.188$ | $D_7 = 5.196$, and so on |

Similarly at $P = 5.43$

$$S = 1.0 + 0.9 \times 5.43 \quad \text{and} \quad D = 12.4 - 1.2 * 5.43$$

$$S = 1.0 + 4.887 = 5.887 \quad \text{and} \quad D = 5.884$$

$$S = 5.88 \quad \text{and} \quad D = 5.88$$

Another way of calculation – from static mathematical model

$$P = \frac{a - c}{b + d}$$

$$P = \frac{12.4 - 1.0}{1.2 + 0.9} = \frac{11.4}{2.1} = 5.428 \quad \text{or} \quad 5.43.$$

**Example 5.** Draw market fluctuations of market price for following values

$P_0 = 5.0 \qquad a = 10.0 \qquad b = 0.9 \qquad c = -2.4 \qquad d = 1.2$

**Solution:** Given equation

$$D = a - bP_1$$
$$S = c + d\,P_0$$
$$S = D$$

$P_0 = 5.0$ then calculation of $P_1$

$$S = D$$
$$c + d\,P_0 = a - b\,P_1$$
$$-2.4 + 1.2 * 5 = 10.0 - 0.9\,P_1$$
$$3.6 = 10.0 - 0.9\,P_1$$

or
$$0.9\,P_1 = 6.4$$
$$P_1 = 7.11$$

Similarly

$P_2 = 4.3$ $\quad P_5 = 9.7038$ $\quad P_8 = 3.1$

$P_3 = 8.04$ $\quad P_6 = 0.839$ $\quad P_9 = 9.6$

$P_4 = 3.055$ $\quad P_7 = 12.65$

**At these Point ($P_0$, $P_1$, $P_7$) the Value of *S* and *D***

$P_0 = 5.0$ $\quad S_0 = 3.6$ $\quad D_0 = 5.5$

$P_1 = 7.11$ $\quad S_1 = 6.13$ $\quad D_1 = 3.60$

$P_2 = 4.3$ $\quad S_2 = 2.76$ $\quad D_2 = 6.13$

$P_3 = 8.04$ $\quad S_3 = 7.25$ $\quad D_3 = 2.76$

$P_4 = 3.055$ $\quad S_4 = 1.26656$ $\quad D_4 = 7.25005$

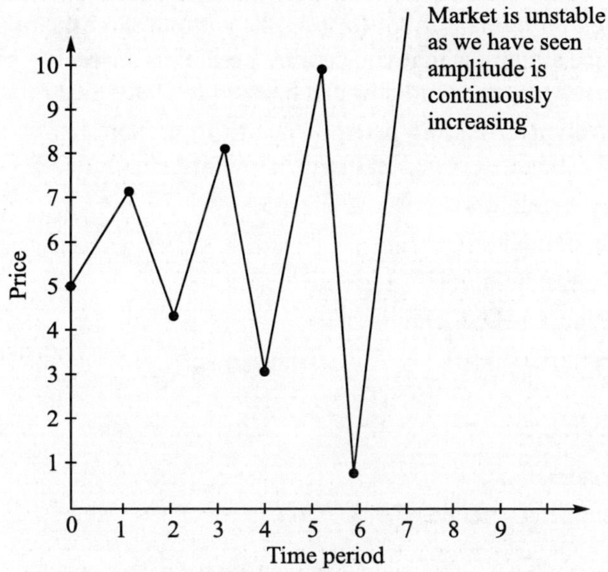


**Fig. 3.8.** Market fluctuations.

In above Fig. 3.8, market is **unstable,** as we have seen amplitude is continuously increasing.

## 3.18 APPLICATION AREAS OF SIMULATION [*UPTU 2005-06*]

There are following areas of simulation:

(*a*) **Simulation in Military System:** Military simulations, also known informally as war games, are models in which theories of warfare can be tested and refined without the need for actual hostilities. They exist in many different forms, with varying degrees of realism. In recent times, their scope has widened to include not only military but also political and social factor.

**(*b*) Simulation in Finance:** In finance, computer simulations are often used for scenario planning. Risk-adjusted net present value, for example is computed from well-defined but not always known (or fixed) inputs. By imitating the performance of the project under evaluation, simulation can provide a distribution of NPV over a range of discount rates and other variables.

**(*c*) Simulation in Engineering, Technology or Process:** Simulation is an important feature in engineering systems or any system that involves many process for example.

(*i*) In electrical engineering delay lines may be used to simulate propagation delay and phase shift caused by an actual transmission line.

(*ii*) In chemical engineering, for example, process simulation are used to give the process parameters immediately used for operating chemical plants, such as oil refineries.

**(*d*) Simulation in Flight System:** A flight simulator is used to train pilots on the ground. It permits a pilot to crash his simulated "aircraft" without being hurt. Flight simulators are often used to train pilots to operate aircraft in extremely hazardous situations, such as landings with no engines, or complete electrical or hydraulic failures. The most advanced simulators have high-fidelity visual systems and hydraulic motion systems. The simulator is normally cheaper to operate than a real trainer aircraft.

**(*e*) Simulation in Robotics:** A robotics simulator is used to create embedded applications for a specific (or not) robot without be dependent of the 'real' robot. In some case, these applications can be transferred on the real robot (or rebuilt) without modifications. Robotics simulators allow to reproduce situations that you cannot 'create' in the reality because cost, time or the 'unique' resource. Also, the simulator can be used to create fast robot prototyping.

**(*f*) Simulation in Marine:** Bearing resemblance to flight simulators, marine simulators train ships' personnel. The most common marine simulators include:

- Ship's bridge simulators
- Engine room simulators
- Cargo handling simulators
- Communication/GMDSS simulators

**Some other popular areas are:**

(*i*) Health care

(*ii*) Transportation systems

(*iii*) Construction systems

(*iv*) Computer system performance.

## 3.19 MONTE-CARLO COMMUTATION VS. STOCHASTIC SIMULATION

***[UPTU 2005-06, 2006-07]***

Monte-Carlo computation or technique was originated during World War-II, when this approach was applied to problems related to the development of the atomic bomb.

**This method is often used when:**

1. Problem is too difficult to solve analytically.
2. The model is complex and non-linear.
3. Problem take considerable amount of time to solve experimentally.

Due to above reasons, the problems is simulated on a high speed computer using random numbers. This technique is based on a gambling principle.

Monte-Carlo method is use to mean any computation that involves random numbers. However, we know that our problem can be **deterministic** or **stochastic.** Stochastic means randomness or dynamic.

So accepted meaning of Monte-Carlo is restricted only to those computations in which random numbers are used to obtain solutions of those problems which are inherently **deterministic.** And if problems are 'stochastic' then term **'stochastic simulation'** is used.

| Sl. No. | Monte-Carlo computation | Stochastic simulation |
|---|---|---|
| 1. | It is use to obtain solutions of problems which are deterministic in nature. | 1. It is use to obtain solutions of problems which are stochastic in nature. |
| 2. | We can determine exact outcome at any time. | 2. We cannot determine exact outcome at any time. |
| 3. | Applies to static model. | 3. Applies to dynamic model |
| 4. | Example: Evaluating the constant $\pi$ by making use of random numbers. | 4. Example: The inventory problem. |

**Deterministic Problem Through Random Numbers (or example of Monte-Carlo Simulation)**

Consider a quadrant of a unit circle (circle of radius one is called unit circle) as shown in Fig. 3.9. All points satisfying the equation.

$$x^2 + y^2 \leq 1 \quad x, y \geq 0 \text{ lie in this quadrant}$$

This equation can be written as:

$$y^2 \leq 1 - x^2$$

or

$$y \leq \sqrt{1 - x^2}$$

Suppose we generate a pair of uniform random numbers $U_1$ and $U_2$ in the range (0, 1) so this pair $(U_1, U_2)$ is acceptable when

$$U_2 \leq \sqrt{1 - U_1^2}$$

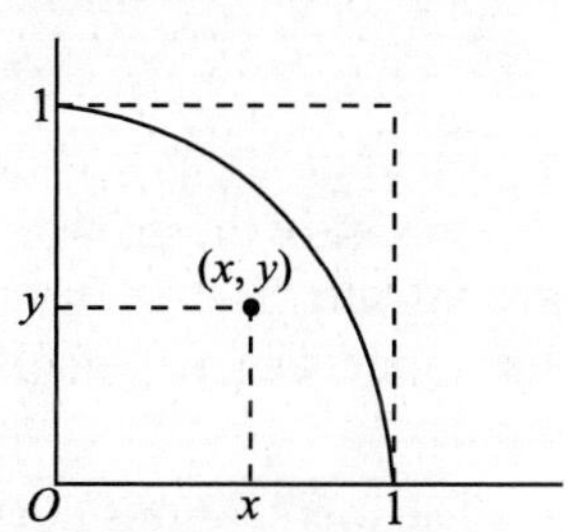


**Fig. 3.9.** Monte-Carlo evaluation of $\pi$.

Otherwise, we have to reject this pair; and generate and test another pair of uniformly distributed random numbers same as in rejection method.

If we have to generate a large number of random pairs and compute the ratio of the number of pairs accepted to those generated, the ratio will approach the area under the curve, which is $\pi/4$. Thus by using random numbers we have solved a completely deterministic problem.

Sometime Monte-Carlo applications are called as being simulation because so many simulation problem involve use of random numbers. However, simulation and Monte-Carlo are both numerical computation techniques. Simulation applies to **dynamic model** whereas Monte-Carlo method is used to **Static Models.**

## 3.20 NUMERICAL COMPUTATION TECHNIQUES FOR CONTINUOUS MODELS

Let we take multi-segment models (see Chapter 8) to explain the general numerical technique of simulation is based on a continuous model. A builder analysis that number of house sold (HS) depends upon the number of families who do not yet have a house. If the number of people without house is decreases then his house selling rate is also decreases. Let HH be the potential number of households, and HS be the number of families with houses.

The situation is represented as in figure:

where:

HH: horizontal line at HH is the total potential market for houses.

HS: HS be the number of families with houses and curve for $y$ indicates the number of houses sold increases with time.

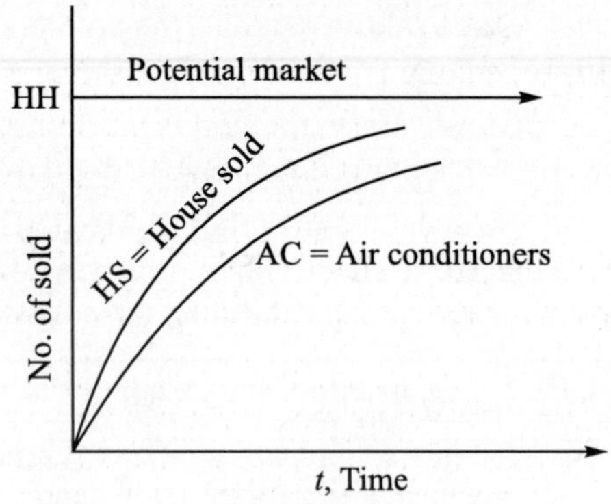


**Fig. 3.10**

The slope of the curve decreases as (HH – HS) gets less, this reflects the slowdown of sales as the market becomes saturated. According modified exponential growth model showing how the number of houses increases, with a given fixed limit HH, is given by following equation

$$\text{HS} = r_1\ (\text{HH} - \text{HS}) \qquad \text{HS} = 0 \text{ at } t = 0. \qquad ...(1)$$

Now consider a air conditioner manufacturer, who designed air conditioner for houses. His rate of sales depends upon the number of houses built. Let $AC$ be the potential number of air conditioners being installed. Then the unfilled market is the difference between the number of house sold (HS) and the number of installed air conditioners. Similarly, we have another equation of modified growth model for the number of air conditioners, but it has a limit of HS.

$$\text{AC} = r_2\ (\text{HS} - \text{AC}) \qquad \text{AC at } t = 0 \qquad ...(2)$$

The change of AC with time is also illustrated in Fig. 3.10. These two equations (1 and 2) represent a model of the growth of air conditioners sales. Due to simplicity, it is possible to solve above model analytically. But if various actual marketing conditions are expanded and included in above model then model can not be solved analytically. Let consider following conditions that make the model more complex:

- The market limit fluctuates upon population growth and economic conditions.
- Coefficient rate (such as $r_1$, $r_2$... etc.) could be influenced by the amount of money spent on advertising.

These conditions make the model complicated and complex and can be solved by continuous simulation techniques. Simulation techniques compute the output step-by-step. Let the computation is made at uniform intervals of time and that the calculation has already progressed to time $t_i$, when the two variables of the problem have the values $\text{HS}_i$ and $\text{AC}_i$. The calculation steps forward an interval $\Delta t$ to $t_{i+1} = t_i + \Delta t$ and the rates are assumed to be constant over the interval. The rates, can be interpreted as the amount of change per unit time. That is,

Rate of change of HS = $\Delta\text{HS}_i / \Delta t$ ...(3)

Rate of change of AC = $\Delta\text{AC}_i / \Delta t$ ...(4)

Apply equation (3) and (4) in equation (1) and (2).

$$\Delta\text{HS}_i / \Delta t = r_1\ (\text{HH} - \text{HS})$$

$$\Delta\text{HS}_i = r_1\ (\text{HH} - \text{HS})^* \ \Delta t \quad \text{and} \quad \Delta\text{AC}_i = r_2\ (\text{HS} - \text{AC})^* \ \Delta t$$

$\text{HS}_i$ and $\text{AC}_i$ are known, it is a simple to get the values of HS and AC at time $t_{i+1}$.

Repetition of the calculation using the new values of HS and AC produces the output at the end of the next interval. Calculation is equivalent to calculating the slope at each point and projecting a short straight line at that slope as shown in Fig. (3.11).

This method is very simple way of integrating differential equations numerically, but it is not very accurate method, unless small steps are used, compared with the rate at which the variable change.

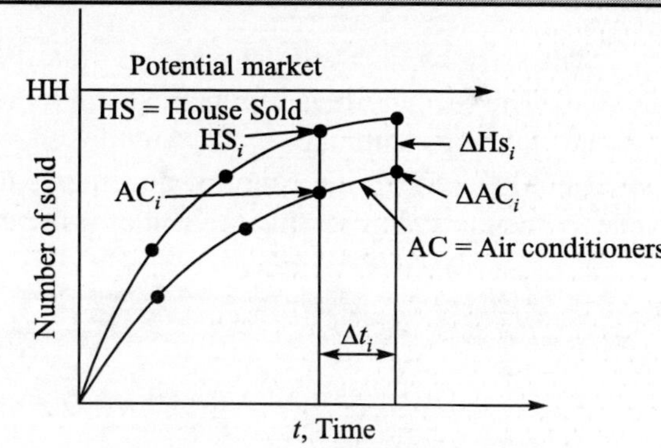


**Fig. 3.11**

Computational interval is generally adjusted in size to match the rate at which the variables are changing.

## 3.21 NUMERICAL COMPUTATION TECHNIQUES FOR DISCRETE MODELS

Let consider a example of a bank to understand the numerical computational technique of simulation with discrete model where a clerk starts his day with bulk of files to be processed. The time taken to process them varies. He works through beginning of each file and takes 5 minutes rest after processing a file in one or more than one hour since he began work or since he last had a break. We assume the times to process the files are given. We can also keep a count of how many files are left. And a count will be initially set to the number of files at the beginning of the day. We also assume that no documents arrive during the day. The count will be decremented for each completed task and work will be stop if value of count becomes zero. The computational involved can be organized as shown in table. The *i*th row corresponds to the *i*th file. The first column numbers the files. There are four columns giving various times in minutes.

**Table 3.7**

| File Number (*i*) | Start Time *Ts* | Work Time *Tw* | Finish Time *Tf* | Cumulative Time *Tc* | Break Flag *Tb* | Number of Jobs *Nj* |
|---|---|---|---|---|---|---|
| 1. | 0 | 40 | 40 | 40 | 0 | 45 |
| 2. | 40 | 20 | 60 | 60 | 5 | 44 |
| 3. | 65 | 10 | 75 | 10 | 0 | 43 |
| 4. | 75 | 25 | 100 | 35 | 0 | 42 |
| 5. | 100 | 35 | 135 | 70 | 5 | 41 |
| 6. | 140 | 15 | 155 | 15 | 0 | 40 |

Clerk started his work from time zero on file number 1. Processing time of file 1 is 40 hence he will complete its task in (0 + 40 = 40 minutes) mention in column four. The fifth column contains the cumulative time (*Tc*) since work started or since the last break. Sixth column represent as time break (*Tb*) that set as 5 minutes as a break after the $i^{th}$ file and if *Tc* is equal to greater than 60 minutes, otherwise set as 0. Or we can say: If $Tc \geq 60$ then $Tb = 5$ minutes else $Tb = 0$.

The file no.1 is finished at 45, with a cumulative time of 40. This is not long enough for a break, so the flag is set 0. The count, which was initialized to 45 jobs is dropped to 44.

The clerk works until there are no more files, or the time he finishes a file goes beyond some time limit. The second file is begun at 40 and requires 20 minutes for processing, which increase *Tc* as 60, so he will takes 5 minutes break. So third file will be starts as 65. The calculation continues in this manner until either, *Nj*, the count of documents to be processed, drops to zero, or the finish time *Tf*, reaches some limit representing the end of task period.

## SUMMARY

- "Simulation" is used to describe any procedure of establishing a model and deriving a solution numerically.
- System that changes with time and involve randomness are good candidate for simulation.
- Simulation can be used to experiment with new designs or policies prior to implementation, so as to prepare for what may happen.
- Analog simulation provides higher speed of simulation and the accuracy of analog simulation depends on the accuracy of component being used.
- Digital simulation provides slow speed of simulation compared to analog simulation and the results from a digital simulation is more accurate.
- There are following type of simulation:
  - ➢ Discrete and continuous simulation
  - ➢ Static and Dynamic simulation
  - ➢ Deterministic and stochastic simulation
  - ➢ Real time simulation hybrid simulation.
- A fighter aircraft wants to destroyed an enemy bomber and how it is possible, discuss in simulation of pure pursuit problem.
- A queueing system consists of one or more servers that provide service of some kind of arriving customers.
- Monte-Carlo simulation is defined to be a scheme emloying random numbers used to obtain solutions of problems which are inherently stochastic or deterministic.
- Advantage of simulation is in understanding how the system operates.
- Disadvantage of simulation is that, it is time consuming and expensive.
- Distributed lag models are the models which change only at fixed intervals of time.
- The cobweb model explains why prices in certain markets are subject to periodical fluctuations.
- Military, finance, flight, health care and transportation systems are various application areas of simulation.
- Monte-Carlo computation is often used when : problem is too difficult, model is complex and non-linear and take considerable amount of time to solve experimently.

## SOLVED QUESTIONS

**Q. 1. What is system simulation?**

**Ans.** System simulation is the technique of solving problems by the observation of the performance over time, of a dynamic model of the system.

**Q. 2. Differentiate between analytic method and numerical methods.**

**Ans.** Analytical methods directly produce general solutions while numerical methods produce solutions in steps; each step gives the solution for one set of conditions, and the calculation must be repeated to expand the range of the solution.

**Q. 3. What is lagged variable?**

**Ans.** If any variable that appears in the form of its current value and one or more previous intervals is said to be a lagged variable.

**Q. 4. Draw the Cobweb model for the following market value:**

$$\mathbf{P_0 = 5.0}$$
$$\mathbf{a = 10.0}$$
$$\mathbf{b = 0.9}$$
$$\mathbf{c = -2.4}$$
$$\mathbf{d = 1.2}$$

Assume the market is always cleared.

**Ans.** Market model in distributed lag form is as follows:

$$D = a - bP$$
$$S = c + dP_{-1}$$
$$D = S$$

or

$$D = a - b\,P_1 \qquad \text{given} \quad P_0 = 5.5$$
$$S = c + dP_0 \qquad a = 10.0$$
$$b = 0.9$$
$$a - b\,P_1 = c + dP_0 \qquad c = -2.4$$
$$a - c = dP_0 + bP_1 \qquad d = 1.2$$

Putting the values

$$10 - (-2.4) = 12.4 \times 5 + 0.9 \times P_1$$
$$12.4 = 6.0 + 0.9\,P_1$$
$$12.4 - 6.0 = 0.9\,P_1$$
$$P_1 = \frac{6.4}{0.9} = 7.11$$
$$\boxed{P_1 = 7.11}$$

Now on the basis of $P_1$ calculate $P_2$

$$12.4 = 1.2\,P_1 + 0.9 * P_2$$
$$12.4 = 1.2 * 7.11 + 0.9\,P_2$$
$$12.4 - 8.532 = 0.9\,P_2$$
$$P_2 = 3.868/0.9 = 4.29$$

Above process is repeated for $P_3$, $P_4$ and $P_7$, and so on.

**Calculation of *D* and *S***

At $P_0 = 5$, $D_0 = 10 - 0.9 \times 5 = 5.5$ and $S_0 = -2.4 + 1.2 * 5 = 3.6$

At $P_1 = 7.11$, $D_1 = 10 - 0.9 * 7.11 = 3.6$ and $S_1 = -2.4 + 1.2 * 7.11 = 6.13$

At $P_2 = 4.29$, $D_2 = 10 - 0.9 * 4.29 = 6.13$ and $S_2 = -2.4 + 1.2 * 4.29 = 2.75$

Other values are:

At $P_3 = 8.04$ $D_3 = 2.76$ $S_3 = 7.25$

At $P_4 = 3.05$ $D_4 = 7.25$ $S_4 = 1.26$

At $P_5 = 9.70$ $D_5 = 1.26$ $S_5 = 9.24$

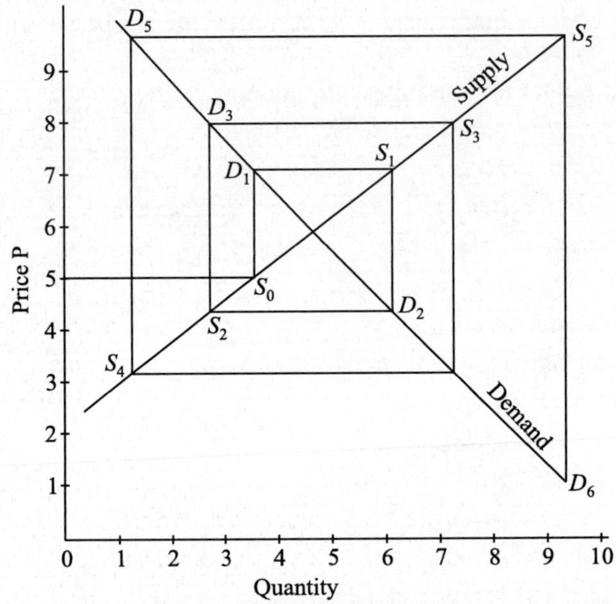


**Fig. 3.12.** Cobweb model.

From the above Fig. 3.12 we can say that there is no point where demand and supply can meet to each other or spiral will never reach the equilibrium price of 5.90 from which time the market will remain unchanged.

or We can say that the given market condition is unstable.

**Q. 5. Derive the mathematical conditions under which the market model of Fig. 3.7 is stable interpret the results graphically.**

**Ans.** We have to take the deviations of price and demand from the equilibrium values.

$$D = a - bP \quad \text{...(1)}$$

$$S = c + dP \quad \text{...(2)}$$

$$D = S \quad \text{...(3)}$$

In equation (3) put the values of equation (1) and (2)

$$a - bP = c + dP$$

$$a - c = dP + bP$$

$$a - c = P(d + b)$$

$$P = \frac{a - c}{b + d}$$

Given values

$$a = 12.4$$

$$b = 1.2$$

$$c = 1.0$$
$$d = 0.9$$
$$P = \frac{12.4 - 1.0}{1.2 + 0.9} = \frac{11.4}{2.1} = 5.428 \approx 5.43$$

At price $P = 5.43$, the graph or spiral will eventually reach the equilibrium price of 5.43, from which time the market will remain unchanged.

**Q. 6. Calculate different parameter for a single server queueing system when inter-arrival time and service times are given by the Table 3.8 below. [UPTU 2005-06]**

**Table 3.8.**

| Customer | Inter-arrival time | Service time |
|---|---|---|
| 1 | 0 | 20 |
| 2 | 10 | 15 |
| 3 | 15 | 10 |
| 4 | 35 | 5 |
| 5 | 30 | 15 |
| 6 | 10 | 15 |
| 7 | 5 | 10 |
| 8 | 5 | 10 |

**Calculate the following parameters:**

**(*a*) Average waiting time of customers (minutes)**
**(*b*) Probability that a customer has to wait in the queue**
**(*c*) Probability of idle time of the server**
**(*d*) Average service time of customer in minutes**
**(*e*) Average time between the arrivals**
**(*f*) Average waiting time of those who wait**
**(*g*) Average time customer spends in the system**
**(*h*) Probability of the server being busy.**

**Ans.** Simulation Table for Q. 6.

**Table 3.9**

| Cust. no. | Interarrival time | Arrival time | Service time | Time service begin | Time wait in queue | End of service | Time cust. spend in server | Idle time of system |
|---|---|---|---|---|---|---|---|---|
| 1 | 0 | 0 | 20 | 0 | 0 – 0 = 0 | 20 | 20 – 0 = 20 | 0 |
| 2 | 10 | 10 | 15 | 20 | 20 – 10 = 10 | 35 | 35 – 10 = 25 | 0 |
| 3 | 15 | 25 | 10 | 35 | 35 – 25 = 10 | 45 | 45 – 25 = 20 | 0 |
| 4 | 35 | 60 | 5 | 60 | 60 – 60 = 0 | 65 | 65 – 60 = 05 | 60 – 45 = 15 |
| 5 | 30 | 90 | 15 | 90 | 90 – 90 = 0 | 105 | 105 – 90 = 15 | 90 – 65 = 25 |
| 6 | 10 | 100 | 15 | 105 | 105 – 100 = 5 | 120 | 120 – 100 = 20 | 0 |
| 7 | 5 | 105 | 10 | 120 | 120 – 105 = 15 | 130 | 130 – 105 = 25 | 0 |
| 8 | 5 | 110 | 10 | 130 | 130 – 110 = 20 | 140 | 140 – 110 = 30 | 0 |
| | 110 | | 100 | | 60 | | 160 | 40 |

(*a*) Average waiting time $= \frac{60}{8} = 7.5$ min

(*b*) Probability that a customer has to wait in queue

$$= \frac{\text{No. of customer who has to wait}}{\text{Total no. of customer}}$$

$$= \frac{5}{8}$$

(*c*) Probability of idle server $= \frac{\text{Total run time of idle server}}{\text{Total run time of simulation}}$

$$= \frac{40}{140} = \frac{2}{7}$$

(*d*) Average service time $= \frac{\text{Total service time (min)}}{\text{Total no. of customer (min)}}$

$$= \frac{100}{8} = \frac{25}{5} = 12.5 \text{ min}$$

(*e*) Average time between arrivals $= \frac{\text{Sum of average time between arrivals}}{\text{Total no. of arrival} - 1}$

$$= \frac{110}{8-1} = \frac{110}{7} = 15.7 \text{ min}$$

(*f*) Average waiting time of those customer who has wait is

$$= \frac{\text{Total time waiting in queue}}{\text{Total no. of customer waiting}}$$

$$= \frac{60}{5} = 12 \text{ min}$$

(*g*) Average time customer spend in the system

$$= \frac{\text{Total time customer spend in system}}{\text{Total no. of customer}}$$

$$= \frac{160}{8} = 20 \text{ min}$$

(*h*) Probability of busy server $= 1 -$ probability of idle server

$$= 1 - \frac{2}{7} = \frac{5}{7}$$

**Q. 7. What is distributed lag models?**

**Ans.** Models that have following two properties:

(*a*) Changing only at fixed intervals of time.

(*b*) Basing current values of the variables on other current values and values that occurred in previous interval are called distributed lag models.

**Q. 8. Discuss simulation for business executive?**

**Ans.** There are many problems faced by business executive that can not be solved by standard operations research tools such as inventory and queueing theory. So business executive had to make decisions based solely on his intuition and experience. Now he can use computer simulation to make better and meaningful decisions, simulation has been used widely for inventory control, facility planning, production scheduling, etc.

**Q. 9. Discuss important steps in simulation and model building.**

**Ans. Important steps in simulation and model building.**

Following steps are there in simulation and model building.

**1. Define an achievable goal**

- Goals change with increasing insight.

**2. Put together a complex mix of skills on the team we need:**

- Knowledge of the system under investigation
- System analyst skills (model formulation)
- Model building skills (model programming)
- Data collection skills
- Statistical skills (Input data representation)
- More statistical skills (output data analysis)
- Even more statistical skills (design of experiment)
- Management skills (to get everyone pulling in the same direction).

**3. Involve the end user**

- Does anyone believe the results?
- Will anyone put the results into action?
- The end-user can do all of the above, but he must be convinced or he must believe it is model.

**4. Choose the appropriate simulation tools 3 alternative tool exist.**

- build model in a general purpose language
- build model in a general simulation language
- use a special purpose simulation package.

**5. Model the appropriate level of detail**

- define the boundaries of the system
- some characteristics of "the environment" may need to be included
- control the tendency to model in great detail will understood elements in the system.

**6. Start early to collect the necessary input data**

- too much data: we need techniques for reducing it to a form usable in our model
- too little data: we need information which can be represented by statistical distributions.

**7. Provide adequate and ongoing documentation**

- use self-documenting language
- insist on built-in user instructions
- set standards for coding cycle.

**8. Develop a plan for adequate model verification**

- Simulation provides something that no other technique does step by step tracing of the model execution.

This provides a very natural way of checking the internal consistency of the model.

**9. Develop a plan for model validation**

- "Doing the right thing."
- How do we know our model represents the system under investigation?

**10. Develop a plan for statistical output analysis:** Techniques for analysis—point estimate, internal estimate.

**Q. 10. What is simulator?**

**Ans.** A simulator is a computer program which contains both

(*a*) The internal behaviour of a real-world system

(*b*) The input processes which drive or control the simulated system.

The simulator output is a set of measurements concerning the observable reactions and performance of the system.

**Q. 11. Prove or disprove cobweb model is a special case of distributed lag model.**

[*UPTU 2005-06*]

**Ans.** In distributed lag model equations are represented as follows:

$$I = 2 + 0.1\, Y_{R-1} \quad \text{...(1)}$$

$$Y_R = 45.45 + 2.27\,(I + G) \quad \text{...(2)}$$

$$T_X = 0.2\, Y_R \quad \text{...(3)}$$

$$C = 20 + 0.7\,(Y_R - T_X) \quad \text{...(4)}$$

In cobweb model, the equations are given as

$$D = a - bP \quad \text{...(5)}$$

$$S = c + dP_{-1} \quad \text{...(6)}$$

where equation (1) and equation (6), both are same

$$I = 2 + 0.1\, Y_{R-1}$$

$$S = c + dP_{-1}$$

**(Note:** We know that $c$ and $d$ are constant which can replace by 2 and 0.1, respectively).

$Y_R$ is national income and $P$ is price, it means both variable $Y_R$ and $P$ is related to money. Cobweb model uses the concept of lag variable and we must have the initial value of $P$, *i.e.*, $P_0$.

This $P_0$ become value of $P_{-1}$ used to calculate the value of next interval and again this new value becomes $P_1$ and again calculate the value of next interval and so on.

So cobweb model is a special case of distributed lag model, also there is some other similarties between cobweb model and distributed lag model.

1. In both the model, initial value of $Y_R$ and $P_0$ are given.
2. Both model consist of the rules of linear algebraic equation example $I = 2.0 + 0.1\, Y_{R-1}$ and $S = c + dP_{-1}$.
3. Both the model consist the concept of lagged variable.
4. Both the model used mainly in econometric studies, in which some economic data are collected over fixed interval of time such as month, or a year.

# EXERCISES

1. Discuss the advantages and disadvantages of system simulation.
2. How is pure-pursuit problem simulated?
3. The pure-pursuit problem can also be simulated in terms of polar coordinates ($R$, $Q$) instead of rectagular coordinates ($x$, $y$). The relevant equations are:

$$\frac{dR}{dt} = VB \cdot \cos\theta - VF$$

$$R\frac{d\theta}{dt} = -VB \cdot \sin\theta$$

where $R$ is the distance between the bomber and fighter ($F$) and $\theta$ is the angle of the line from the fighter to the bomber. Write a C++ program using polar coordinates.
4. Use a cobweb model to investigate a market in which the supply and demand functions are:

$$D = \frac{17.91}{P^{\frac{1}{2}}} - 4.66$$

$$9S = 5.0\,(P_{-1} - 1)$$

Assume the market is always cleared.
5. Calculate different parameter for a single server quencing system when arrival time and service times are given by the Table 3.10 below:

**Table 3.10**

| Customer | Arrival time | Service time |
|---|---|---|
| 1 | 0 | 3 |
| 2 | 3 | 2 |
| 3 | 8 | 4 |
| 4 | 2 | 3 |
| 5 | 9 | 2 |
| 6 | 10 | 5 |
| 7 | 16 | 6 |

(*a*) Average waiting time of customers (minutes)
(*b*) Probability that a customer has to wait in the queue
(*c*) Probability of idle time of server
(*d*) Probability of the server being busy
(*e*) Average time between the arrivals
(*f*) Average waiting time of those who wait
(*g*) Average time customer spends in the system
(*h*) Average service time of customer in minutes.
6. In a field there are four animals—a dog, a mongoose, a snake, and a mouse. Dogs kill mongooses, mongooses kill snakes, and snakes kill mice. The speeds of the mouse, snake, mongoose and

dog are respectively, 7, 11, 17 and 30 km/hour. Simulate the chase with different starting positions to see which animal gets killed first.

7. How is inventory system is simulated?
8. Discuss following term:
   (*a*) Web based simulation
   (*b*) Real time simulation
   (*c*) Hybrid simulation
   (*d*) Monte-Carlo simulation
9. What are the application areas of simulation?
10. Compare the analytical method and simulation method of study.
11. What are the types of system simulation? Explain each with example.
12. Explain cobweb model with an example. Demonstrate its use.
13. Write the steps in a sound simulation study.
14. Prove or disprove cobweb model is a special case of distributed lag model.
15. What is single-server queueing problem? Simulate it with a suitable example.

■■

# 4

# Input Data Model

**INSIDE THIS CHAPTER**



## 4.1 INTRODUCTION

Input data provide the driving force for a simulation model. In the simulation of a queueing system, typical input data are the distributions of time between arrivals and service times. For an inventory system simulation, input data include the distributions of demand and lend time. For the simulation of a reliability system, the distribution of time-to-failure of a component is an example of input data. The development of system model has two distinct but closely related components.

- **Structural Modeling:** This is first component which describe the structure and operation of the system in term of entities, location, resources and processes. The result is a mathematical or logical representation, such as a queuing system model, that describes how the system behaves when executed.
- **Data Modeling:** This is second component. In this component, descriptive data required to execute the structural model are developed. These data include interarrival times, process service times, resource schedules and failure rates, entity travel times, and all of the other system data needed to describe the operation of the system quantitatively. The data model is use to generate the many random variate strems that must be supplied during the simulation of system model.

A key distinction in input modeling problems is the:

- **Presence of data:** When we have data, then we *fit* a model to the data. Good software is available for this.
- **Absence of data:** When no data are available then we have to creatively use what we can get to construct an input model.

The next subsection of 4.2.2 describes input modeling when data is available for modeling.

## 4.2 STEPS IN DEVELOPING INPUT DATA MODELS

Input models represent the uncertainty in a stochastic simulation. Random variates are generated based on the input models. Simulation outputs are determined by the input models. The quality of the output is no better than the quality of inputs. The quality of input models depends on the raw data, which is often collected by people who don't know simulation. There are following four steps in developing input data models.

**Step 1: Collect data** from the real system of interest. This often requires a substantial time and resource commitment. Unfortunately, in some situations it is not possible to collect data. When data are not available, expert opinion and knowledge must be used to make educated guesses.

**Step 2: Identify a probability distribution** to represent the input process. When data are available, this step typically begins by developing a frequency distribution, or histogram, of the data. Based on the frequency distribution and structural knowledge of the process, a family of distributions is chosen.

**Step 3: Choose parameters** that determine a specific instance of the distribution family, When data are available, these parameters may be estimated from the data.

**Step 4:** Evaluate the chosen distribution and the associated parameters **for goodness-of-fit**. The chi-square and the Kolmogorov-Smirnov tests are standard goodness of-fit tests. If not satisfied that the chosen distribution is a good approximation of the data, then the analyst returns is to the second step, chooses a different family of distributions, and repeats the procedure.

### 4.2.1 Collect Data

"GIGO," or "garbage-in, garbage-out" is a basic concept in computer science and it applies equally in the area of discrete system simulation. Many are fooled by a pile of computer output or a sophisticated animation, as if these were the absolute truth. Even if the model structure is valid, if the input data are inaccurately collected, inappropriately analyzed, or not representative of the environment, the simulation output data will be misleading and possibly damaging or costly when used for policy or decision-making garbage-in-garbage-out. Suggestions that may enhance and facilitate data collection:

1. **Plan ahead:** Begin by a practice or pre-observing session, watch for unusual circumstances. Planning is important even if data will be collected automatically (*e.g.*, via computer data collection) to ensure that the appropriate data are available. When data have already been collected by someone else, be sure to allow plenty of time for converting the data into a usable format.
2. **Analyze the collected data** to check adequacy. There is no need to collect superfluous data.
3. **Combine homogeneous data sets,** *e.g.*, successive time periods, during the same time period on successive days.
4. **Be aware of data censoring.** The quantity is not observed in its entirety, danger of leaving out long process times.
5. To determine whether there is a relationship between two variables, build a scatter diagram.
6. **Check for autocorrelation.** Autocorrelation may exist in successive time periods or for successive customers.
7. **Collect input data, not performance data.** Input data typically represent the uncertain quantities that are largely beyond the control of the system and will not be altered by changes made to improve the system. Output data, on the other hand, represent the performance of the system when subjected to the inputs, performance that we may be trying to improve.

### 4.2.2 Identifying the Underlying Statistical Distribution with Data

Identify a probability distribution to represent the input process. When data are available, this step typically begins by developing a frequency distribution, or histogram, of the data. Based on the frequency distribution and structural knowledge of the process, a family of distributions is chosen. The specific distribution within a family is specified by estimating its parameters.

#### Histograms

A frequency distribution or histogram is useful in determining the shape of a distribution. A histogram is constructed as follows:

(*a*) Divide the range of the data into equal intervals.

(*b*) Label the horizontal axis to conform to the intervals selected.

(*c*) Determine frequency of occurrences within each interval.

(*d*) Label the vertical axis so that the total occurrences can be plotted for each interval.

(*e*) Plot the frequencies on the vertical axis.

The number of class intervals depends on:

- The number of observations
- The dispersion of the data
- Suggested: the square root of the sample size.

#### Histogram for Continuous Data

Corresponds to the probability density function of a theoretical distribution and **for discrete data** Corresponds to the probability mass function. If there are few data points then it may be necessary to combine adjacent cells to eliminate the ragged appearance of the histogram.

**Example 4.1: For Discrete Data**

The number of vehicles arriving at the north-west corner of an intersection in a 5-minute period between 7:00 a.m. and 7:05 a.m. was monitored for five workdays over a 20-week period. Table 4.1 shows the resulting data. The first entry in the table indicates that there were 12. 5-minute periods during which zero vehicles arrived, 10 periods during which one vehicle arrived, and so on.

**Table 4.1**

| Arrivals per period | Frequency | Arrivals per period | Frequency |
|---|---|---|---|
| 0 | 12 | 6 | 7 |
| 1 | 10 | 7 | 5 |
| 2 | 19 | 8 | 5 |
| 3 | 17 | 9 | 3 |
| 4 | 10 | 10 | 3 |
| 5 | 8 | 11 | 1 |

Since the number of automobiles is a discrete variable, and since there are ample data, the histogram can have a cell for each possible value in the range of data. The resulting histogram is shown in Fig. 4.1.

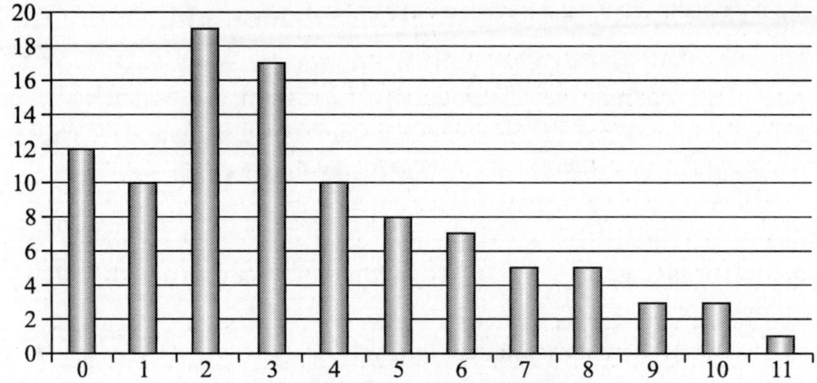


**Fig. 4.1** Histogram of number of arrivals per period.

**Example 4.2: For Continuous Data**

Life tests were performed on a random sample of 50 PDP-11 electronic chips at 1.5 times the normal voltage, and their lifetime (or time to failure) in day was recorded:

| | | | | | | | | | |
|---|---|---|---|---|---|---|---|---|---|
| 79.919 | 3.081 | 0.062 | 1.961 | 5.845 | 3.027 | 6.505 | 0.021 | 0.012 | 0.123 |
| 6.769 | 59.899 | 1.192 | 34.760 | 5.009 | 18.387 | 0.141 | 43.565 | 24.420 | 0.433 |
| 144.695 | 2.663 | 17.967 | 0.091 | 9.003 | 0.941 | 0.878 | 3.371 | 2.157 | 7.579 |
| 0.624 | 5.380 | 3.148 | 7.078 | 23.960 | 0.590 | 1.928 | 0.300 | 0.002 | 0.543 |
| 7.004 | 31.764 | 1.005 | 1.147 | 0.219 | 3.217 | 14.382 | 1.008 | 2.336 | 4.562 |

**Table 4.2**

| Chip life (Days) | Frequency | Chip life (Days) | Frequency |
|---|---|---|---|
| $0 <= x_i < 3$ | 23 | $30 <= x_i < 33$ | 1 |
| $3 <= x_i < 6$ | 10 | $33 <= x_i < 36$ | 1 |
| $6 <= x_i < 9$ | 5 | ... | ... |
| $9 <= x_i < 12$ | 1 | $42 <= x_i < 45$ | 1 |
| $12 <= x_i < 15$ | 1 | ... | ... |

Life time, usually considered a continuous variable. The histogram is prepared by placing the data in class intervals. The range of the data is rather large, from 0.002 day to 144.695 days. However, most of the values (30 of 60) are in the zero-to-5 day range. Using intervals of width three results in Table 4.2 and histogram prepared in Fig. 4.2.

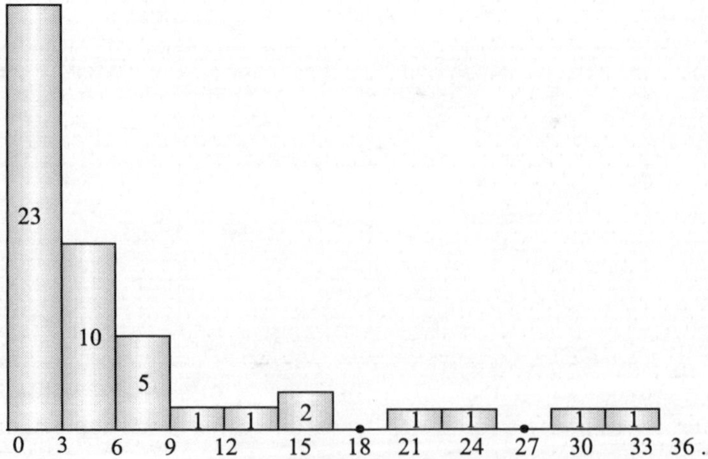


**Fig. 4.2** Histogram of chip life.

#### 4.2.2.2 Selecting the Family of Distributions

Now another important task is how to select family of distribution. A family of distribution is selected based on:

1. The context of the input variable.
2. Shape of the histogram.

**Frequently encountered distributions:**

**Easier to analyze:** Exponential, normal and poisson distribution.

**Harder to analyze:** Beta, gamma and Weibull.

Most probability distributions were invented to represent a particular physical situation. If we know the physical basis for a distribution, then we can match it to the situation we have to model. A number of examples are follows:

| Distribution | Basis of Selection |
|---|---|
| Binomial | No. of successes in $n$ trials. |
| Poisson | No. of independent events that occur in a fixed amount of time or space. |
| Normal | Distribution of a process that is the sum of a number of component processes. |
| Exponential | Time between independent events, or a process time that is memoryless. |
| Weibull | Time to failure for components. |
| Gamma | An extremely flexible distribution used to model nonnegative random variable. |
| Beta | An extremely flexible distribution used to model bounded (fixed upper and lower limits) random variable. |
| Discrete or continuous uniform | Models complete uncertainty, since all outcomes are equally likely. This is often overused when there are no data. |
| Triangular | A process for which only the minimum, most likely, and maximum values are known. |
| Empirical | Resamples from the actual data collected. |

Keep in mind that there is no true distribution for any stochastic input process. An input model is an approximation of reality, so the goal is to obtain a good approximation that yields useful from the simulation experiment.

#### 4.2.2.3 Quantile-Quantile Plots

Histogram is not as useful for evaluating the fit of the chosen distribution. When there are a small number of data points, say 30 or fewer, a histogram can be rather ragged. Our perception of fit depends on the widths of the histogram intervals. But even if the intervals are well chosen, grouping data into cell makes it difficult to compare a histogram to a continuous probability density function.

*A quantile-quantile (Q-Q)* plot is a useful tool for evaluating distribution fit that does not suffer from these problems.

If $X$ is a random variable with cdf $F$, then the $q$-quantile of $X$ is the $\gamma$ such that

$$F(\gamma) = P(X \le \gamma) = q, \text{ for } 0 < q < 1$$

When $F$ has an inverse, we can write $\gamma = F^{-1}(q)$.

Let $\{x_i,\ i = 1, 2, \ldots.., n\}$ be a sample of data from $X$ and $\{y_j, j = 1,2, \ldots, n\}$ be the observations in ascending order: where $y_1 \le y_2 \le \ldots \le y_n$. Let $j$ denote the ranking number and hence $j = 1$ for the smallest and $j = n$ for the largest. The *Q-Q* plot is based on the fact that $y_j$, is an estimate of the $\left(j - \frac{1}{2}\right)/n$ quantile of *X*. Or we can say,

$$y_j \text{ is approximately } F^{-1}\,(j - 1/2)/n$$

Now suppose that we have chosen a distribution with cdf *F* as a possible representation of the distribution of *X*. The plot of $y_j$ versus $F^{-1}\,(j - 1/2)\,/n)$ is

- Approximately a straight line if *F* is a member of an appropriate family of distributions
- The line has slope 1 if *F* is a member of an appropriate family of distributions with appropriate parameter values.

**Example 4.3:** Check whether the door installation times follows a normal distribution.

The observations are now ordered from smallest to largest: $y_j$ are plotted versus $F^{-1}\,((j - 0.5)\,n)$ where *F* has a normal distribution with the sample mean (99.99 sec) and sample variance $(0.2832)^2$ sec$^2$.

| j | Value | j | Value | j | Value | j | Value |
|---|---|---|---|---|---|---|---|
| 1. | 99.55 | 6. | 99.82 | 11. | 99.98 | 16. | 100.26 |
| 2. | 99.56 | 7. | 99.83 | 12. | 100.02 | 17. | 100.27 |
| 3. | 99.62 | 8. | 99.85 | 13. | 100.06 | 18. | 100.33 |
| 4. | 99.65 | 9. | 99.90 | 14. | 100.17 | 19. | 100.41 |
| 5. | 99.79 | 10. | 99.96 | 15. | 100.23 | 20. | 100.47 |

The ordered observed are then plotted versus $F^{-1}\,((j - 1/2)/20)$, for $j = 1, 2, ..., 20$ where *F* is the cdf of the normal distribution with mean 99.99 and variance $(0.2832)^2$ sec$^2$. To obtain a *Q-Q* plot. The plotted values are shown in Fig. 4.3.

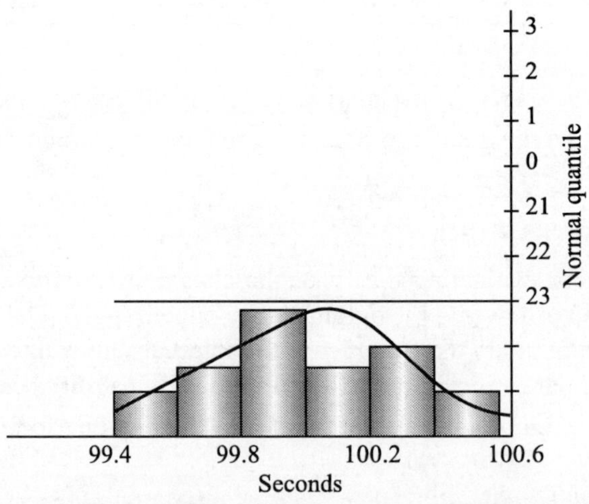


**Fig. 4.3** Histogram and *q-q* plot of the installation.

**Consider the following while evaluating the linearity of a Q-Q plot:**

- The observed values never fall exactly on a straight line.
- The ordered values are ranked and hence not independent, unlikely for the points to be scattered about the line.

- Variance of the extremes (largest and smallest values) are higher than the middle of the plots. Linearity of the points in the middle of the plot is more important than the linearity at the extremes.

**Q-Q plot can also be used to check homogeneity**

- It can be used to check whether a single distribution can represent two sample sets
- Given two random variables $X = (x_1, x_2 \ldots, x_n)$ and $Z = (z_1, z_2 \ldots, z_n)$, then plotting the ordered values of $X$ and $Z$ against each other reveals approximately a straight line if $X$ and $Z$ are well represented by the same distribution.

### 4.2.3 Parameter Estimation

Next step after selecting a family of distributions is choose parameters that determine a specific instance of the distribution family. When data are available, these parameters may be estimated from the data.

In a number of instances the sample mean , or the sample mean and sample variance, are used to estimate the parameters of the hypothesized distribution. Equations (4.1) and (4.2) can be used when discrete or continuous raw data are available.

The sample mean $X$, is defined by

$$\overline{X} = \left( \sum_{i=1}^{n} X_i \right) \Big/ n \qquad \ldots(4.1)$$

and the sample variance, $S^2$, is defined by

$$S^2 = \left( \sum_{i=1}^{n} X_i^2 - n\overline{X}^2 \right) \Big/ (n-1) \qquad \ldots(4.2)$$

If the data are discrete and grouped in a frequency distribution, Eq. (4.1) and Eq. (4.2) can be modified to provide for much greater computational efficiency. The sample mean can be computed by

$$\overline{X} = \left( \sum_{j=1}^{k} f_j X_j \right) \Big/ n \qquad \ldots(4.3)$$

And the sample variance, $S^2$, is defined by

$$S^2 = \left( \sum_{j=1}^{k} f_j X_j^2 - n\overline{X}^2 \right) \Big/ (n-1) \qquad \ldots(4.4)$$

where $k$ is the number of distinct values of $X$ and $f_i$ is the observed frequency of the value $X_i$ of $X$.

When raw data are unavailable (data are grouped into class intervals) the approximate sample mean and variance are:

$$\overline{X} = \frac{\sum_{j=1}^{c} f_j m_j}{n} \qquad \ldots(4.5)$$

$$S^2 = \frac{\sum_{j=1}^{c} f_j m_j^2 - n\overline{X}^2}{n-1} \qquad \ldots(4.6)$$

where $f_j$, is the observed frequency in the $j^{th}$ class interval, $m_j$ is the midpoint of the $j^{th}$ interval and $c$ is the number of class interval.

**Suggested Estimators**

Numerical estimates of the distribution parameters are needed to reduce the family of distribution to a specific distribution and to test the resulting hypothesis, common methods for fitting distributions are *maximum likelihood, method of moments,* and *least squares.*

**Remember:** There is no "true distribution" just waiting to be found.

Table 4.3 contains suggested estimators for distribution often used in simulation.

**Table 4.3**

| Distribution | Parameter(s) | Suggested Estimator(s) |
|---|---|---|
| Poisson | $\alpha$ | $\hat{\alpha} = \overline{X}$ |
| Exponential | $\lambda$ | $\hat{\lambda} = 1/\overline{X}$ |
| Gamma | $\beta, \theta$ | $\hat{\beta}$<br>$\hat{\theta} = 1/\overline{X}$ |
| Uniform on (0, $b$) | $b$ | $\hat{b} = \{(n+1)/n\} \times [\max(X)]$ (unbiased) |
| Normal | $\mu, \sigma^2$ | $\hat{\mu} = \overline{X}$<br>$\hat{\sigma}^2 = S^2$ (unbiased) |
| Weibull with $v = 0$ | $\alpha, \beta$ | $\hat{\beta}_0 = \overline{X}/S$<br>$\hat{\beta}_j = \hat{\beta}_{j-1} - f(\hat{\beta}_{j-1})/f'(\hat{\beta}_{j-1})$<br>Iterate untill convergence<br>$\alpha = \left\{(1/n)\sum_{i=1}^{n} X_i^{\hat{\beta}}\right\}^{1/\hat{\beta}}$ |

**Note:** If a parameter is denoted by $\alpha$, the estimator will be denoted by $\hat{\alpha}$.

### 4.2.4 Testing for Goodness of Fit

Evaluate the chosen distribution and the associated parameters for goodness-of-fit. The **chi-square** and **Kolmogorov-Smirnov** tests are standard goodness-of-fit tests. If not satisfied that the chosen distribution is a good approximation of the data, then the analyst returns to the second step, chooses a different family of distributions, and repeats the procedure.

No single correct distribution in a real application exists. If very little data are available, it is unlikely to reject any candidate distributions if a lot of data are available, it is likely to reject all candidate distributions.

### Chi-Square Test

This test is valid for large sample sizes, for both discrete and continuous distributional assumptions when parameters are estimated by maximum likelyhood. The test procedure begins by arranging the $n$ observations into a set of $k$ class intervals or cells. The test statistic is given by

$$\chi_0^2 = \sum_{i=1}^{k} (O_i - E_i)^2 / E_i$$

where $O_i$ is the observed frequency in the $i$th class interval and $E_i$ is the expected frequency in that class interval.

The hypotheses are:

$H_0$: The random variable, $X$, conforms to the distributional assumption with the parameter(s) given by the parameter estimate(s).

$H_1$: The random variable, $X$, does not conform.

The critical value $\chi^2_{\alpha,\, k-s-1}$ is found in standard table. The null hypothesis, $H_0$ is rejected if

$$\chi_0^2 > \chi^2_{\alpha,\, k-s-1}$$

If an $E_i$ value is too small, it can be combined with expected frequencies in adjacent class intervals. The corresponding O$i$ values should also be combined and $k$ should be reduced by one for each cell that is combined.

Although there is no general agreement regarding the minimum size of $E_i$, values of 3, 4 and 5 have been widely used. In general, minimum expected frequency of 5 was suggested.

The $\chi_0^2$ approximately follows the chi-square distribution with $k$-$s$-1 degrees of freedom, where $s$ represents the number of parameters of the estimated distribution. For uniform distribution has $s = 0$, for Poisson distribution has $s = 1$, and normal distribution has $s = 2$.

**Table 4.4:** Recommendations for number of class intervals for continuous data

| Sample size ($n$) | Number of class intervals ($k$) |
|---|---|
| 20 | Do not use the chi-square test |
| 50 | 5 to 10 |
| 100 | 10 to 20 |
| > 100 | $\sqrt{n}$ to $n/5$ |

**Example 4.4: (Chi-square test applied to Poisson Assumption)**

In the previous example, the vehicle arrival data were analyzed. Since the histogram of the data, shown in Fig. 4.2, appeared to follow a Poisson distribution, the parameter $\hat{\alpha} = 3.64$ was determined. Thus, the following hypotheses are formed:

$H_0$: The random variable is Poisson distributed

$H_1$: The random variable is not Poisson distributed

The pmf for the Poisson distribution was given:

$$P(x) = \begin{cases} (e^{-\alpha}\, \alpha^x)/x\,!, & x = 0, 1, 2 \ldots \\ 0, & \text{otherwise} \end{cases} \qquad \ldots(4.7)$$

For $\alpha = 3.64$, the probabilities associated with various values of $x$ are obtained using equation (4.7) with the following results.

$p(0) = 0.026$ $\quad p(3) = 0.211$ $\quad p(6) = 0.085$ $\quad p(9) = 0.008$
$p(1) = 0.096$ $\quad p(4) = 0.192$ $\quad p(7) = 0.044$ $\quad p(10) = 0.003$
$p(2) = 0.174$ $\quad p(5) = 0.140$ $\quad p(8) = 0.020$ $\quad p(11) = 0.001$

the value of $E_1$ is given by $np_0 = 100(0.26) = 2.6$ and so on for $E_i$.

Since $E = 2.6 < 5$, $E_1$ and $E_2$ are combined. In that case $O_1$ and $O_2$ are also combined and $k$ is reduced by one. The last five class intervals are also combined for the same reason, and $k$ is further reduced by four.

**Table 4.5**

| $x_i$ | Observed frequency $O_i$ | Expected frequency $E_i$ | $(O_i - E_i)^2/E_i$ |
|---|---|---|---|
| 0 | 12 } 22 | 2.6 } 12.2 | 7.87 |
| 1 | 10 | 9.16 | |
| 2 | 19 | 17.4 | 0.15 |
| 3 | 17 | 21.1 | 0.80 |
| 4 | 10 | 19.2 | 4.41 |
| 5 | 8 | 14.0 | 2.57 |
| 6 | 7 | 8.5 | 0.26 |
| 7 | 5 } 17 | 4.4 } 7.6 | 11.62 |
| 8 | 5 | 2.0 | |
| 9 | 3 | 0.8 | |
| 10 | 3 | 0.3 | |
| 11 | 1 | 0.1 | |
| | 100 | 100.0 | 27.68 |

The calculated $\chi_0^2$ is 27.68. The degrees of freedom for the tabulated value of $\chi_0^2$ is $k - s - 1 = 7 - 11 = 5$. Here, $s = 1$, since one parameter was estimated from the data. At $\alpha = 0.05$ the critical value $\chi^2_{0.05,\,5}$ is 11.1. Thus, $H_0$ would be rejected at level of significance 0.05. The analyst must now search for a better fitting model or use the empirical distribution of the data.

**Chi-Square Test with Equal Probabilities**

If continuous distributional assumption is being used, class intervals that are equal in probability rather than equal in width of interval should be used. If using equal probability then $p_i = \dfrac{1}{k}$.

Since $$E_i = np_i \geq 5$$

$$\Rightarrow \quad n/k \geq 5 \quad \text{(substitution of } p_i = \frac{1}{k}\text{)}$$

and solve for $k$ yields

$$k \leq n/5 \qquad \ldots(4.8)$$

Equation (4.8) was used in determining the recommended maximum number of class intervals in Table 4.4.

**Example 4.5:** Since the histogram of the data, shown in Fig. 4.2 (histogram of chip life) appeared to follow an exponential distribution, the parameter $\hat{\lambda} = 1\backslash\overline{X} = 0.084$ was determined. Thus, the following hypotheses are formed:

$H_0$: The random variable is exponentially distributed

$H_1$: The random variable is not exponentially distributed.

In order to perform the chi-square test with intervals of equal probability the endpoints of the class intervals must be determined. The number of intervals should be less then or equal to $n/5$. Here $n = 50$, so that $k \leq 10$. In Table 4.4, it is recommended that 7 to 10 class intervals be used. Let $k = 8$, then each interval will have probability $p = 0.125$. The endpoints for each interval are computed from the *cdf* for the exponential distribution, as follows:

$$F(a_i) = 1 - e^{-\lambda a_i} \qquad \text{...(4.9)}$$

where $a_i$ represents the endpoint of the $i$th interval, $i = 1, 2, ..., k$. Since $F(a_i)$ is the cumulative area from zero to $a_i$, $F(a_i) = ip$, so Equation 4.9 can be written as

$$ip = 1 - e^{-\lambda q_i}$$

or

$$e^{-\lambda a_i} = 1 - ip$$

Taking the logarithm of both sides and solving for $a_i$ gives a result for the endpoints of $k$ equiprobable intervals for the exponential distribution, namely,

$$a_i = \{-1/\lambda\} \ln (1 - ip),\ i = 0, 1, ..., k \qquad \text{...(4.10)}$$

Regardless of the value of $\lambda$, equation (4.10) will always result in $a_0 = 0$ and $a_k = \infty$.

With $\lambda = 0.084$ and $k = 8$, $a_1$ is determined from equation (4.10) as

$$a_1 = \{-1/0.084\} \ln (1 - 0.125) = 1.590$$

Continued application of equation (4.10) for $i = 2, 3, ... 7$ results in $a_2, ..., a_7$ as 3.425, 5.595, 8.252, 11.377, 16.503, and 24.755. Since $k = 8$, $a_8 = \infty$. The first interval is [0, 1.590], the second interval is [1.590, 3.425], and so on. The expectation is that 0.125 of the observations will fall in each interval. The observations, expectations, and the contributions to the calculated value of $\chi_0^2$ are shown in Table 4.6.

**Table 4.6**

| Class Interval | Observed Frequency ($O_i$) | Expected Frequency ($E_i$) | $(O_i - E_i)^2/E_i$ |
|---|---|---|---|
| [0, 1.590] | 19 | 6.25 | 26.01 |
| [1.590, 3.425] | 10 | 6.25 | 2.25 |
| [3.425, 5.595] | 3 | 6.25 | 1.69 |
| [5.595, 8.252] | 6 | 6.25 | 0.01 |
| [8.252, 11.677] | 1 | 6.25 | 4.41 |
| [11.677, 16.503] | 1 | 6.25 | 4.41 |
| [6.503, 24.755] | 4 | 6.25 | 0.81 |
| [24.755, ∞] | 6 | 6.25 | 0.01 |
| | 50 | 50 | 39.6 |

The calculated value of $\chi^2$ is 39.6. The degree of freedom are given by $k - s - l = 8 - 1 - 1 = 6$. At $\alpha = 0.05$, the tabulated value of $\chi^2_{0.05,\,6}$ is 12.6. Since $\chi_0^2 > \chi^2_{0.05,\,6}$, the null hypothesis is rejected. (The value at $\chi^2_{0.01,\,6}$ is 16.8, so the null hypothesis would also be rejected at level of significance $\alpha = 0.01$).

### Kolmogorov-Smirnov Test

The chi square test requires that the data be placed in class interval in case of continuous distribution, this grouping is arbitrary. A hypothesis may be accepted when data are grouped in one way but rejected when grouped another way. Also, distribution of the chi-square test statistics is known only approximately. As a result of these considerations, goodness-of-fit tests, other than the chi-square are desired. This test is based on:

- Examining a *q-q* plot.
- Very useful when you don't want to estimate any parameter from data.
- When the sample size is small.

The Kolmogorov-Smirnov test is defined by:

$H_0$: The data follow a specified distribution.

$H_a$: The data do not follow the specified distribution.

**Test:** The Kolmogorov-Smirnov test statistic is defined as statistics

$$D = \max_{l \le i \le n} [i/N - U_i, U_i - (i-1)/N]$$

The distribution being tested must be a continuous distribution (*i.e.,* no discrete distributions such as the binomial or Poisson), and it must be fully specified (*i.e.,* the location, scale, and shape parameters cannot be estimated from the data).

The hypothesis regarding the distributional form rejected if the test statistic, *D*, is greater than the critical value obtained from a standard table.

### p-Values and "Best Fits"

The outcome of the tests depends on the significance level. Traditional significance level are 0.01, 0.05, 0.1. Sometimes you reject a hypothesis for some value but the test valid for another one to that case it is important to find out the significance level.

The *p*-value is the significance level at which one should just rejected for the given value of the test. Therefore, a large value for *p-value* indicates a **good fit** and small *p*-value suggests a **poor-fit**.

With respect to testing distributions:

- Test many possible distributions
- Select the one with the largest *p*-value

Many software use *p*-value as the ranking measure to automatically determine the "best fit". There are several things to keep in mind:

1. Software may not know about the physical basis of the data distribution families it suggests may be inappropriate.
2. Close conformance to the data does not always lead to the most appropriate input model.
3. *p-value* does not say much about where the lack of fit occur
4. Recommended: always inspect the automatic selection using graphical methods.

## 4.3 SELECTING INPUT MODELS WITHOUT DATA

In this case the modeler must be resourceful in choosing input model and must carefully check the sensitivity of results to the chosen models. There are a number of ways to obtain information about a process even if no data is available:

1. Use references that describes model for the input data.
2. Get expert opinion-help to characterize the inputs.

3. Use physical limitations to bound the problem (*e.g.,* computer data entry cannot be faster than a person can type).
4. Exploit the nature of the process-choose a distribution that is most closely related to the underlying input process.

In general, uniform, triangular and beta distributions are used when nothing else is known. A useful refinement is obtained when a minimum, maximum and one or more "breakpoint" can be given. A breakpoint is an intermediate value and a probability of being less than or equal to that value. The following example explains how breakpoint used.

**Example 4.6:** Production planning simulation, Input of sales volume of various products is required. Salesperson of product XYZ-123 says that: No fewer than 1000 units and no more than 5,000 units will be sold. According her experience she believes there is 90% chances of selling more than 2,000 units, a 25% chance of selling more than 3500 units and only 1% chance of selling more than 4500 units.

Following table 4.7 summarizes this information. Translating these information into a cumulative probability of being less than or equal to those goals for simulation output.

**Table 4.7**

| *i* | Interval (Sales) | Cumulative Frequency, *ci* |
|---|---|---|
| 1. | $1000 \le x \le 2000$ | 0.10 |
| 2. | $2000 < x \le 3500$ | 0.75 |
| 3. | $3500 < x \le 4500$ | 0.99 |
| 4. | $4500 < x \le 5000$ | 1.00 |

## 4.4 MULTIVARIATE AND TIME-SERIES INPUT MODELS

Before describing the multivariate and time-series input models, we have to understand the covariance and correlation.

### 4.4.1 Convariance and Correlation

Let $X_1$ and $X_2$ be two random variables, and let $\mu_i = E(X_i)$ $\sigma_i^2 = \text{Var}(X_i)$ be the mean and variance of $X_i$, respectively. The covariance and correlation are measures of the linear dependence between $X_1$ and $X_2$. Relationship between $X_1$ and $X_2$ is described by the model.

$$(X_1 - \mu_1) = \beta(X_2 - \mu_2) + \varepsilon$$

**Some important points:**

- $\varepsilon$ is a random variable with mean 0 that is independent of $X_2$.
- If $(X_1 - \mu_1) = \beta(X_2 - \mu_2)$, then this model is perfect.
- If $X_1$ and $X_2$ are statistically independent, then $\beta = 0$ and the model is of no value.
- A positive value of $\beta$ indicates that $X_1$ and $X_2$ tend to be above or below their means together, while a negative value of $\beta$-indicates that they tend to be on opposite sides of their means.

The covariance between $X_1$ and $X_2$ is defined to be:

$$\text{Cov}(X_1, X_2) = E[(X_1 - \mu_1)(X_2 - \mu_2)] = E(X_1X_2) - \mu_1\mu_2 \qquad \text{...(4.11)}$$

A value of Cov $(X_1, X_2) = 0$ implies $\beta = 0$ in our model of dependence, while cov $(X_1, X_2) < 0$ $(> 0)$ implies $\beta < 0$ $(> 0)$.

The covariance can take any value between $-\infty$ and $\infty$. The correlation standardizes the covariance to be between $-1$ and 1.

$$\rho = \text{corr}\,(X_1, X_2) = \frac{\text{cov}\,(X_1, X_2)}{\sigma_1\sigma_2} \quad \text{...(4.12)}$$

Again, a value of $\text{corr}(X_1, X_2) = 0$ implies $\beta = 0$ in our model, while $\text{corr}(X_1, X_2) < 0\ (> 0)$ implies $\beta < 0\ (> 0)$. The closer $\rho$ is to $-1$ or 1, the stronger the linear relationship is between $X_1$ and $X_2$.

### 4.4.2 Multivariate Input Models

If $X_1$ and $X_2$ each are normally distributed, then dependence between them can be modeled by the bivariate normal distribution with parameters $\mu_1$, $\mu_2$, $\sigma_1^2$, $\sigma_2^2$ and $\rho = \text{corr}\,(X_1, X_2)$. Estimation of $\mu_1$, $\mu_2$, $\sigma_1^2$ and $\sigma_2^2$ was already described. To estimate $\rho$, suppose that we have $n$ independent and identically distributed pairs $(X_{11}, X_{21})$, $(X_{12}, X_{22})$ ... $(X_{1n}, X_{2n})$. Then the sample convariance is

$$\widehat{\text{cov}}\,(X_1, X_2) = \frac{1}{n-1}\sum_{j=1}^{n} (X_{1j} - \overline{X}_1)(X_{2j} - \overline{X}_2)$$

$$= \frac{1}{n-1}\left(\sum_{j=1}^{n} X_{1j}\, X_{2j} - n\overline{X}_1\overline{X}_2\right) \quad \text{...(4.13)}$$

where $\overline{X}_1$ and $\overline{X}_2$ are the sample means. The correlation is estimated by

$$\hat{\rho} = \frac{\widehat{\text{cov}}\,(X_1, X_2)}{\hat{\sigma}_1\hat{\sigma}_2} \quad \text{...(4.14)}$$

where $\hat{\sigma}_1$ and $\hat{\sigma}_2$ are the sample standard deviations.

The following alogorithm can be used to generate bivariate normal random variables:

1. Generate $Z_1$ and $Z_2$, independent standard normal random variables.
2. Set $X_1 = \mu_1 + \sigma_1 Z_1$
3. Set $X_2 = \mu_2 + \sigma_2\left(\rho Z_1 + \sqrt{1-\rho^2}\, Z_2\right)$.

The bivariate normal distribution will not be appropriate for all multivariate input modeling problem.

**Example 4.7:** An inventory simulation includes the lead time and annual demand for LCD. An increase in demand results in an increase in lead time, since the final assembly of the LCD must be made according to the specifications of the purchaser. Therefore, rather than treat lead time and demands as independent random variables a multivariate input model should be developed.

Let $X_1$ represent average lead time to deliver (in months), and $X_2$ the annual demand, for LCD. The following data were available on demand and lead time for the last 10 years:

| **Lead Time** | 6.5 | 4.3 | 6.9 | 6.0 | 6.9 | 6.9 | 5.8 | 7.3 | 4.5 | 6.3 |
|---|---|---|---|---|---|---|---|---|---|---|
| **Demand** | 103 | 83 | 116 | 97 | 112 | 104 | 106 | 109 | 92 | 96 |

Following table represent the way to calculate sample mean and standard deviation.

| Lead Time | $(X_i - \overline{X})$ | $(X_i - \overline{X})^2$ | Demand |
|---|---|---|---|
| 6.5 | 0.36 | .1296 | 103 |
| 4.3 | –1.84 | 3.386 | 83 |
| 6.9 | 0.76 | .5776 | 116 |
| 6.0 | –0.14 | .0196 | 97 |
| 6.9 | 0.76 | .5776 | 112 |
| 6.9 | 0.76 | .5776 | 104 |
| 5.8 | –0.34 | .1156 | 106 |
| 7.3 | 1.16 | 1.346 | 109 |
| 4.5 | –1.64 | 2.689 | 92 |
| 6.3 | 0.16 | 256 | 96 |
| $\sum_{i=1}^{10} X_1 = 61.4$ | | 9.442 | $\Sigma X_2 = 1018$ |

$n = 10$

Sample mean of lead time

$$\overline{X}_1 = \frac{\sum_{i=1}^{n} X_i}{n} \Rightarrow \frac{61.4}{10} = 6.14$$

Sample mean of demand

$$\overline{X}_2 = \frac{1018}{10} = 101.8$$

For standard deviation of lead time $\hat{\sigma}_1$:

$$\hat{\sigma}_1 = S = \sqrt{\frac{1}{n-1}\sum_{i=1}^{n}(X_i - \overline{X})^2} \text{ or } \sqrt{\frac{1}{9}\times 9.442} \Rightarrow \text{ or } \sqrt{1.049}$$

$$\hat{\sigma}_1 = 1.02$$

Similarly calculate standard deviation for demand $\hat{\sigma}_2$ which is 9.93.

To estimate the correlation we need

$$\sum_{j=1}^{10} X_{1j}\, x_{2j} = 6328.5$$

Therefore, $\widehat{\text{Cov}}$ = [6328.5 – (10) (6.14) (101.80)]/(10 – 1) = 8.66 and

$$\hat{\rho} = \frac{8.66}{(1.02)(9.93)} = 0.86$$

Clearly, lead time and demand are strongly dependent. Before accepting this model, lead time and demand should be checked individually to see if they are well represented by normal distribution.

### 4.4.3 Time-Series Input Models

If $X_1, X_2, X_3, \ldots$ is a sequence of identically distributed, but dependent and covariance-stationary random variables, then there are a number of time series models that can be used to represent the process. We will describe two models that have the characteristic that the auto correlations take the form

$$\rho_h = \text{corr}\,(X_t, X_{t+h}) = \rho^h$$

for $h = 1, 2, \ldots$ here the lag-$h$ autocorrelation decreases geometrically as the lag increases so that observations for apart in time are nearly independent.

**AR(1) Model**

Consider the time-series model $X_t$ where each $X_t$ is normally distributed.

$$X_t = \mu + \phi(X_{t-1} - \mu) + \varepsilon \qquad \ldots(4.15)$$

for $t = 2, 3, \ldots$, where $\varepsilon_2, \varepsilon_3 \ldots$ are independent and identically distributed with mean 0 and variance $\sigma_\varepsilon^2$, and $-1 < \phi < 1$. If the initial value $X_t$ is chosen correctly, then $X_1, X_2, \ldots$ are all normally distributed with mean $\mu$, variance $\sigma_\varepsilon^2/(1 - \phi^2)$, and

$$\rho_h = \phi^h$$

for $h = 1, 2, \ldots$

This time series model is called the auto regressive order –1 model, or AR(1) for short Estimation of the parameter $\phi$ can be obtained from the fact that

$$\phi = \rho' = \text{corr}\,(X_t, X_{t+1})$$

the lag-1 autocorrelation. Therefore to estimate $\phi$, we first estimate the lag-1 auto covariance by

$$\widehat{\text{cov}}\,(X_t, X_{t+1}) = \frac{1}{n-1}\sum_{t=1}^{n-1}(X_t - \overline{X})(X_{t+1} - \overline{X})$$

$$= \frac{1}{n-1}\left(\sum_{t=1}^{n-1} X_t\, X_{t+1} - (n-1)\overline{X}^2\right) \qquad \ldots(4.16)$$

and the variance $\sigma^2 = \text{var}\,(X)$ by the usual estimator $\hat{\sigma}^2$.

Then $$\hat{\phi} = \frac{\widehat{\text{cov}}\,(X_t, X_{t+1})}{\hat{\sigma}^2}$$

Finally, estimate $\mu$ and $\sigma_\varepsilon^2$ by $\hat{\mu} = \overline{X}$ and

$$\hat{\sigma}_\varepsilon^2 = \hat{\sigma}^2(1 - \hat{\phi}^2)$$

respectively.

The following algorithm generates a stationary AR(1) time series, given values of the parameters $\phi$, $\mu$ and $\sigma_\varepsilon^2$.

1. Generate $X_t$ from the normal distribution with mean $\mu$ and variance $\sigma_\varepsilon^2/(1 - \phi^2)$. Set $t = 2$.
2. Generate $\varepsilon_t$ from the normal distribution with mean 0 and variance $\sigma_\varepsilon^2$.
3. Set $X_t = \mu + \phi(X_{t-1} - \mu) + \varepsilon_t$
4. Set $t = t + 1$ and go to 2.

### EAR(1) Model

For this model each $X_t$ is exponentially distributed. This time services model is called the exponential auto regressive order-1 model, or EAR(1) for short. Only auto correlations greater than 0 can be represented by this model. Consider the time-series model

$$X_t = \begin{cases} \phi X_{t-1}, & \text{with probability } \phi \\ \phi X_{t-1} + \varepsilon_t, & \text{with probability } 1-\phi \end{cases}$$

for $t = 2, 3 ...$, where $\varepsilon_2, \varepsilon_3, ...$ are independent and identically exponentially distributed with mean $1/\lambda$ and $0 \le \phi < 1$. If the initial value $X_1$ is chosen appropriately, then $X_1, X_2, ...$ are all exponentially distributed with mean $1/\lambda$ and

$$\rho_h = \phi^h \quad \text{for } h = 1, 2, ...$$

Estimation of the parameters proceeds as for the AR(1) by setting $\hat{\phi} = \hat{\rho}$, the estimated lag-1 autocorrelation, and setting $\hat{\lambda} = 1/\overline{X}$.

The following algorithm generates values of the parameter $\phi$ and $\lambda$.

1. Generate $X_t$ from the exponential distribution with mean $\frac{1}{\lambda}$. Set $t = 2$.
2. Generate $U$ from the uniform distribution on [0, 1]. If $U \le \phi$, then set $X_t = \phi X_{t-1}$ otherwise, generate $\varepsilon_t$ from the exponential distribution with mean $\frac{1}{\lambda}$ and set $X_t = \phi X_{t-1} + \varepsilon_t$.
3. Set $t = t + 1$ and go to 2.

**Difference between AR(1) and EAR(1)**

| AR(1) | EAR(1) |
|---|---|
| 1. Auto regressive order-1 model | 1. Exponential autoregressive order-1 model |
| 2. Each $X_t$ is normally distributed | 2. Each $X_t$ is exponentially distributed. |
| 3. Model is represented by following equation: $X_t = \mu + \phi(X_{t-1} - \mu) + \varepsilon_t$ | 3. Model is represented by following equation: $X_t = \begin{cases} \phi X_{t-1}, & \text{with probability } \phi \\ \phi X_{t-1} + \varepsilon_t, & \text{with probability } 1-\phi \end{cases}$ |

## SUMMARY

- Input models provide the driving force for a simulation model. The quality of the output no better than the quality of inputs.
- Following steps in input data modeling:
  1. Collecting the raw data
  2. Identifying the underlying statistical distribution
  3. Estimating the parameter
  4. Testing for goodness of fit
- Suggested estimators were given for distributions of ten used in simulation. In a number of instances, there are function of the sample mean and sample variance.

- The $q$–$q$ plot is a useful graphical method in the testing of distributional hypotheses. The K-S, chi$^2$, and Anderson–Darling goodness of fit tests can be applied to many distributional assumptions.
- Multivariate input models:
  1. Fixed, finite number of random variables.
  2. For example, lead time and annual demand for an inventory model.
  3. An increase in demand results in lead time increase, hence variable are dependent.
- Time-series input models.
  1. Infinite sequence of random variables.
  2. For example time between arrivals of orders to buy and sell stock.
  3. Buy and sell orders tend to arrive in bursts, hence, times between arrivals are dependent.

## SOLVED QUESTIONS

**Q. 1. Discuss advantages and limitation of Kolmogorov-Smirnov test.**

**Ans.** An attractive feature of this test is that the distribution of the K-S test statistic itself does not depend on the underlying cumulative distribution function being tested. Another advantage is that it is an exact test (the chi-square goodness) of fit test depends on an adequate sample size for the approximation to be validity. Despite these advantages, the K-S test has several important limitations.

1. It only applies to continuous distributions.
2. It tends to be more sensitive near the center of the distribution all the than at the tails.
3. Perhaps the most serious limitation is that the distribution must be fully specified. That is, if location, scale, and shape parameters are estimated from the data, the critical region of the K-S test is no long valid. It typically must be determined by simulation.

Due to limitations 2 and 3 above, many analysts prefer to use Anderson-Darling goodness-of-fit test. However, the Anderson-Darling test is only available for a few specified distributions.

**Q. 2. How do you perform three Chi-square test (the test goodness-of-fit the test of independence, and the test of homogeneity) that are used to analyzed categorical responses.**

**Ans.** There are three **Chi-Square tests** presented the tests of goodness of fit, independence and **homogeneity**. All three tests are based on the **Chi-Square statistic.**

$$\chi_0^2 = \sum \frac{(O_i - E_i)^2}{E_i}$$ where $O_i$ is the observed count and

$E_i$ is the expected count under the corresponding null hypothesis.

The **goodness-of-fit test** answers the questions. "Do the data fit well compared to a specified distribution for one categorical veriable" This test considers one categorical variable and assesses whether the protortion of sampled observations falling into each category matches well enough to the null distribution the given problem. For instance, the null distribution might be specified a manufacturer, a product label, or the results of a previous study. The null hypothesis for the goodness of fit test specifies this null distribution which describes the population proportion of observation in each category.

The **test of homogeneity** answers the question, Do two or more populations have the same distribution for one categorical variable?" This test considers one categorical variable and assesser whether this variable is distributed the same in two (or more) difference populations. The null hypothesis for the test of homogeneity is that the distribution of the categorical variable is the same for the two or more) populations.

The **test of independence** answers the question, "Are two factors (or variables) independent for a population under study?" This test considers two categorical variable and assesses whether there is a relationship between these two variables for a single population. The null hypothesis for the test of independence is that the two categorical variables are independent (that is, they are not related) for the population of interest.

**Q. 3. How to identify distribution?**

**Ans.** There are three ways to identify distributions:

1. Histograms
2. Selecting a family of distributions
3. Quantile-quantileplots.

**Q. 4. Assumes that the raw data on chip like shown in example 4.2 were lost. However, the data shown in Table 4.2 are still available determine approximate values of $\overline{X}$ and $S^2$.**

**Ans.** Equation 4.5 and 4.6 are used as this is the case when no data is available

$$f_i = 23, \quad m_i = 1.5, \quad f_2 = 10, \quad m_2 = 4.5, ..., c = 49 \text{ (no. of class interval)}$$

$$\sum_{j=1}^{49} f_j m_j = 614$$

or

$$\sum_{j=1}^{49} f_j m_j^2 = 37{,}226.5 \text{ with } n = 50$$

$$\overline{X} = \frac{614}{50} = 12.28 \qquad \text{(By equation 4.5)}$$

$$S^2 = \frac{37{,}226.5 - 50(12.28)^2}{49} = 605.849 \qquad \text{(By equation 4.6)}$$

$$S = 24.614$$

However, applying equation 4.1 and 4.2 to the original data in example 4.2 results in $\overline{X} = 11.894$ and $S = 24.953$. Thus, if the raw data are either discarded or lost, some inaccuracies may result.

**Q. 5. Consider following table of Arrivals per periods and frequence (see Table-4.1) and compute sample mean $\overline{X}$ and sample variance $S^2$ and standard deviation $S$.**

**Ans.** From table 4.1, $n = 100$ (Addition of total no. of frequency)

$$f_1 = 12, \quad X_1 = 0$$
$$f_2 = 10, \quad X_2 = 1 \ldots$$

$$\sum_{j=1}^{k} f_j X_j = f_1X_1 + f_2X_2 \ldots f_{12}X_{12}$$

$$= 12 \times 0 + 10 \times 1 + 19 \times 2 + 17 \times 3 + 10 \times 4 + 8 \times 5 + 7 \times 6 + 5 \times 7 + 5 \times 8 + 3 \times 9 + 3 \times 10 + 1 \times 11$$

$$= 0 + 10 + 38 + 51 + 40 + 40 + 42 + 35 + 40 + 27 + 30 + 11 = 364$$

Similarly

$$\sum_{j=1}^{k} f_j X_j^2 = 2080$$

From equation (4.3), $\overline{X} = \frac{364}{100} = 3.64$

From equation (4.4), $S^2 = \frac{2080 - 100\,(3.64)^2}{99} = 7.63$

The sample standard deviation is = $\sqrt{S^2} = \sqrt{7.63} = 2.76$

**Q. 6. The following data are randomly generated from a gamma distribution:**

| | | | |
|---|---|---|---|
| 1.691 | 1.437 | 8.221 | 5.976 |
| 1.116 | 4.435 | 2.345 | 1.782 |
| 3.810 | 4.589 | 5.313 | 10.90 |
| 2.649 | 2.432 | 1.581 | 2.432 |
| 1.843 | 2.466 | 2.833 | 2.361 |

**Determine the maximum-likelihood estimation β.**

**Ans.**

| Data (X) | ln (X) |
|---|---|
| 1.691 | 0.52532007 |
| 1.116 | 0.109750864 |
| 3.81 | 1.337629189 |
| 2.649 | 0.97418221 |
| 1.843 | 0.411394679 |
| 1.437 | 0.362557607 |
| 4.435 | 1.489527616 |
| 4.589 | 1.523662135 |
| 2.432 | 0.888713964 |
| 2.4466 | 1.902597405 |
| 8.221 | 2.106691856 |
| 2.345 | 0.852285402 |
| 5.313 | 1.670156647 |
| 1.581 | 0.458057558 |
| 2.833 | 1.041336221 |
| 5.976 | 1.787751448 |
| 1.782 | 0.577736329 |
| 10.9 | 2.388762789 |
| 2.432 | 0.888713964 |
| 2.361 | 0.859085258 |
| Sum = 70.212/20 | Sum = 21.35591321 |

$$\overline{X} = 70.212/20 = 3.5106$$

$$\ln \overline{X} = 1.25578696$$

$$\text{Sum of } \ln X = 21.35591$$

$$1/20 \ln X = 21.35591/20 = 1.067795$$

$$M = 1.25578696 - 1.067795661$$
$$= 0.1879913$$

$$1/M = 5.31939503$$

By interpolation in standard table

$$\beta = 2.815$$

$$\theta = 0.3848516$$

**Q. 7. Discuss some method of estimation.**

**Ans. Common approaches**

1. Lest squares (LS)
   - Minimize the sum of square of the deviation
   - Used in linear regression
2. Maximum likelihood estimation (MLE)
   - Estimate parameter that is most consistent with the data
   - Widely used
3. Minimum Chi-squared
   - Variable made of frequencies
   - Compare observed with expected frequencies.

**Q. 8. The percentage rates of return on 10 items are 18.8, 27.9, 21.0, 6.1, 37.4, 5, 22.9, 1.0, 3.1, and 8.3. Estimate the parameters of lognormal model of this data.**

**Ans.** It is known that distribution is lognormal from table 4.3 parameters of lognormal data are $\mu$ and $\sigma^2$ and suggested estimators are

$$\hat{\mu} = \overline{X} \text{ (After taking ln of the data)}$$

$$\hat{\sigma}^2 = S^2 \text{ (After taking ln of the data)}$$

$$n = 10$$

See following table based on above question.

| Item No. | Rate of return ($x_i$) | ln($x_i$) | ln ($X_i$) − $\overline{X}$ | (ln ($X_i$) − $\overline{X}$)$^2$ |
|---|---|---|---|---|
| 1. | 18.8 | 2.9 | 0.6 | .36 |
| 2. | 27.9 | 3.3 | 1.0 | 1 |
| 3. | 21.0 | 3.0 | 0.7 | .49 |
| 4. | 6.1 | 1.8 | −0.5 | .25 |
| 5. | 37.4 | 3.6 | −1.3 | 1.69 |
| 6. | 5 | 1.6 | −0.7 | .49 |
| 7. | 22.9 | 3.1 | 0.8 | .64 |
| 8. | 1 | 0 | −2.3 | 5.29 |
| 9. | 3.1 | 1.1 | −1.2 | 1.44 |
| 10. | 8.3 | 2.1 | −0.2 | .04 |
| | $\sum_{i=1}^{10} X_i$ | 22.5 | | 11.64 |

$$\overline{X} = \frac{\sum_{i=1}^{10} x_i}{n} = \frac{22.5}{10} = 2.25$$

$$= 2.3 \text{ (decimal one place)}$$

$$S^2 = \frac{\sum_{i=1}^{n} \left(\ln(x_i) - \overline{X}\right)^2}{n-1} \Rightarrow \frac{11.64}{9} = 1.29 \approx 1.3$$

Hence $\hat{\mu} = 2.3$ and $\hat{\sigma}^2 = 1.3$

## EXERCISES

1. Records pertaining to the montly number of job-related at an underground coal mine were being studied by a federal agency. The values for the past 100 months were as follows:

| Injuries per month | Frequency of occurrence |
|---|---|
| 0 | 35 |
| 1 | 40 |
| 2 | 13 |
| 3 | 6 |
| 4 | 4 |
| 5 | 1 |
| 6 | 1 |

Apply the chi-square test to these data to test the hypothesis that the underlying distribution is Poisson. Use a level of significance of $\alpha = 0.05$.

2. Write the steps in developing input data model.
3. Differentiate between AR(1) and EAR(1) model.
4. We have an 80% chance of landing the contract with Big corp. If we land the contract, then there is only a 10% chance that the initial orders will arrive first quarter; a 50% chance they will start the second quarter; and a 40% chance they will start in quarter 3 Sales will be 250K in each quarter we produce. How to model the sales for the year related to Big corp?
5. Why do we need input models at all?
6. What is a "Good-Fit"?
7. How would you collect data which is to be used as input to simulation model?
8. Discuss the methods which chosen distribution is evaluated for goodness of fit test.
9. In case there is no data but the distribution is known then, how would you genorate samples to be used as input to simulationi model? Generate one set of samples inputs using of the method.

5

# Process and Validation of Modeling and Simulation

INSIDE THIS CHAPTER



The most important and difficult tasks facing a model developer is the verification and validation of the simulation model. It is the responsibility of the model developer to work closely with the end users throughout the period of development and validation.

This chapter describes process of model building and verification and validation of simulation model. It also describes the process of simulation.

## 5.1 PROCESS OF MODEL BUILDING

The facility or process of interest is usually called a system, and in order to study it scientifically we often have to make a set of assumptions about how it works. These assumptions, which usually take the form of mathematical or logical relationships, constitute a model that is used to try to gain some understanding of how corresponding system behaves. There are following steps to build a model of the system.

### Step 1

The first step in model building consists of observing the real system and the interactions among its various components and collecting data on its behaviour. Persons familiar with the system, or any subsystem, should be questioned to take advantage of their special knowledge. Operators, technicians, repair and maintenance personnel, engineers, supervisors, and managers understand certain aspects of the system which may be unfamiliar to others. As model development proceeds, new questions may arise, and the model developers will return to this step of learning true system structure and behaviour.

### Step 2

The second step is the construction of a conceptual model—a collection of assumptions on the components and the structure of the system, plus hypotheses on the values of model input parameters.

Figure 5.1 shows comparison between real system to the conceptual model.

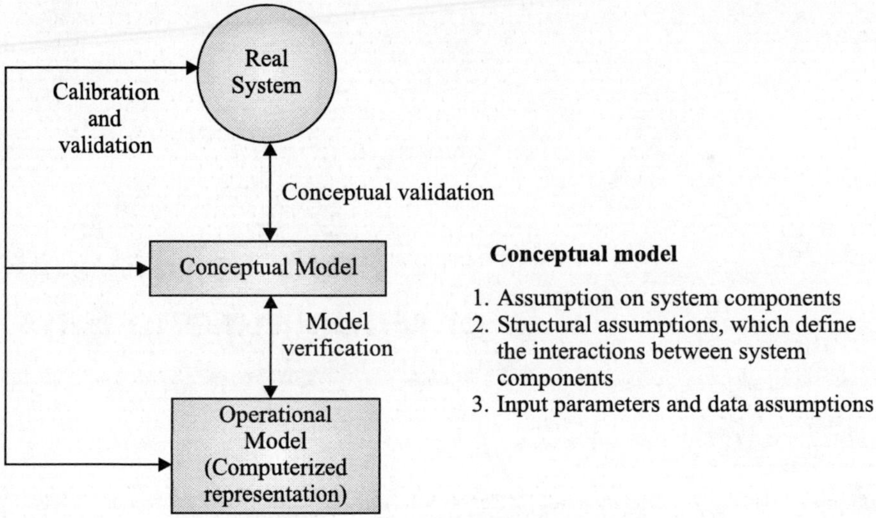


**Fig. 5.1.** Model building, verification, and validation.

### Step 3

Step 3 is the translation of the operational model into a computer recognizable form—the computerized model.

A model building is not a linear process with three steps. The model builder will return to each of these steps many times while building, verifying, and validating the model.

Above Figure 5.1 shows model building process in which the need for verification and validation causes continual comparison of the real system to the conceptual model and to the operational model, and repeated modification of the model to improve its accuracy.

## 5.2 GOODNESS-OF-FIT TESTS

Goodness-of-fit tests provide helpful guidance for evaluating the suitability of a potential input model. It is especially important to understand the effect of sample size. If very little data are available, then a goodness-of-fits test is unlikely to reject any candidate distribution; but if a lot of are available, then a goodness-of-fit test will likely reject all candidate distributions. Therefore, failing to reject a candidate distribution should be taken as one piece of evidence in favour of that choice, while rejecting an input model is only one piece of evidence against the choice.

Chi-square test and Kolmogorov-Smirnov test are introduced in this section to hypothesis about distributional forms of input data.

### Chi-Square Test

One procedure for testing the hypothesis that a random sample of size $N$ of the random variable $X$-follows a specific distributional form is the chi-square goodness-of-fit test.

This test is valid for large sample sizes, for both discrete and continuous distributional assumptions. When parameters are estimated by maximum likelihood. The test procedure begins by arranging the N observations into a set of $n$ class intervals or cells. Formula of this test statistic is given by

$$\rho^2 = \sum_{i=1}^{n} \frac{(0_i - E_i)^2}{E_i}$$

where $O_i$ = Observed frequency in the $i$th class interval.

$E_t$ = Expected frequency in that class interval [for example, Please refer Section 9.5.1.]

### Kolmogorov-Smirnov Goodness-of-Fit Test

The chi-square goodness-of-fit test can accommodate the estimation of parameters from the data with a resultant decrease in the degrees of freedom. The chi-square test requires that the data be placed in class intervals, and in the case of a continuous distributional assumption, this grouping is arbitrary. Changing the number of classes and the interval width affects the value of the calculated and tabulated chi-square. It is also possible that a hypothesis may be accepted when the data are grouped one way but rejected when grouped another way. As a result of these considerations, goodness-of-fit tests, other than the chi-square, are desired.

The Kolmogorov-Smirnov test was presented in Section 9.5.1 to test for the uniformity of numbers and again it is also use to perform the gap test. Both of these uses fall into the category of testing for goodness-of-fit. The Kolmogorov-Smirnov test is particularly useful when sample sizes are small and when no parameters have been estimated from the data.

The Kolmogorov-Smirnov test has been modified so that it can be used in several situations when the parameters are estimated from the data. The computation of the test statistic is the same, but different tables of critical values are used. Different tables of critical values are required for different distributional assumptions.

## 5.3 VARIANCE REDUCTION TECHNIQUES

Most simulation experiments are conducted for obtaining some important characteristic of a population, for example, average waiting time per customer in a queue.

Suppose we make $m$ observations, $x_1$ $x_2$,... $x_m$ and obtain the mean $\bar{x}$.

Standard deviation $\sigma_{\bar{x}}$ is represented by closeness parameter of sample mean $\bar{x}$ to population mean $\mu$. If sample size $m$ is larger, then $\bar{x}$ will be closer to $\mu$. This phenomenon is called stochastic convergence. We used a single long run of length $m$, or $b$ replication of $m/b$ observations each. The standard deviation of the sample mean could always be expressed as

$$\sigma_{\bar{x}} = \frac{1}{m}$$

If we have to reduce the standard deviation by a factor $R$, then we have to make $R^2$ as many observations. Number of techniques have been developed for reducing the variance without increasing the sample-size. The following are some of the better-known variance-reduction technique.

Variance Reduction Technique →
- → Antithetic Sampling
- → Correlated Sampling
- → Importance Sampling
- → Control Variates
- → Stratified Sampling
- → Russian Roulette and Splitting

1. **Antithetic Sampling:** Suppose we have made two simulation runs, each of size $m/2$ to estimate the mean value of some random variables of our interest (let assume $m$, as an even no). Suppose the values of the observations in two runs are

$$p_1, p_2 \ldots, p_{m/2}$$

and $$q_1, q_2 \ldots, q_{m/2}$$

The averages of these two replications are

$$\bar{p} = \frac{2}{m}\sum_{i=1}^{m/2} p_i$$

$$\bar{q} = \frac{2}{m}\sum_{i=1}^{m/2} z_i$$

and their variances are

$$\sigma_{\bar{p}}^2 = \sigma_{\bar{q}}^2 = \frac{2\sigma^2}{m}$$

$\sigma^2$: The variance of the parent population.

The grand mean of the two replications is

$$\bar{x} = \frac{1}{m}\sum_{i=1}^{m/2} (p_i + q_i) = \frac{\bar{p}+\bar{q}}{2}$$

*i.e.*, the average of the two averages.

The variance of this grand mean $\bar{x}$ is

$$\sigma_{\bar{x}}^2 = \frac{1}{m}\sum_{i=1}^{m/2} [(p_i - \bar{p})^2 + (q_i - \bar{q})^2],$$

which turns out to be

$$\sigma_{\bar{x}}^2 = \frac{\sigma_{\bar{p}}^2 + \sigma_{\bar{q}}^2}{4} + \frac{1}{2}\text{Cov}(\bar{p}, \bar{q}), \qquad \ldots(1)$$

Cov $(u, v)$ denotes the covariance of random variables $u$, $v$.

If we introduce a negative correlation between each of the $m/2$ pairs of the observations (*i.e.*, when the value of $p_i$ is high and the value of $q_i$ will be low, and vice versa), the second term in equation (1), will be negative, and the variance of the mean of the $m/2$ pairs of observations

$$\frac{p_1 + q_1}{2}, \frac{p_2 + q_2}{2}, \ldots \frac{p_{m/2} + q_{m/2}}{2}$$

will be less than the variance of the mean of one continuous run of $m$ observations. This method of purposely introducing a negative correlation between pairs of observations, for obtaining a more statistically reliable answer, is called *antithetic sampling.*

2. **Correlated Sampling:** This method is useful when we are interested in estimating the differences in average characteristics of two alternate designs or strategies.

   Let $\bar{X}$ and $\bar{Y}$ are samples which estimate, say, the average waiting time per customer using two different queue disciplines. We will estimate the difference between the two performances. If completely random samplings wear used to obtain $\bar{X}$ and $\bar{Y}$, then the difference is given by

$$\bar{D} = \bar{X} - \bar{Y}$$

and the confidence we have in this value of the difference is determined by its variance, which is

$$\sigma_{\bar{D}}^2 = \sigma_{\bar{X}}^2 + \sigma_{\bar{Y}}^2$$

If $X$ and $Y$ were statistically independent. If the same set of random numbers are used to generate the stochastic variates in the two experiments, $\bar{X}$ and $\bar{Y}$ are likely to be positively correlated. In that case variance of the difference could be much smaller because

$$\sigma_{\bar{D}}^2 = \sigma_{\bar{X}}^2 + \sigma_{\bar{Y}}^2 - 2\,\text{Cov}\,(X, Y)$$

So we can reduce the variance at no increase in computer running time. This method is opposite to the antithetic sampling technique.

3. **Importance Sampling:** This method is use in a sampling procedure which deliberately distorts the original distribution so that a larger proportion of samples is drawn from the interval of importance. An appropriate weighting function is used to multiply the sampled data which makes correction for the distortion introduced.
4. **Control Variates:** When simulating a complex system then we have seen often that simplified version of the system can be handled analytically and the value of parameter can be calculated, corresponding to the parameter of interest in simulation.

   **Example:** Suppose we want to simulate a multi-server queuing system with a complicated priority scheme, and the parameter of our interest is the mean queue length. We find a simplified model and find that a FCFS, single-server queueing system is mathematically handled. The average queue length is given by following equation (1), if the inter arrival and the service times follow the negative exponential distribution.

$$\text{Average queue length} = \frac{\rho^2}{1-\rho} \qquad \text{...(2)}$$

**Note:** $\rho$ is an important ratio. It is called the utilization factor of the service facility

$$\rho = \frac{\text{Average service time}}{\text{Average arrival gap}}$$

In next step we simulate the simpler model using a sequence of random numbers $u_1\ u_2, \ldots.\ u_m$.

Let $\bar{P}$ be the estimate of the parameter of interest (mean queue length) obtained through simulation.

Let $\bar{P}'$ be the value of the same parameter obtained analytically. The difference

$$E = \bar{P} - \bar{P}'$$

give us the error due to sampling when the sequence of random numbers $u_1\ u_2, \ldots.\ u_n$. This error $E$ may also be used an as estimate of the sampling error in the more complex simulation.

In the last step, we simulate the given complex system using the random number sequence, $u_1$ $u_2 \ldots u_n$ and obtain the value of the parameter corresponding to $\bar{P}$. Let this value $b\ \bar{Q}$. Now we correct this observed value $\bar{Q}$ by subtracting the estimated sampling error $E$ from it. Thus the estimated value of the parameter

$$\bar{R} = \bar{Q} - E = \bar{Q} - \bar{P} + \bar{P}'$$

The variance $\sigma^2$ of this new estimate will be smaller than the variance of $\bar{Q}$ be cause $\bar{P}$ and $\bar{Q}$ have strong positive correlation. Since they both use the same sequence of random numbers as input.

This technique of using a control variate, which clearly affects the response in both systems to introduce a positive correlation between their responses is easy to apply.

5. **Stratified Sampling:** We can also reduce the variance without having to increase the number of sample points; by breaking the parent population into classes (of strata) such that there is a minimum variation within each class. We take then an appropriately small number of samples from each strata. These samples are then combined to obtain an overall estimate of the mean. Clearly such sampling would be more efficient than a simple random sampling.
6. **Russian Roulette and Splitting:** This method is used by classifying experiments with partial results at each stage as 'interesting' or 'uninteresting' and then proceeding further with only the interesting one. This method can be extremely effective in variance reduction. If good definitions of interesting on uninteresting are available at each stage of the experiment. Although this basic approach has been used for a long time in sequential sampling schemes by quality control engineers it is not easily applicable to a typical simulation experiment.

## 5.4 VERIFICATION AND VALIDATION OF SIMULATION MODEL

One of the most important task in model development is verification and validation of the simulation model. The verification and validation process consists of the following components.

1. Verification is concerned with building the model right. It is utilized in the comparison of the conceptual model to the computer representation that implements that conception. It asks the following question:
   — Is the model implemented correctly in computer?
   — Are the input parameters and logical structure of the model correctly represented?
2. Validation is concerned with building the right model. It is utilized to determine that a model is an accurate representation of the real system. Validation is usually achieved through the calibration of the model, an iterative process of comparing the model to actual system behaviour and using the discrepancies between the two and the insights gained, to improve the model.

**Verification of Simulation Models**

The purpose of model verification is to assure that the conceptual model is reflected accurately in the computerized representation. Number of suggestions can be given for use in the verification process as:

1. Have the computerized representation checked by some one other than its developer.
2. Closely examine the model output for reasonableness under a variety of settings of the input parameters.
3. Have the computerized representation print the input parameters at the end of the simulation, to be sure that these parameter values have not been changed inadvertently.
4. Make the computerized representation as self-documenting as possible.
5. Give a precise definition of every variable used and a general description of the purpose of each major section of code.
6. If the computerized representation is animated, verify that what is seen in the animation imitates the actual system.

7. Use interactive run controller (IRC) or debugger in successful simulation model building. Because IRC assists in finding and correcting errors.
8. Graphical interfaces are recommended for accomplishing verification and validation. The graphical representation of the model is essentially a form of self-documentation. It simplifies the task of model understanding. These suggestions are basically the same ones any software engineer would follow.

For certain models, it is possible to consider more than whether a particular statistic is reasonable. It is possible to compute certain long-run measures of performance.

Another important way to aid the verification process is the off-neglected documentation phase. If a model builder writes brief comments in the computerized model, plus definitions of all variables and parameters, and descriptions of each major section of the computerized model, it becomes much simpler for someone else. Documentation is also important as a means of clarifying the logic of a model and verifying its completeness.

A more sophisticated technique is the use of a trace. A trace designed specifically for use in a simulation program. Would give the value of selected variable each time the simulation clock was incremented. Thus, a simulation trace is nothing more than a detailed printout of the state of the simulation model as it changes over time.

Most simulation software has built-in capability to conduct a trace without the programmer having to do any extensive programming. Some software allows a selective trace. For example, a trace could be set for specific location in the computerized model. Any time an entity goes through that location or those locations, a message is written.

## 5.5 CALIBRATION AND VALIDATION OF MODELS

Verification and validation are conceptually different and usually are conducted simultaneously by the modeler. Validation is the overall process of comparing the model and its behaviour to the real system and its behaviour.

**Calibration:** Calibration is the iterative process of comparing the model to the real system making adjustments to the model comparing the revised model to reality, comparing again, and so on.

Figure. 5.2 shows iterative process of calibrating a model.

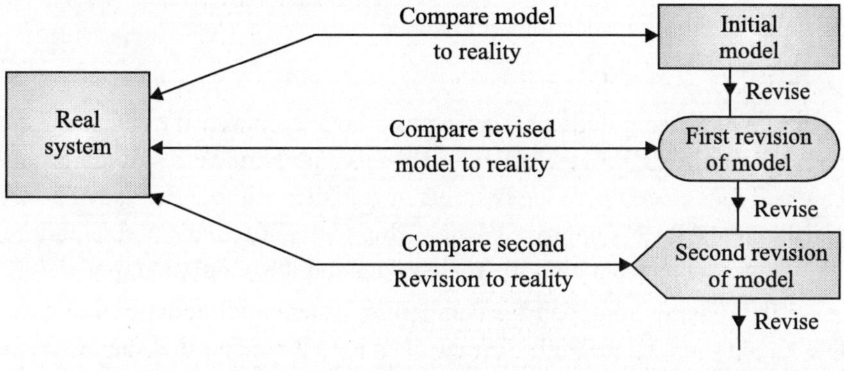


**Fig. 5.2**

The comparison of model to reality is carried out by a variety of tests— like—subjective tests, objective tests.

**Subjective Tests**

This test involve people who are knowledgeable about one or more aspects of the system making judgement about the model and its output.

**Objective Tests**

This test always require data on the system's behaviour plus the corresponding data produced by the model. Then one or more statistical tests are performed to compare some aspect of the system data set to the same aspect of the model data set.

No model is ever totally representative of the system under study each revision of the model, as shown in Figure 5.3, involves some cost, time, and effort. If level of accuracy cannot be obtained within the budget constraints, either expectations of model accuracy must be lowered, or the model must be abandoned. Yucesan and Jacobon indicate that verifying simulation model is so difficult as to be intractable. They offer theorems to confirm this intractibility.

Naylor and Finger formulated a following step approach in validation which has been widely followed:

1. Build a model that has high face validity
2. Validate model assumptions
3. Validating input-output transformations
4. Input-output validation: using historical input data
5. Input-output validation: using a turing test.

## 5.6 COMPARISON BETWEEN MODEL DATA WITH REAL SYSTEM DATA

The idea of comparing the model and system output data for the existing system might be called results validation. If the two sets of data compare "closely", then the model of the existing system is considered "valid". Here we present statistical procedures that might be useful for carrying out the comparison of model and system output data. We can have three approaches to this comparison problem.

1. Inspection approach
2. Confidence-interval approach
3. Time-series approach

We will discuss here only inspection approach.

**Inspection Approach**

This approach is used by most simulation practitioners who attempt the aforementioned comparison is to compute one or more statistics from the real world observations and corresponding statistics from the model output data, and then to compare the two sets of statistics without the use of a formal statistical procedure. The difficulty with this approach is that this idea particularly vulnerable to the inherent randomness of the observations from both the real system and the simulation model.

We can describe a better approach for comparing system and model output data if the system data are complete enough and in the right format. It is recommended that the system and model be compared by "driving" the model with historical system input data, rather than sample from the input probability distributions, and then comparing the model and system outputs.

Figure 5.3 show the system outputs are those corresponding to the historical system input data.

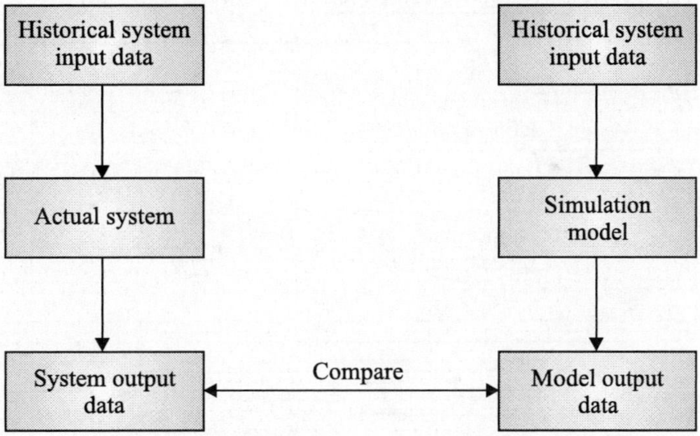


**Fig. 5.3.** The correlated inspection approach.

Thus, the system and model experience exactly the same observations from the input random variables, which should result in a statistically more precise comparison.

We call this idea the correlated inspection approach, since it generally results in comparable model and system output statistics being positively correlated. This approach is a more definitive way to validate the assumptions of the simulation model other than the probability distributions.

## 5.7 SIMULATION PROCESS

A set of steps to guide a model builder in a thorough and sound simulation process has shown in Fig. 5.4 on next page.

**1. Formulate the Problem and Plan the Study**

Every study should begin with a statement of the problem. The simulation analyst must ensure that the problem being described is clearly understood. If a problem statement is being developed by the analyst, it is important that the policy makers understand and agree with the formulation.

Overall plan should include a statement of the alternative systems to be considered, and a method for evaluating the effectiveness of these alternatives. It should also include the plans for the study in terms of—

- No. of people involved
- The cost of the study
- No. of days required to accomplish each phase of the work
- What are the anticipated results at the end of each phase
- Scope of the model
- Software to be used
- Time frame for the study and the required resources.

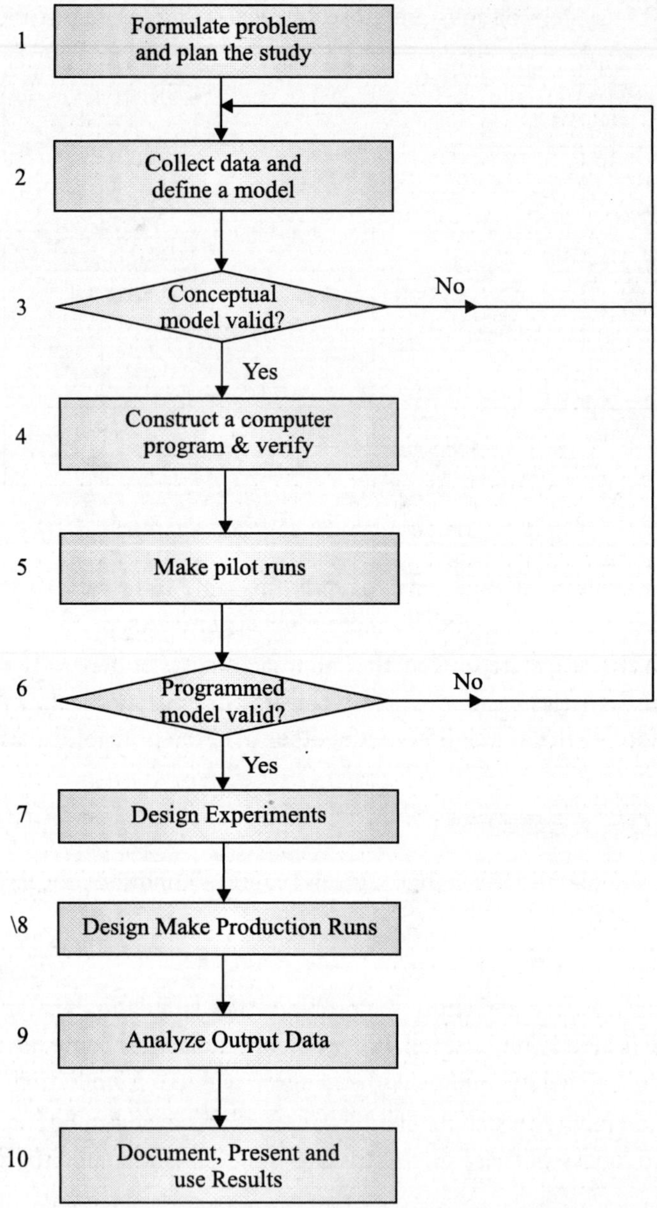


**Fig. 5.4.** Simulation process.

**2. Collect Data and Define the Model**

There is a constant interplay between the collection of the data and define the model. As the complexity of the model changes, the required data elements may also change. Since data collection takes such a large portion of the total time required to perform a simulation, it is necessary to begin it as early as possible, usually together with the early stages of model building.

The level of model detail should depend on the following:

- Project objectives
- Performance measures

- Data availability
- Credibility concerns
- Computer constraints
- Time and money constraints.

**3. Is the Conceptual Model Valid?**

Perform a structured walk-through of the conceptual model using the assumption document before an audience of managers, analysts, and subject-matter experts.

- It helps ensure that the model's assumptions are correct and complete.
- It promotes ownership of the model.
- It takes place before programming begins to avoid significant reprogramming later.

**4. Construct a Computer Program and Verify**

Program the model in a programming language (*e.g.*, C or FORTRAN) or in simulation software (*e.g.*, Arena, Automod, External, Promodel WITNESS). Benefits of using a programming language are that one is often known, they have a low purchase cost and they may result in a smaller model execution time.

Verify the simulation computer program. If the input parameters and logical structure of the model are correctly represented in the computer, verification has been completed.

**5. Make Pilot Runs**

Make pilot runs for validation purposes in step 6.

**6. Is the Programmed Model Valid?**

If there is an existing system, then compare model and system performance measures for the existing system, the simulation analysts and subject matter experts should review the model results for correctness.

**Note:** Validation is the determination that a model is an accurate representation of the real system.

**7. Design Experiments**

Specify the following for each system configuration of interest

- Length of each run
- Number of independent simulation runs using different random numbers
- The number of replications to be made for each run.

**8. Make Production Runs**

Production runs are made for use in step 9.

**9. Analyze Output Data**

Two major objectives in analyzing output data are:

1. Determining the absolute performance of certain system configuration.
2. Comparing alternative system configurations in a relative sense.

**More Runs?**

Based on the analysis of runs that have been completed, the analyst determines if additional runs are needed and what design those additional experiments should follow.

**10. Document, Present, and use Results**

**Document**

Program documentation is necessary for numerous reasons. If the program is going to be used again by the same or different analysts, it may be necessary to understand how the program operates. This will build confidence in the program, so that model users and policy-makers can make decisions based on analysis. If the program is to be modified by the same or a different analyst, this can be greatly facilitated by adequate documentation.

- It presents result of study
  → Use animation to communicate model to managers and other people who are not familiar with all the model detail.
- Results are used in decision-making process if they are both valid and credible.

## SUMMARY

- Process of model building involve three step:
  (*i*) Analysis of real system
  (*ii*) Construction of a conceptual model
  (*iii*) Operational model or computerized model
- Estimation methods are used to estimate parameters from observations on random variables. This method uses confidence interval.
- Goodness-of-fit tests provide helpful guidance for evaluating the suitability of a potential input model. It uses two method chi-square test and Kolmogorov-Smirnov test that tests the hypothesis.
- Antithetic sampling, correlated sampling, importance sampling and control variates, etc., are some variance-reduction technique that is developed for reducing the variance without increasing the sample-size.
- Verification is concerned with building the model right.
- Validation is concerned with building the right model.
- Calibration is the iterative process of comparing the model to the real system.
- The correlated inspection approach, generally results better while we are comparing model data with real system.
- Process of simulation involve formulate problem and plan the study, define a model and validate, construct computer program and verify, make pilot runs and validate, design experiments make production runs, analyze output data and document, present results.

## SOLVED QUESTIONS

**Q. 1. Discuss some general perspective on validation of simulation models.**

**Ans.** The following are some general perspectives on validation:

(*a*) If a simulation model is "Valid", then it can be used to make decisions about the system similar to those that would be made if it were feasible and cost-effective to experiment with the system itself.

(*b*) The ease or difficulty of the validation process depends on the complexity of the system being modeled and on whether a version of the system currently exists.

(*c*) A model that is valid for one purpose may not be for another.

(*d*) The measure of performance used to validate a model should include those that the decision maker will actually use for evaluating system designs.

**Q. 2. What are six classes of techniques for increasing the validity and credibility of a simulation model?**

**Ans.** Six classes of techniques are following:

(*i*) Collect high-quality information and data on the system.

(*ii*) Interact with the manager on a regular basis.

(*iii*) Maintain on assumptions document and perform a structured walk-through.

(*iv*) Validate components of the model by using quantitative techniques.

(*v*) Validate the output from the overall simulation model.

(*vi*) Animation.

**Q. 3. Write some basic concepts of modeling and simulation.**

**Ans. Basic concepts:**

In the following, an introduction to the basic concepts of modeling and simulation is given.

Object — Object is some entity in the real world. Such an object can exhibit widely varying behaviour depending on the context in which it is studied, as well as the aspects of its behaviour which are under study.

Base model — Base model is a hypothetical, abstract representation of the object's properties, in particular, its behaviour, which is valid in all possible contexts, and describes all the object's facts.

System — System is well defined object in the real world under specific conditions, only considering specific aspects of its structure and behaviour.

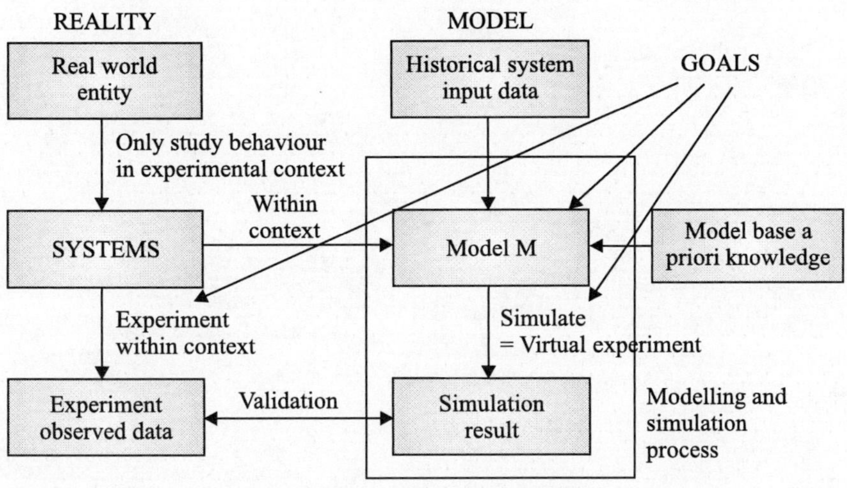


**Fig. 5.5.** modeling and simulation.

## EXERCISES

1. Differentiate verification and validation of simulation model.
2. Explain simulation process in detail.
3. Discuss, process of modeling and simulation in detail.
4. Compare between model data and real system data.
5. Explain following term:
   (*a*) Calibration, (*b*) Modeling and Simulation.

6

# Basic of Probability

**INSIDE THIS CHAPTER**



## 6.1 INTRODUCTION

In our day-to-day life, we may have many situations where uncertainty plays a vital role. We usually use statements like "there is a chance for rain today" or " probably I will get A grade in university examination", etc. In all these contexts the term chance or probably is used to indicate uncertainty. The word Probability is related with the occurrence of uncertainty, and Probability theory is the discipline which tries to quantify the concept of chance or likelihood.

**Probability theory is applied to situations where uncertainty exists.**

The chance of rain tomorrow is 40%. In such cases they are saying that the probability of a rain even is 0.4. Other probabilistic character.

## 6.2 EXPERIMENT

An experiment is an activity which can be repeated in more or less same condition and will have some specific outcome or outcomes. By combining hydrogen with oxygen under certain conditions results in formation of $H_2O$ or tossing a coin and checking whether the face value is Head or Tail are examples of experiments. Generally, experiment are of two types:

1. **Deterministic experiment:** In deterministic experiment, if we repeat the experiment, outcome or outcomes will not change; provided the initial conditions of the experiment remains same. Experiments like combining hydrogen and oxygen are examples of deterministic experiments.
2. **Random experiment or stochastic experiment:** A Random experiment may be defined as an experiment with more than one possible outcome, which may be repeated any number of times under more or less similar conditions and the outcomes of which vary irregularly from repetition to repetition. For example, throwing a die and looking on the face value is a random experiment as the face value will be any value from the set {1, 2, 3, 4, 5, 6}, which cannot be predicted before conducting the experiment. Random experiments are common in all kinds of research activities.

The set of all possible outcomes of an experiment is called the sample space and is denoted by $S$. The outcomes themselves are called the sample points in the sample space.

## 6.3 SAMPLE SPACE, EVENTS

### 6.3.1 Sample Space

The set of all sample outcomes or sample points of a random experiment is called its sample space, which is denoted by the letter $S$ or $\Omega$. For example: Sample space of the random experiment of throwing a die is $S = \{1, 2, 3, 4, 5, 6\}$ and sample space of the random experiment of tossing a coin is, $S = \{H, T\}$.

### 6.3.2 Sample Points

Every elements in the set of all outcomes of the random experiments are called sample points.

### 6.3.3 Event

Those subsets of the sample space to which probabilities are assigned by probability set function under consideration are called an Event. For example: Consider the random experiment of throwing a die and the sample space will be, $S = \{1, 2, 3, 4, 5, 6\}$. One can define an event $E$ as getting odd number. Then $E = \{1, 3, 5\}$. Note that the event $E$ is a subset of sample space $S$.

**Types of Events**

- **Simple and Compound Event:** If an event has only one outcome, that is, only one sample point it is called simple event. For example: The event of getting head in the experiment of tossing a coin.

  If an event has more than one sample points then it is called compound event. *E.g.*: The event of getting odd or even when rolling a die.
- **Mutually Exclusive Event:** Two events $A$ and $B$ are said to be mutually exclusive if the occurrence of any one of the events exclude the occurrence of the other for example $A \cap B$ is empty.
- **Exhaustive Event:** Let $E_1, E_2, E_3, \ldots E_n$ are $n$ events of sample space then $E_0$ is said to be exhaustive if $[n_{i=1} E_i = S]$.

## 6.4 WHAT IS PROBABILITY?

There is really no answer to this question. Some people think of it as 'limiting frequency'. That is, to say that the probability of getting heads when a coin is tossed means that, if the coin is tossed many times, it is likely to come down heads about half the time. But if you toss a coin 1000 times, you are not likely to get exactly 500 heads. You wouldn't be surprised to get only 495. But what about 450, or 100? Some people would say that you can work out probability by physical arguments, like the one we used for a fair coin. But this argument doesn't work in all cases, and it doesn't explain what probability means. Some people say it is subjective. You say that the probability of heads in a coin toss is 1/2 because you have no reason for thinking either heads or tails more likely; you might change your view if you knew that the owner of the coin was a magician or a con man. But we can't build a theory on something subjective. We regard probability as a mathematical construction satisfying some axioms (devised by the Russian mathematician A.N. Kolmogorov). We develop ways of doing calculations with probability, so that (for example) we can calculate how unlikely it is to get 480 or fewer heads in 1000 tosses of a fair coin. The answer agrees well with experiment.

## 6.5 KOLMOGOROV'S AXIOMS

Remember that an event is a subset of the sample space $S$. $E$ number of events, say $E_1$, $E_2$, ..., are called *mutually disjoint* or *pairwise disjoint* if $E_i / E_j = \phi$ for any two of the events $E_i$ and $E_j$; that is, no two of the events overlap. According to Kolmogorov's axioms, each event $E$ has a probability $P(E)$, which is a number.

These numbers satisfy three axioms:

**Axiom 1:** For any event $E$, we have $P(E) \geq 0$.

**Axiom 2:** $P(S) = 1$.

**Axiom 3:** If the events $E_1$, $E_2$,... are pairwise disjoint, then

$$P(E_1 \cup E_2 \cup ...) = P(E_1) + P(E_2) + ...$$

Note that in Axiom 3, we have the union of events and the sum of numbers.

Don't mix these up; never write $P(E_1) \cup P(E_2)$, for example. Sometimes we separate Axiom 3 into two parts:

**Axiom 3*a*** if there are only finitely many events $E_1$, $E_2$, ..., $E_n$, so that we have

$$P(E_1 \cup ... \cup E_n) = \sum_{i=1}^{n} P(E_i),$$

**Axiom 3*b*** for infinitely many. We will only use Axiom 3*a*, but 3*b* is important later on.

Notice that we write

$$\sum_{i=1}^{n} P(E_i)$$

for

$$P(E_1) + P(E_2) + ... + P(E_n)$$

## 6.6 TYPES OF PROBABILITY

**1. Relative frequency or estimated frequency:**

Relative frequency is based on observation or actual measurements. If an experiment is performed $N$ times, and the event $E$ occurs $fr(E)$ times, then the ratio is called the relative frequency or estimated probability of $E$. $N$, the number of times that the experiment is performed, is called the number of trials or the sample size. If $E$ consists of a single outcomes, then we refer to $P(E)$ as the estimated probability of the outcomes, and write $P(s)$.

**Example:** A die is rolled 80 times.

$$P(E) = \frac{fr(E)}{N}$$

The number 3 is rolled 6 times. The relative frequency of rolling a 3 is 6/80.

**Some Properties of Estimated Probability**

Let $S = \{s_1, s_2, ..., s_n\}$ be a sample space and let $P(s_i)$ be the estimated probability of the event $\{s_i\}$. Then

(*a*) $0 \leq P(s_i) \leq 1$

(*b*) $P(s_1) + P(s_2) + ... + P(s_n) = 1$

(*c*) If $E = \{e_1, e_2, e_r\}$, then $P(E) = P(e_1) + P(e_2) + ... + P(e_r)$.

**In words:**

(*a*) The estimated probability of each outcome is a number between 0 and 1.

(*b*) The estimated probabilities of all the outcomes add up to 1.

(*c*) The estimated probability of an event $E$ is the sum of the estimated probabilities of the individual outcomes in $E$.

**2. Theoretical Probability:** The **theoretical probability,** $P(E)$, of an event $E$ is a probability determined from the nature of the experiment rather than through actual experimentation.

The estimated probability approaches the theoretical probability as the number of trials gets larger and larger.

**Important:**

1. We write $P(E)$ for both estimated and theoretical probability. Which one we are referring to should always be clear from the context.
2. Theoretical probability satisfies the same properties (shown above) as estimated probability.

**Computing Theoretical Probability for Equally Likely Outcomes**

In an experiment in which all outcomes are equally likely, the probability of an event $E$ is given by number of favourable outcomes

$$P(E) = \frac{\text{Number of favourable outcomes}}{\text{Total number of out comes}}$$

$$= \frac{n(E)}{n(S)}$$

where $n(E)$ is the number of elements in $E$, and $n(S)$ is the number of elements in $S$. Example: The theoretical probability of rolling a 3 on a regular 6 sided die is 1/6.

**3. Personal or subjective probability:** These are values (between 0 and 1 or 0 and 100%) assigned by individuals based on how likely they think events are to occur. Example: The probability of my being asked on a date for this weekend is 10%.

## 6.7 PROBABILITY DISTRIBUTIONS

The usefulness of probability theory comes in understanding probability distributions (also called probability functions and probability densities or masses). Probability distributions list or describe probabilities for all possible occurrences of a random variable. There are two types of probability distributions:

- Discrete distributions
- Continuous distributions.

### 6.7.1 Probability Mass Function

Let $X$ be a random variable. If number of possible values of $X$ is finite, then $X$ is called discrete variable. A finite sample space is just a finite set $X$. A probability distribution is an assignment of a number $P(x_i)$ to each outcome $s_i$ in a sample space $X = \{x_1, x_2,..., x_n\}$ so that

(*a*) $0 \leq P(x_i) \leq 1$

(*b*) $P(x_1) + P(x_2) + ... + P(x_n) = 1$.

The collection pairs $(x_i, P(x_i))$, $i = 1, 2, \ldots$, is called the probability distribution of $S$, and $P(x_i)$ is called the *probability mass function.* Given a probability distribution, we obtain the probability of an event $E$ by adding up the probabilities of the outcomes in $E$.

### 6.7.2 Probability Density Function

The probability density function (PDF) is the probability distribution of a continuous variable. Since continuous variables are uncountable, it is difficult to write down the probabilities of all possible events. Therefore, the PDF is always a function which gives the probability of one event, $x$. For a continuous variable $X$ the probability that $X$ lies in the interval $[a, b]$ is given by:

$$P(a \le X \le b) = \int_a^b f(x)\,dx$$

The function $f(x)$ is called probability density function (pdf) of the random variable $X$. The pdf satisfies the following conditions.

(a) $f(x) \ge 0\ \forall\ x$ in $R_X$

(b) $\int_{R_X} f(x)\,dx = 1$

(c) $f(x) = 0$, if $x$ is not in $R_X$

where $R_X$ is range space of the random variable $X$.

### 6.7.3 Cumulative Distribution Function

The distribution function is sometimes called the **cumulative distribution function (*cdf*),** denoted by $F(x)$, measures the probability that random variable $X$ is defined for each real numbers as follows:

$$F(x) = P(X \le x) \quad \text{for} \quad -\infty < x < \infty$$

where $P(X \le x)$ means the probability associated with the event $\{X \le x\}$. Thus, $F(x)$ is the probability that, when the experiment is done, the random variable $X$ will have taken on a value no larger than the number $x$. All random variable (discrete & continuous) have a cumulative distribution function.

If $X$ is discrete, then $F(x) = \sum_{\text{all } x_i \le x} p(x_i)$ [*cdf* is summing up the probabilities]

If $X$ is continuous, then $F(x) = \int_{-\infty}^{x} f(t)\,dt$ [*cdf* is integral of its probability density function]

**Properties of distribution function $F(x)$:**

1. $0 \le F(x) \le 1$ for all $x$.
2. $F(x)$ is non-decreasing [*i.e.*, if $x_1 < x_2$ $F(x_1) \le F(x_2)$]
3. $\lim_{x \to \infty} F(x) = 1$ and $\lim_{x \to -\infty} F(x) = 0$ (since $X$ takes on only finite values)

**Important point**

- A table of the probabilities cumulated over the events.

| Value | Probability | Value | Probability |
|---|---|---|---|
| $X \le 1$ | 1/6 | $X \le 4$ | 4/6 |
| $X \le 2$ | 2/6 | $X \le 5$ | 5/6 |
| $X \le 3$ | 3/6 | $X \le 6$ | 1 |

- The Cumulative Distribution (CD) is a monotonically increasing set of numbers.
- The *CD* always ends with at the highest value of 1.

## 6.8 DISCRETE AND CONTINUOUS PROBABILITY DISTRIBUTIONS

If a variable can take on any value between two specified values, it is called a continuous variable; otherwise, it is called a discrete variable.

Following examples will clarify the difference between discrete and continuous variables.

- Suppose the fire department mandates that all fire fighters must weigh between 250 and 350 pounds. The weight of a fire tighter would be an example of a continuous variable; since a fire fighter's weight could take on any value between 250 and 350 pounds.
- Suppose we flip a coin and count the number of heads. The number of heads could be any integer value between 0 and plus infinity. However, it could not be any number between 0 and plus infinity. We could not, for example, get 2.5 heads. Therefore, the number of heads must be a discrete variable.

Just like variables, probability distribution can be classified as discrete or continuous.

**Discrete Probability Distributions**

If a random variable is a discrete variable, its probability distribution is called a **discrete probability distribution.**

With a discrete probability distribution, each possible value of the discrete random variable can be associated with a non-zero probability. Thus, a discrete probability distribution can always be presented in tabular form.

An example will make this clear. Suppose you flip a coin two times then you can have four possible outcomes: HH, HT, TH, and TT. Now, let the random variable $X$ represent the number of heads that result from this experiment. The random variable $X$ can only take on the values 0, 1, or 2, so it is a discrete random variable.

The probability distribution for this statistical experiment appears below.

| Number of heads | Probability |
|---|---|
| 0 | 0.25 |
| 1 | 0.50 |
| 2 | 0.25 |

The above table represents a *discrete* probability distribution because it relates each value of a discrete random variable with its probability of occurrence. There are following discrete probability distributions.

- Binomial probability distribution
- Hyper geometric probability distribution
- Bernoulli probability distribution
- Poisson probability distribution
- Geometric Distribution

And many more.

**Continuous Probability Distributions**

If a random variable is a continuous variable, its probability distribution is called a **continuous probability distribution.** A continuous probability distribution differs from a discrete probability distribution in several ways.

- The probability that a continuous random variable will assume a particular value is zero.
- As a result, a continuous probability distribution cannot be expressed in tabular form.
- Instead, an equation or formula is used to describe a continuous probability distribution.

Most often, the equation used to describe a continuous probability distribution is called a **probability density function.** Sometimes, it is referred to as a **density function,** a **PDF,** or a **pdf.** For a continuous probability distribution, the density function has the following properties:

- Since the continuous random variable is defined over a continuous range of values (called the **domain** of the variable), the graph of the density function will also be continuous over that range.
- The area bounded by the curve of the density function and the $x$-axis is equal to 1, when computed over the domain of the variable.
- The probability that a random variable assumes a value between $a$ and $b$ is equal to the area under the density function bounded by $a$ and $b$.

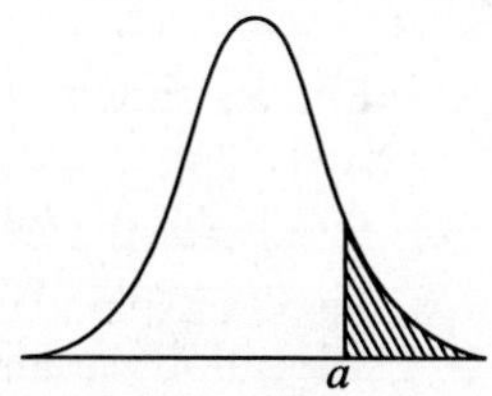


Fig. 6.1

For example, consider the probability density function shown in the graph below. Suppose we wanted to know the probability that the random variable $X$ was less than or equal to $a$. The probability that $X$ is less than or equal to $a$ is equal to the area under the curve bounded by $a$ and minus infinity–as indicated by the shaded area.

The shaded area in the graph represents the probability that the random variable $X$ is less than or equal to $a$. This is a cumulative probability. However, the probability that $X$ is *exactly* equal to $a$ would be zero. A continuous random variable can take on an infinite number of values. The probability that it will equal a specific value (such as $a$) is always zero. There are following continuous probability distributions.

- Normal probability distribution
- Uniform distribution
- Exponential distribution
- Weibull distribution
- Student's t distribution
- Chi-square distribution
- $F$ distribution

And many more.

## 6.9 PARAMETERIZATION OF CONTINUOUS DISTRIBUTION

For continuous distribution such as normal, exponential, etc., there are various ways to define, or parameterize, the probability density function. Parameters are classified, on the basis of their physical or geometric interpretation. They usually come in three types:

- Location parameter
- Scale parameter
- Shape parameters

1. **The location parameter:** Location determines the mean value of the distribution. In some cases, the mean of the distribution is the location parameter, but this need not be the case. When available, the location parameter can take on any finite value.

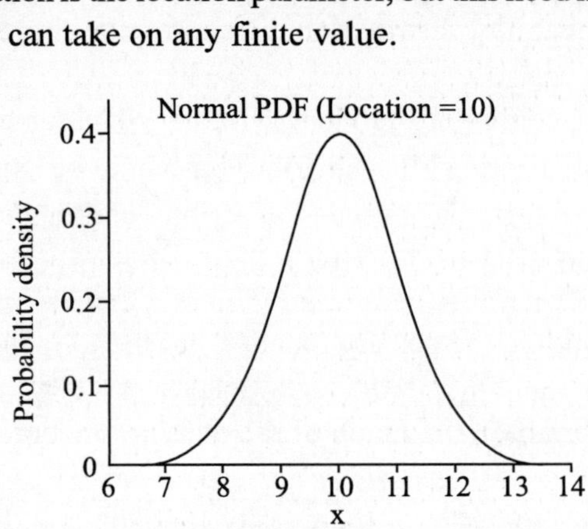


**Fig. 6.2 (*a*)**

Fig. 6.1 shows that probability density function for a normal distribution with a location parameter of 10 and a scale parameter of 1. The effect of the location parameter is to translate the graph, relative to the standard normal distribution, 10 units to the right on the horizontal axis. A location parameter of –10 would have shifted the graph 10 units to the left on the horizontal axis. That is, a location parameter simply shifts the graph left or right on the horizontal axis.

2. **The scale parameter:** Scale parameter determines the horizontal scale of the distribution. A large scale parameter means that the graph is the distribution is stretched out. For some distributions, the standard deviation serves as the scale parameter, but this need not be the case. When available, the scale parameter can take on any finite, strictly positive value.

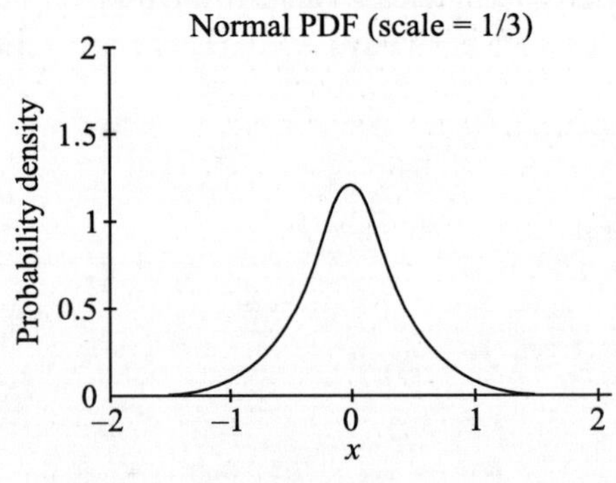


**Fig. 6.2 (*b*)**

Fig. 6.2 (*b*) has a scale parameter of 3 (and a location parameter of zero). The effect of the scale parameter is to stretch out the graph. The maximum $y$ value is approximately 0.13 as opposed 0.4 in the previous graphs. The $y$ value, *i.e.*, the vertical axis value, approaches zero at about (+/–) 9 as opposed to (+/–) 3 with the first graph.

The effect of a scale parameter greater than one is to stretch the *pdf*. The greater the magnitude, the greater the stretching. The effect of a scale parameter less than one is to compress the *pdf*. The compressing approaches a spike as the scale parameter goes to zero. A scale parameter of 1 leaves the *pdf* unchanged (if the scale parameter is 1 to begin with) and non-positive scale parameters are not allowed.

**Location and Scale Together:** The following graph shows the effect of both a location and a scale parameter. The plot has been shifted right 10 units and stretched by a factor of 3.

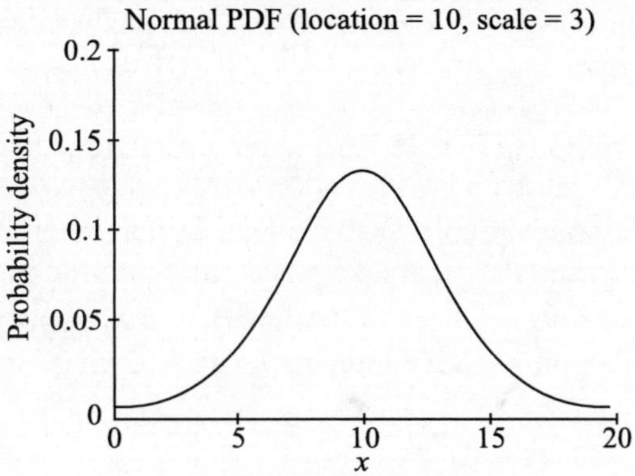


**Fig. 6.3**

**Standard Form:** The standard form of any distribution is the form that has location parameter zero and scale parameter one. It is common in statistical software packages to only compute the standard form of the distribution. There are formulas for converting from the standard form to the form with other location and scale parameters. These formulas are independent of the particular probability distribution.

**Formulas for Location and Scale based on the Standard Form:** The following are the formulas for computing various probability functions based on the standard form of the distribution. The parameter *a* refers to the location parameter and the parameter *b* refers to the scale parameter. Shape parameters are not included.

| | |
|---|---|
| Cumulative Distribution Function | $F(x; a, b) = F((x - a)/b; 0, 1)$ |
| Probability Density Function | $f(x; a, b) = (1/b) f((x - a)/b; 0,1)$ |
| Per cent Point Function | $G(x; a, b) = a + bG(x; 0, 1)$ |
| Hazard Function | $h(x; a, b) = (1/b)\, h((x - a)/b; 0, 1)$ |
| Cumulative Hazard Function | $H(x; a, b) = H((x - a)/b; 0, 1)$ |
| Survival Function | $S(x; a, b) = S((x - a)/b; 0, 1)$ |
| Inverse Survival Function | $Z(x; a, b) = a + bZ(x; 0, 1)$ |
| Random Numbers | $Y(a, b) = a + bY(0, 1)$ |

3. **Shape Parameter:** One or more *shape parameters* may further determine the exact form of the distribution. Shape parameters usually affect the skewness and kurtosis of a distribution.

   A type of distribution may have one or more of these parameters, depending on the nature and common usage of the distribution.

   *For example:*

   - The exponential distribution only has a scale parameter.
   - The normal distribution has a location and a scale parameter.
   - The beta distribution has two shape parameters.

## 6.10 IMPORTANT DISTRIBUTION FUNCTION

Here are some explain some important distribution function.

### 6.10.1 Poisson Distribution

In statistics and probabilistic theory , the **Poisson distribution** is a discrete probability distribution that expresses the *probability of a number of events occurring in a fixed period of time if these events occur with a known average rate and independently of the time since the last event.* The distribution was first introduced by Simeon-Denis Poisson (1781–1840) and published, together with his probability theory, in 1838 in his research work "Research on the Probability of Judgments in Criminal and Civil Matters". The work focused on certain random variables $N$ that count, among other things, the number of discrete occurrences (sometimes called "arrivals") that take place during a time-interval of given length.

If the expected number of occurrences in this interval is $\lambda$, then the probability that there are exactly $n$ occurrences ($n$ being a non-negative integer, $n = 0, 1,2, ...$) is equal to

$$f(n; \lambda) = \frac{\lambda^n e^{-\lambda}}{n!},$$

where

- $e$ is the base of the natural logarithm ($e = 2.71828...$)
- $n$ is the number of occurrences of an event—the probability of which is given by the function
- $n$ ! is the factorial of $n$
- $\lambda$ is a positive real number, equal to the expected number of occurrences that occur during the given interval. For instance, if the events occur on average 4 times per minute, and you are interested in the number of events occurring in a 10 minute interval, you would use as your model a Poisson distribution with $\lambda = 10 \times 4 = 40$.

As a function of $n$, this is the probability mass function. The Poisson distribution can be derived as a limiting case of the binomial distribution.

**Possible Application**

The Poisson distribution can be applied to systems with a large number of possible events, each of which is rare. A classic example is the nuclear decay of atoms.

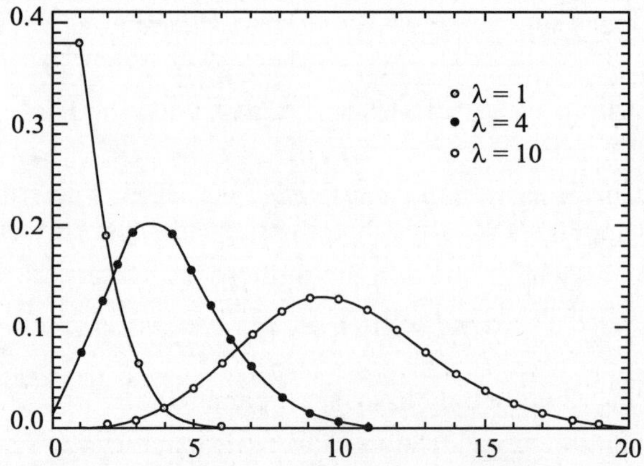


**Fig. 6.4.** A diagram showing the Poisson distribution.

**Parameter:** $\lambda > 0$

**Range:** $\{0, 1, ...\}$

**Mean:** $\lambda$

**Variance:** $\lambda$

### 6.10.2 Exponential Distribution

It describes the time between events in a Poisson process, *i.e.*, a process in which events occur continuously and independently at a constant average rate. This distribution is sometime called a negative exponential, occurs in applications such as reliability theory and queuing theory. Following random variables will often be modeled as exponential:

(*i*) Time between two successive job arrivals or interarrival time.

(*ii*) Service time at a server in a queuing network: server can be resource such as CPU, I/O device.

(*iii*) Time required to repair a component that has malfunctioned.

A continuous random variable $X$ is said to have an Exponential ($\lambda$) distribution if it has probability density function.

$$f(x; \lambda) = \begin{cases} \lambda e^{-\lambda x}, & x \geq 0 \\ 0, & x < 0 \end{cases}$$

where $\lambda > 0$ is called the rate of the distribution.

The distribution is supported on the interval $[0, \infty)$.

In the study of continuous-time stochastic processes, the exponential distribution is usually used to model the time until something happens in the process.

The cumulative distribution function is given by

$$F(X; \lambda) = 1 - e^{-\lambda X} \quad \text{if} \quad X \geq 0 \text{ otherwise 0 for } X \leq 0$$

There are following properties of exponential distribution function:

1. **Mean, variance, and median:**

(*a*) The mean or expected value of an exponentially distributed random variable $X$ with rate parameter $\lambda$ is given by

$$E[X] = 1/\lambda$$

(*b*) The variance of $X$ is given by:

$$\text{VAR}[X] = 1/\lambda^2$$

(*c*) The median of $X$ is given by

$$M[X] = \ln^2/\lambda < E[X]$$

where ln = Natural logarithm

2. **Memoryless:** An important property of the exponential distribution is that it is memoryless. It is memoryless because each subsequent event is completely independent from the previous events.

$$P\,[T > s + t \mid T > s] = P[T > t] \quad \text{for all} \quad s, t \geq 0$$

This says that the conditional probability that we need to wait, for example, more than another 5 seconds before the first arrival, given that the first arrival has not yet happened after 15 seconds, is equal to the initial probability that we need to wait more than 5 seconds for the first arrival. So, if we waited for 15 seconds and the first arrival didn't happen ($T > 15$), probability that we'll need to wait another 5 seconds for the first arrival ($T > 15 + 5$) is the same as the initial probability that we need to wait more than 5 seconds for the first arrival ($T > 5$). The fact that $\Pr(T > 20 \mid T > 15) = \Pr(T > 5)$ does *not* mean that the events $T > 40$ and $T > 30$ are independent. It means that the probability of waiting an additional $t$ units of time for the next event given that you've already waited $s$ units of time is the same as the probability of waiting $t$ units of time for an event starting immediately.

The exponential distributions and the geometric distributions are the only memoryless probability distributions.

3. **Quantiles:** The quantile function (inverse cumulative distribution function) for exponential ($\lambda$) is:

$$F - 1(p; \lambda) = -\ln(1 - p)/\lambda$$

for $0 \le p < 1$. The quartiles are therefore: first quartile ln $(4/3)/\lambda$ , median $\ln(2)/\lambda$ and third quartile $\ln(4)/\lambda$.

### 6.10.3 Normal Distribution

A normal distribution is a very important statistical data distribution pattern occurring in many natural phenomena, such as height, blood pressure, lengths of objects produced by machines, etc. Certain data, when graphed as a histogram, creates a bell-shaped curve known as a normal curve, or normal distribution.

The **normal distribution** refers to a family of continuous probability distributions described by the normal equation. The normal distribution is defined by the following equation:

$$P(x) = \frac{1}{\sigma\sqrt{2\pi}} e^{-(x-\mu)^2/(2\sigma^2)}$$

where $X$ = normal random variable

$\mu$ = mean

$\sigma$ = standard deviation

$\pi$ = approximately 3.14159

$e$ = approximately 2.71828.

The random variable $X$ in the normal equation is called the **normal random variable.** The normal equation is the probability density function for the normal distribution.

If the transformation $z = (x - \mu)/\sigma$ is made, the distribution is transformed to a form in which the mean is 0 and the standard deviation is 1.

$$z = (x - \mu)/\sigma$$

or

$$x = z\sigma + \mu$$

The variable z is usually called the **standard normal variate.**

**Parameters:**

Location parameter $\mu \in (-\infty, \infty)$

Scale parameter $\sigma > 0$.

**Mean:** $\mu$

**Variance:** $\sigma^2$

**Mode:** $\mu$

**The Normal Curve**

The graph of the normal distribution depends on two factors—the mean and the standard deviation. The mean of the distribution determines the location of the center of the graph, and the standard deviation determines the height and width of the graph. When the standard deviation is large, the curve is short and wide; when the standard deviation is small, the curve is tall and narrow. All normal distributions look like a symmetric, bell-shaped curve, as shown below.

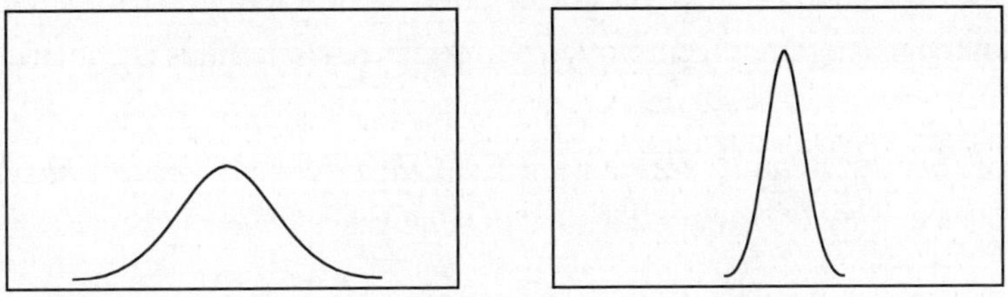

(*a*) (*b*)

**Fig. 6.5 (a), (b)**

The curve on the left (Fig. 6.5*a*) is shorter and wider than the curve on the right (Fig. 6.5*b*), because the curve on the left has a bigger standard deviation.

**Probability and the Normal Curve**

The normal distribution is a continuous probability distribution. This has several implications for probability.

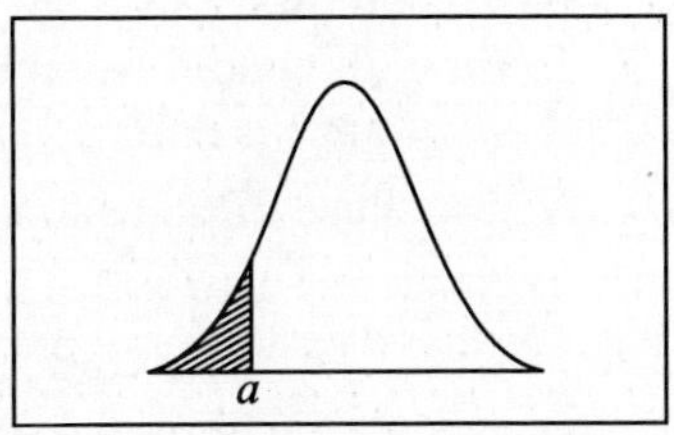

**Fig. 6.6**

- The total area under the normal curve is equal to 1.
- The probability that a normal random variable $X$ equals any particular value is 0.
- The probability that $X$ is greater than $a$ equals the area under the normal curve bounded by $a$ and plus infinity (as indicated by the *non-shaded area* in the Fig. 6.6).
- The probability that $X$ is less than $a$ equals the area under the normal curve bounded by $a$ and minus infinity (as indicated by the *shaded* area in the Fig. 6.6).

Additionally, every normal curve (regardless of its mean or standard deviation) conforms to the following "rule".

- About 68% of the area under the curve falls within 1 standard deviation of the mean.
- About 95% of the area under the curve falls within 2 standard deviations of the mean.
- About 99.7% of the area under the curve falls within 3 standard deviations of the mean.

Collectively, these points are known as the **empirical rule** or the **68-95-99.7 rule.** Clearly, given a normal distribution, most outcomes will be within 3 standard deviations of the mean.

**Characteristics of Normal Distribution**

1. Normal distributions are symmetrical with a single central peak at the mean (average) of the data. 50 per cent of the distribution lies to the left of the mean and 50 per cent lies to the right of the mean.
2. The spread of a normal distribution is controlled by the standard deviation, $\sigma$. If value of standard deviation is smaller then data are more concentrated.
3. The mean and the median are the same in a normal distribution.
4. The total area under the normal curve is equal to 1.
5. Continuous random variable.

**Applications of Normal Distribution**

1. Modern portfolio theory commonly assumes that the returns of a diversified asset portfolio follow a normal distribution.
2. In operation management, process variations often are normally distributed.
3. In human resource management, employee performance sometimes is considered to be normally distributed.

**Example 6.1:** *Suppose scores on an IQ test are normally distributed. If the test has a mean of 100 and a standard deviation of 10, what is the probability that a person who takes the test will score between 90 and 110?*

**Solution:** Here, we want to know the probability that the test score falls between 90 and 110. The "trick" to solving this problem is to realize the following:

$$P(90 < X < 110) = P(X < 110) - P(X < 90)$$

We use the normal distribution calculator to compute both probabilities on the right side of the above equation.

- To compute $P(X < 110)$, we enter the following inputs into the calculator: The value of the normal random variable is 110, the mean is 100, and the standard deviation is 10. We find that $P(X < 110)$ is 0.84.
- To compute $P(X < 90)$, we enter the following inputs into the calculator: The value of the normal random variable is 90, the mean is 100, and the standard deviation is 10. We find that $P(X < 90)$ is 0.16.

We use these findings to compute our final answer as follows:

$$P(90 < X < 110) = P(X < 110) - P(X < 90)$$

$$P(90 < X < 110) = 0.84 - 0.16$$

$$P(90 < X < 110) = 0.68$$

Thus, about 68% of the test scores will fall between 90 and 110.

## 6.11 DESCRIPTIVE CHARACTERISTICS

Descriptive characteristics are numerical values that express a particular property of statistical attributes. They are divided as follows:

1. Measures of central tendency: mean, mode and median
2. Measures of variability: range, variance, standard deviation variability coefficient
3. Measures of skewness: coefficient of skewness
4. Measures of kurtosis: coefficient of kurtosis.

### 6.11.1 Central Tendency

Measure of variability for the collection of data values is number that is meant to convey the idea of centeralness for the data set.

Number such as the mean and median give, in some sense, the central or middle values of the data. The data we discuss will come either from a random sample of a larger population or from the larger population itself.

Consider following notation:

| Random Sample | Large Population |
|---|---|
| $N$ = No. of items in the sample<br>$\overline{x}$ = sample mean<br>$s^2$ = sample mean | $N$ = No. of elements in the population<br>$\mu$ = population mean<br>$\sigma^2$ = population variance |

**Note:** (*i*) Greek letters are used with the population and are called parameters.
(*ii*) Latin letters are used with the samples and are called statistics.

**Mean (Arithmetic mean)**

This is the average of the data values:

$$\mu = \frac{\sum_{i=1}^{N} x_i}{N}$$

**Sample mean**

$$\overline{x} = \frac{\sum_{i=1}^{n} x_i}{n}$$

**Geometric Mean**

The geometric mean is a useful measure of central tendency for samples that are log-normally distributed (*i.e.*, the logarithms of the observations are from an approximately normal distribution). The geometric mean can be calculated only from positive values.

$$\overline{x}_G = \sqrt[n]{\prod_{i=1}^{n} x_i}$$

**Harmonic Mean**

The harmonic mean is used to average rates. For example, suppose we want the average speed of a bus that travels a fixed distance every day at speeds $s_1$, $s_2$, and $s_3$. The average speed, found by dividing the total distance by the total time, is equal to the harmonic mean of the three speeds. The harmonic mean is appropriate when the distance is constant from trial to trial and the time required was variable. However, if the times were constant and the distances were variable, the arithmetic mean would have been appropriate.

$$\overline{x}_H = \frac{n}{\sum_{i=1}^{n} \frac{1}{x_i}}$$

**Median**

The median is the 50th percentile of the data set. It is the point that splits the database in half. Unlike mean, median is not affected by extreme values.

**Mode**

This is the most frequently occurring value in a data.

**Example 6.1:** *Find the mode of a given set of data.*

*Set of data: 34, 43, 81, 106, 106, 115*

**Solution:** The mode in a set of data is the number with the highest frequency. Hence, 106 is the mode.

If single value has highest frequency then it is called as unimodal. If two different value has same highest frequency then is called as bimodal and similarly trimodal etc.

### 6.11.2 Measures of Variability

Measure of variability for the collection of data values is number that is meant to convey the idea of spread for the data set.

**Range**

The difference between the largest and smallest values for a variable.

**Interquartile Range**

It is the difference between the third quartile and the first quartile (between the 75th percentile and the 25th percentile). This represents the range of the middle 50 per cent of the distribution.

**Variance**

It is an average of the squared deviations from the mean.

$$\sigma^2 = \frac{\sum_{i=1}^{N} (x_i - \mu)^2}{N}$$

**Sample variance**

$$\sigma^2 = \frac{\sum_{i=1}^{n} (x_i - \overline{x})^2}{n-1}$$

**Standard Deviation**

It measures the average distance between a single observation and its mean.

$$s = \sqrt{\frac{\Sigma(x_i - \overline{x})^2}{n-1}}$$

**Standard Error of Mean (SEM)**

$$\text{SEM} = \frac{s}{\sqrt{n}}$$

**Coefficient of Variation**

The coefficient of variation is a relative measure of dispersion. It is most often used to compare the amount of variation in two samples.

$$k = \frac{s}{\overline{x}}$$

**Example 6.2:** *Find the sample mean $\overline{x}$, variance $s^2$, and standard deviation s for the data:*

*4, 6, 6, 7, 9, 10*

**Solution:** Data sets: 4, 6, 6, 7, 9, 10

$$n = 6$$

$$\overline{x} = \frac{4+6+6+7+9+10}{6} = \frac{42}{6} = 7$$

$$s^2 = \frac{(4-7)^2+(6-7)^2+(6-7)^2+(7-7)^2+(9-7)^2+(10-7^2}{6-1}$$

$$s^2 = \frac{9+1+1+0+4+9}{5}$$

Variance: $s^2 = \dfrac{24}{5} = 4.8$

Standard deviation $s = \sqrt{4.8} \approx 2.19$

### 6.11.3 Measures of Skewness

This statistic measures the direction and degree of asymmetry. A value of zero indicates a symmetrical distribution. A positive value indicates skewness (longtailedness) to the right while a negative value indicates skewness to the left. In a perfectly symmetrical, non-skewed, distribution the mean, median and mode are equal, positive skewness indicates that mean is greater than median, which means that most of values are less than the mean. Negative skewness indicates that median is greater than mean, which means that most of values is greater than the average. Histograms (Figure 6.7–6.9) present triangular distributions with the same mean (10), variance (2) and kurtosis (2.4) with different skewness.

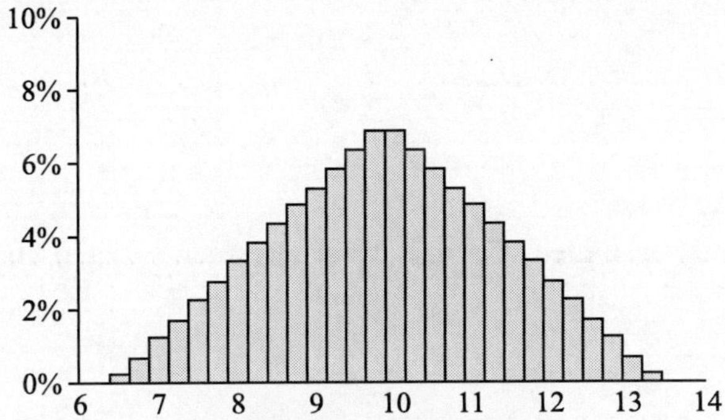


**Fig. 6.7.** Symmetrical distribution.

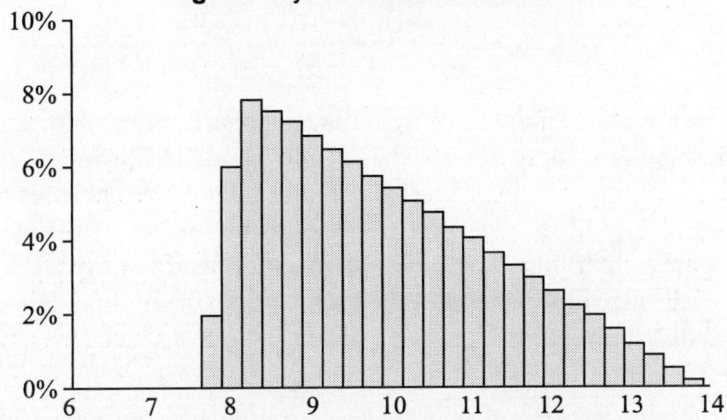


**Fig. 6.8.** Distribution with positive skewness (0.55).

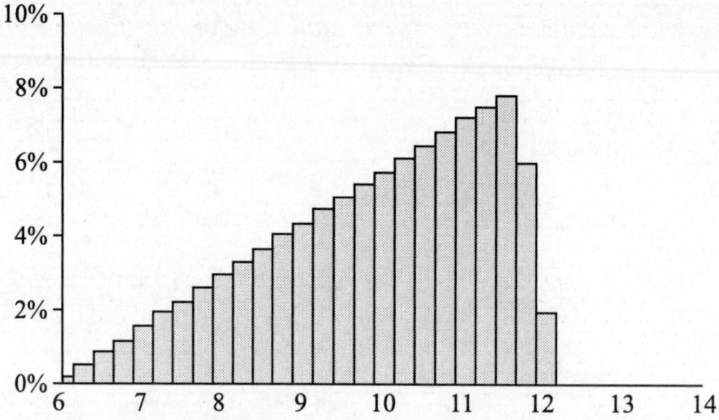


**Fig. 6.9.** Distribution with negative skewness (–0.55).

There are two most frequently used measures of skewness. Pearson's $b_1$ and Fisher's $g_1$ are computed as follows:

$$b_1 = \frac{m_3}{m_2^{3/2}}$$

$$g_1 = \frac{\sqrt{n(n-1)}\, b_1}{n-2}$$

where $m_r$, are central moments:

$$m_r = \frac{\sum_{i=1}^{n} (x_i - \overline{x})^r}{n}$$

### 6.11.4 Measures of Kurtosis

There are two most frequently used measures of kurtosis. Pearson's $b_2$ and Fisher's $g_2$ are computed as follows:

$$b_2 = \frac{m_4}{m_2^2}$$

$$g_2 = \frac{(n+1)(n-1)}{(n-2)(n-3)}\left[b_2 - \frac{3(n-1)}{n+1}\right]$$

This statistic measures the heaviness of the tails of a distribution. The usual reference point in kurtosis is the normal distribution. If the $b_2$ equals 3 ($g_2$ zero) and the skewness is zero, the distribution is normal. Unimodal distributions that have kurtosis greater than 3 have heavier tails than the normal. These same distributions also tend to have higher peaks in the center of the distribution. Unimodal distributions whose tails are lighter than the normal distribution tend to have a kurtosis that is less than 3. In this case, the peak of the distribution tends to be broader than the normal. Histograms (Figure 6.10) present four symmetrical distributions with the same mean (0) and variance (5/3) with different kurtosis.

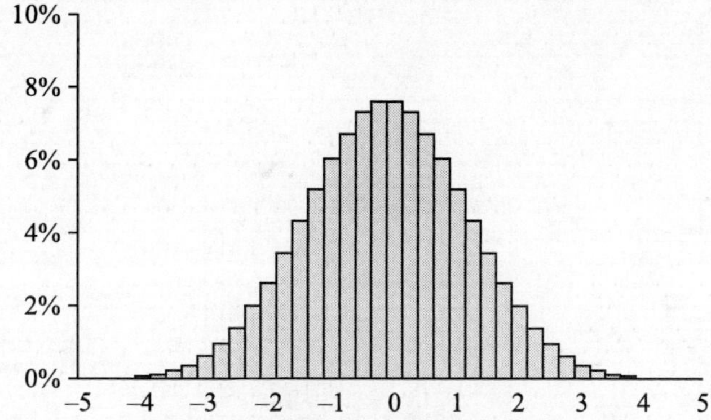


**Fig. 6.10.** Normal distribution ($b_2$= 3).

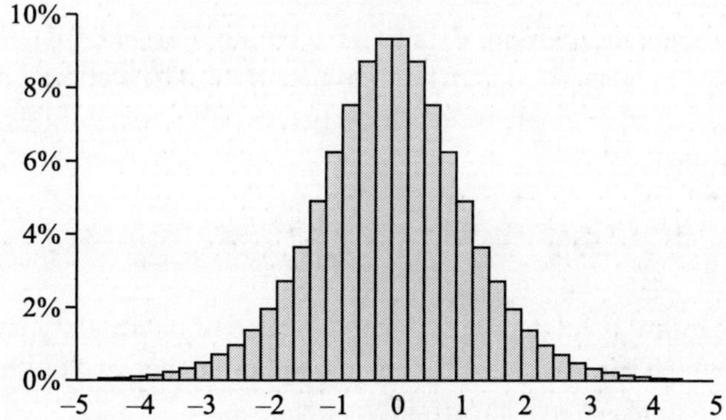


**Fig. 6.11.** Student's t distribution with 5 degrees of freedom ($b_2$ = 9).

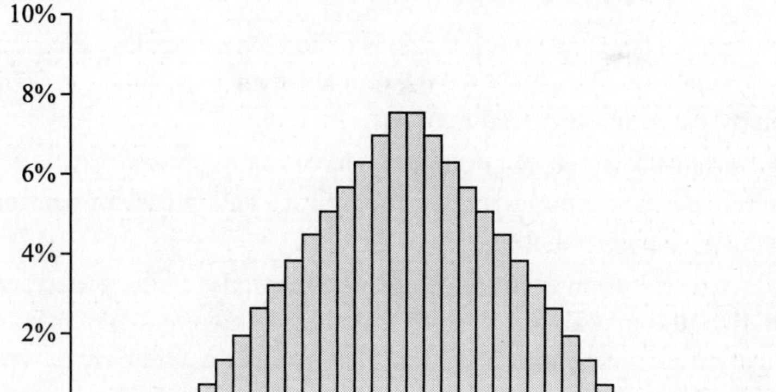


**Fig. 6.12** Triangular distribution ($b_2$ = 2.4).

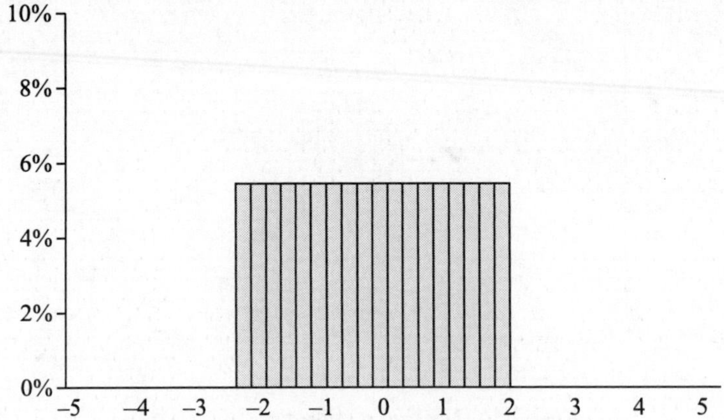


**Fig. 6.13.** Uniform distribution ($b_2$ = 1.8).

The measures of shape require more data to be accurate. For example, a reasonable estimate of the mean may require only ten observations in a random sample. The standard deviation will require at least thirty. A reasonably detailed estimate of the shape (especially if the tails are important) will require several hundred observations.

## SUMMARY

- The word Probability is related with the occurrence of uncertainty, and Probability theory is the discipline which tries to quantify the concept of chance or likelihood. Probability theory is applied to situations where uncertainty exists.
- The set of all possible outcomes of an experiment is called the sample space and is denoted by *S*. The outcomes themselves are called the sample points in the sample space. The event *E* is a subset of sample space *S*.
- The collection pairs ($x_i$, $P(x_i)$), $i = 1, 2, \ldots$, is called the probability distribution of *S*, and $P(x_i)$ is called the probability mass function. the probability density function (PDF) is the probability distribution of a continuous variable.
- All random variable (discrete & continuous) have a cumulative distribution function (*cdf*). In case of discrete cdf is summing up the probabilities and in case of continuous cdf is integral of its probability density function.
- For continuous distribution such as normal, exponential etc., there are various ways to define, or parameterize, the probability density function. Parameters are classified, on the basis of their physical or geometric interpretation. they usually come in three types:

  1. location parameter  2. scale parameter  3. shape parameters
- Poisson distribution is a discrete probability distribution that expresses the probability of a number of events occurring in a fixed period of time if these events occur with a known average rate and independently of the time since the last event.

$$f(n; \lambda) = \frac{\lambda^n e^{-\lambda}}{n!},$$

- Exponential distribution describes the time between events in a Poisson process, *i.e.* a process in which events occur continuously and independently at a constant average rate. a continuous random variable $X$ is said to have an Exponential ($\lambda$) distribution if it has probability density function.

$$f(x;\lambda) = \begin{cases} \lambda e^{-\lambda x}, & x \geq 0, \\ 0, & x < 0 \end{cases}.$$

- A normal distribution is a very important statistical data distribution pattern occurring in many natural phenomena, such as height, blood pressure, lengths of objects produced by machines, etc.

$$P(x) = \frac{1}{\sigma\sqrt{2\pi}} e^{-(x-\mu)^2/(2\sigma^2)}$$

- Descriptive characterstics are numerical values that express a particular property of statistical attributes. they are divided as follows:
  1. Measures of Central tendency-Mean, mode and median
  2. Measures of variability-range, variance, standard deviation, variability coefficient
  3. Measures of skewness-coefficient of skewness
  4. Measures of kurtosis-coefficient of kurtosis.

## SOLVED QUESTIONS

**Q. 1. What is the difference between probability distribution functions and probability density functions?**

**Ans.** Probability density function (PDF) of a continuous random variable is a function that describes the relative likelihood for this random variable to occur at a point in the observation space.

The PDF is the derivative of the probability distribution (also known as cummulative distribution function (CDF)) which described the entire range of values (distribution) a continuous random variable takes in a domain. The CDF is used to determine the probability a continuous random variable occurs any (measurable) subset of that range. This is performed by integrating the PDF over some range (*i.e.*, taking the area under of CDF curve between two values).

**Note 1:** Over the entire domain the total area under the CDF curve is equal to 1.

**Note 2:** A continuous random variable can take on an infinite number of values. The probability that it will equal a specific value is always zero.

*For Example, Example of CDF of a normal distribution:*

If test scores are normal distributed with mean 100 and standard deviation 10. The probability a score is between 90 and 110 is:

$$\begin{aligned} P(90 < X < 110) &= P(X < 110) - P(X < 90) \\ &= 0.84 - 0.16 = 0.68. \end{aligned}$$

*i.e.*, Aproximately 68%.

**Q. 2. Discuss some rule of probability.**

**Ans.** 1. The probability of an event is between 0 and 1. A probability of 1 is equivalent to 100% certainty. Probabilities can be expressed at fractions, decimals, or percents.

$$0 \leq pr(A) \leq 1$$

2. The sum of the probabilities of all possible outcomes is 1 or 100%. If *A, B,* and *C* are the only possible outcomes, then $pr(A) + pr(B) + pr(C) = 1$.

   **Example:** A bag contains 5 red marbles, 3 blue marbles, and 2 green marbles,

   $$pr(\text{red}) + pr(\text{blue}) + pr(\text{green}) = 1 \frac{5}{10} + \frac{3}{10} + \frac{2}{10} = 1$$

3. The sum of the probability of an event occurring and it not occurring is 1. $pr(A) + pr(\text{not } A) = 1$ or $pr(\text{not } A) = 1 - pr(A)$

   **Example:** A bag contains 5 red marbles, 3 blue marbles, and 2 green marbles.

   $$pr\ (\text{red}) + pr(\text{not red}) = 1 \qquad \frac{3}{10} + pr(\text{not red}) = 1 \qquad pr\ (\text{not red}) = \frac{7}{10}$$

4. If two events *A* and *B* are independent (this means that the occurrence of *A* has no impact at all on whether *B* occurs and vice versa), then the probability of *A* and *B* occurring is the product of their individual probabilities.

   $$pr(A \text{ and } B) = pr(A) \cdot pr(B)$$

   **Example:** Roll a die and flip a coin. $pr(\text{heads and roll a 3}) = pr(H)$ and $pr(3)$ $\frac{1}{2} \cdot \frac{1}{6} = \frac{1}{12}$

5. If two events *A* and *B* are mutually exclusive (meaning *A* cannot occur at the same time as *B* occurs), then the probability of either *A* or *B* occurring is the sum of their individual probabilities. $Pr(A \text{ or } B) = pr(A) + pr(B)$

   **Example:** A bag contains 5 red marbles, 3 blue marbles, and 2 green marbles. *pr*(red or green)

   $$pr(\text{red}) + pr\ (\text{green})\ \frac{5}{10} + \frac{2}{10} = \frac{7}{10}$$

6. If two events *A* and *B* are not mutually exclusive (meaning it is possible that *A* and *B* occur at the same time), then the probability of either *A* or *B* occurring is the sum of their individual probabilities minus the probability of both *A* and *B* occurring.

   $$Pr(A \text{ or } B) = pr(A) + pr(B) - pr(A \text{ and } B).$$

   **Example:** There are 20 people in the room: 12 girls (5 with blond hair and 7 with brown hair) and 8 boys (4 with blond hair and 4 with brown hair). There are a total of 9 blonds and 11 with brown hair. One person from the group is chosen randomly.

   $$pr(\text{girl or blond}) = pr(\text{girl}) + pr(\text{blond}) = pr(\text{girl and blond})\ \frac{12}{20} + \frac{9}{20} - \frac{5}{20} = \frac{16}{20}.$$

7. The probability of at least one event occurring out of multiple events is equal to one minus the probability of none of the events occurring. $pr(\text{at least one}) = 1 - pr(\text{none})$.

   **Example:** Roll a die 4 times. What is the probability of getting at least one head on the 4 rolls.

   $$pr(\text{at least one } H) = 1 - pr(\text{no } H) = 1 - pr(TTTT) = 1 - \frac{1}{2} \cdot \frac{1}{2} \cdot \frac{1}{2} \cdot \frac{1}{2} = 1 - \frac{1}{16} = \frac{15}{16}$$

8. If event *B* is a subset of event *A*, then the probability of *B* is less than or equal to the probability of *A*. $pr(B) \leq pr(A)$.

   **Example:** There are 20 people in the room: 12 girls (5 with blond hair and 7 with brown hair) and 8 boys (4 with blond hair and 4 with brown hair), *pr* (girl with brown hair) ≤ *pr*(girl)

   $$\frac{7}{20} \leq \frac{12}{20}.$$

**Q. 3. How is a normal distribution related to probability distribution?**

**Ans.** Normal distributions are one type of continuous probability distribution.

**Q. 4. If a normal distribution is always a probability distribution, is a probability distribution always a normal distribution?**

**Ans.** No, not every probability distribution is a normal distribution.

**Q. 5. If a normal distribution has mean 200 and standard deviation 20, find *K* so that the probability that a sample value is less than *K* is 0.975.**

**Ans.**
$$(K - 200)/20 = 1.96$$

Therefore $K = 239$.

**Q. 6. In making switches, it has been determined by empirical studies that there is, on average, one defect per switch. What is the probability of selecting a sample of five switches that contains zero defects?**

**Ans.** There are two methods to solve this problem. The first method is to use the above formula where $x = 0$, $n = 5$, and $p = 1$, therefore $np = 5 \times 1 = 5$.

$$P(0) = e^{-5}\,(5)^0/0!$$
$$= 0.00674/1 = 0.00674 \quad \text{or} \quad 0.674\ \%$$

The second and most widely used method is to use the Poisson tables that are published in most statistics books. To use the tables, find the value of *x* in the left most column, then find the value of np on the top row and read $P(x)$ at the intersection of the two values.

The Poisson table value for $P(0) = 0.006738$ or $0.674\%$

**Q. 7. In manufacturing the Que model car, a study determined that on average there are three defects per car. What is the probability of buying a Que with less than three defects?**

**Ans.** $P(\text{less than 3 defects}) = P(0) + P(1) + P(2)$

Use the Poisson tables and find $P(0)$, $P(1)$ and $P(2)$ where $np = 3$

$P(\text{less than 3 defects}) = 0.049 + 0.149 + 0.224 = 0.422$.

**Q. 8. What do you mean by conditional probability?**

**Ans.** Conditional probability is defined as the probability of an event occurring if another has occurred or has been specified to occur simultaneously, and the outcome of the first event affects the probability of the second event. Conditional events are not independent.

The probability of *B* occurring given that *A* has already occurred is stated as $P(B/A)$, where the symbol/means "given that."

The formulas for conditional probability are shown below. These are known as *Bayes Formulas*.

$$P(B/A) = \frac{P(A\ \&\ B)}{P(A)}$$

$$P(A/B) = \frac{P(A\ \&\ B)}{P(B)}$$

Since the two formulas have a common term $P(A\ \&\ B)$, they may be used together to solve many problems involving conditional probability.

Conditional events are not independent so $P(A\ \&\ B)$ is not equal to $P(A)\ X\ P(B)$. From Bayes formulas:

$$P(A\ \&\ B) = P(B/A)\ P(A)$$
$$P(A\ \&\ B) = P(A/B)\ P(B).$$

**Q. 9. The number of items bought in a store for a sample of 250 customers are given as follows:**

| Number of Items ($x_i$) | Number of custome ($n_i$) |
|---|---|
| 1 | 25 |
| 2 | 128 |
| 3 | 47 |
| 4 | 38 |
| 5 | 12 |

**Calculate:**

**(*i*) Probability distribution for each item.**

**(*ii*) Cumulative distribution for each item.**

**Ans.**

| Number of Items ($x_i$) | Number of Customers ($n_i$) | Probability Distribution $n_i/N$ | Cumulative distribution |
|---|---|---|---|
| 1 | 25 | 0.1 | 0.1 |
| 2 | 128 | 0.51 | 0.61 |
| 3 | 47 | 0.19 | 0.80 |
| 4 | 38 | 0.15 | 0.95 |
| 5 | 12 | 0.05 | 1.00 |
| Total (N) | 250 | | |

(*i*) By using formula:

$$p(x_i) = n_i/N \quad \text{for each } i = 1, 2 \ldots 5$$

$$p(x_1) = 25/250 = 0.1 \text{ and so on (see in above table)}$$

(*ii*) See in table.

**Q. 10. The life of a laser-ray device used to inspect cracks in aircraft wings is given by X, a continuous random variable assuming all values in the range $x \geq 0$. The pdf of the life time, in years is as follows:**

$$f(x) = \begin{cases} \frac{1}{2}e^{-x/2}, & x \geq 0 \\ 0 & \text{otherwise} \end{cases}$$

**calculate the probability that the life of the laser-ray device is between 2 and 3 years.**

**Ans.**

$$P(2 \leq X \leq 3) = \frac{1}{2}\int_2^3 e^{-x/2}\, dx$$

$$= e^{-3/2} + e^{-1}$$

$$= -0.223 + 0.368$$

$$= 0.145$$

**11. Discuss some properties of distribution.**

**Ans.** Each distribution object has one or more following properties:

- The **Mean** property returns the mean of the distribution.
- The **Standard Deviation** property returns the standard deviation of the distribution.

- The **Variance** property returns the variance.
- The **InterQuartileRange** property returns the difference between the third and the first quartile.
- The **Skewness** property returns the skewness of the distribution.
- The **Kurtosis** property returns the kurtosis supplement of the distribution.
- The **IsSymmetrical** property indicates whether the distribution is symmetrical about the mean. When a distribution is symmetrical, the skewness is zero. However, for most distributions, symmetry can the calculated much more efficiently from the distribution parameters.

**12. Explain population, mean and variance of discrete probability distribution.**

**Ans.** **Population:** The discrete probability distribution describes a **population**. since we have **populations**, we can describe them by computing various **parameters**. Two of the population parameters we studied previously are: **Population mean and population variance.**

**Population mean:** Our general definition of the population mean is:

$$\mu = \frac{\sum_{i=1}^{N} x_i}{N}$$

- If we know that the random variable X is **discrete**, wecan **re-express m in terms of the probability distribution of *X*.** We have:

$$\mu = \sum_{\text{all } x} xP(x)$$

- This parameter is also called the **expected value of *X*** and is **represented by *E*(*X*).**

**Population Variance:** The population variance of a discrete random variable can be expressed similarly. It is the weighted average of the **squared deviations** from the mean:

$$V(X) = \sigma^2 = \sum_{\text{all } x} (x - \mu)^2 \, P(x)$$

- As before, there is a "short-cut" formulation...

$$V(X) = \sigma^2 = \sum_{\text{all } x} x^2 P(x) - \mu^2$$

- The standard deviation is the same as before:

$$\sigma = \sqrt{\sigma^2}$$

**13. A statistics instructor has observed that the number of typographical errors in new editions of textbooks varies considerably from book to book. After some analysis he concludes that the number of errors is Poisson distributed with a mean of 1.5 per 100 pages. the instructor randomly selects 100 pages of a new book. What is the probability that there are no typos?**

**Ans.**

$$P(0) = \frac{e^{-\mu}\mu^x}{x!} = \frac{e^{-1.5}1.5^0}{0!} = .2231$$

**There is about a 22% chance of finding zero errors.**

**14.** ***In the continuation of Q-12*, Suppose that the instructor has just received a copy of a new statistics book. He notices that there are 40 pages.**

**(*a*) What is the probability that there are no typos?**

**(*b*) What is the probability that there are five or fewer typos?**

**Ans.** (*a*) First, note that we are now talking about an interval of 400 pages. In the original statement of the problem, we were told that the expected number of typos in an interval of 100 pages was $\mu = 1.5$. therefore, the expected number of typos in an interval of 400 pages is 4*1.5 = 6. Thus, when we deal with an interval of 400 pages, we must use $\mu = 6$ in the POisson distribution formula.

$$P(0) = \frac{e^{-\mu}\mu^x}{x!} = \frac{e^{-6}6^0}{0!} = .002479$$

Hence there is a very small chance there are no typos.

(*b*) For a 400 page book, what is the probability that there are five or less typos?

$$P(x \le 5) = P(0) + P(1) + \ldots + P(5)$$

$$= \frac{e^{-6}6^0}{0!} + \frac{e^{-6}6^1}{1!} + \ldots + \frac{e^{-6}6^5}{5!} = 0.446$$

There is about a 45% chance there are 5 or less typos.

## EXERCISES

1. Discuss normal distribution.
2. Differentiate between continuous random variable and discrete random variable.
3. What are deterministic and random variable.
4. What are properties of exponential distribution function.
5. What are the applications of normal distribution?
6. What do you mean by central tendency?
7. Given a normal distribution of values for which the mean is 70 and the standard deviation is 4.5. Find:

   (*a*) The probability that a value is between 65 and 80, inclusive.

   (*b*) The probability that a value is greater than or equal to 75.

   (*c*) The probability that a value is less than 62.

   (*d*) The 80th percentile for this distribution.
8. The lifetime of a battery is normally distributed with a mean life of 40 hours and a standard deviation of 1.2 hours. Find the probability that a randomly selected battery lasts longer than 42 hours.
9. Let $X: S \to (0; 1]$ have the uniform distribution, and let $Y = 1 = X$. Find the probability density function $f_Y: [1, \infty) \to [0, \infty]$ for $Y$.

# 7

# Continuous and Discrete System Simulation

**INSIDE THIS CHAPTER**



## 7.1 SIMULATION OF CONTINUOUS SYSTEM *[UPTU 2006-07]*

There are two fundamentally different types of system:

(*i*) Continuous system

(*ii*) Discrete system

In continuous system, state changes smoothly or continuously with time are called continuous systems for example the movement of the aircraft occurs smoothly.

In discrete system changes are predominantly discontinuous or random.

For example, in factory system, changes are predominantly discontinuous. The inventory system is also a discrete system because demand is discrete number. In this chapter, we will deal with continuous systems only.

Continuous dynamic systems, are those system where the state or the variables vary continuously with time. It generally described in term of differential equations.

If the set of differential equations describing a system are ordinary, linear, and time-invariant then any analytic solution is easy to obtain. Generally differential equations of a more difficult nature can only be solved numerically. So we need simulation to get required numerical solution.

The methods of applying simulation to continuous models where the differential equations are linear and have constant coefficients, and then generalizing to more complex equations.

## 7.2 DIFFERENTIAL EQUATIONS

Consider the example of a linear differential equation with constant coefficients was given in Section 1.10.2, to describe the wheel suspension system of an automobile. The equation was

$$M\ddot{x} + D\dot{x} + Kx = KF(t)$$

dependent variable $x$ appears together with its first and second derivatives $\dot{x}$ and $\ddot{x}$ and that the terms involving these quantities are multiplied by constant coefficient added. The quantity $F(t)$ is an input to the system, depending upon the independent variable $t$. A linear differential equation with constant coefficients is always of this form, although derivatives of any order may enter the equation.

If the dependent variable or any of its derivatives appear in any other form, such as being raised to a power, or are combined in any other way (*e.g.*, multiplied together). The differential equation is said to be non-linear.

### 7.2.1 Partial Equation

When more than one independent variable occurs in a differential equation, the equation is said to be a *partial differential equation.* It can involve the derivatives of the same dependent variable with respect to each of the independent variables.

The general method of solving such equations numerically is to use finite differences to convert the equations into a set of ordinary differential equations, which can be solved by the described methods.

**Note:** Differential equations, both linear and non-linear, occur repeatedly in scientific and engineering studies.

## 7.3 ANALOG VS. DIGITAL SIMULATION

For most studies the model is clearly either of a continuous or discrete nature, and that is the determining factor in deciding whether to use an analog or digital computer for system simulation with an analog computer, however, is more properly described as being based on a mathematical model than as being a physical model. The most widely used form of analog computer is the electronic analog computer, based on the use of high gain direct current amplifiers, called operational amplifiers. It consists special purpose elements such as integrators, multipliers, adders, function generation and non-linear elements. These variables and their derivatives are represented by means of voltages.

Historically analog computers came into being years before digital computers did, and have played a major role in simulation of continuous dynamic systems. Although analog computers are giving way to digital computers at an increasing pace, they are still in occasional use.

### 7.3.1 Disadvantages of Analog Computers

1. **Inadequate Accuracy:** Digital simulation is more accurate than that from an analog simulation. The accuracy of analog simulation depends on the accuracy of the components being used.
2. **Scaling Needed:**
   - The magnitudes of dependent variables are represented in an analog computer by voltages. This magnitude scaling is a tedius task in an analog simulation.
   - While in case of digital computer, magnitude scaling problem does not arise.
   - In magnitude scaling, analog computer is also require time scaling.
3. **Hardware Set up Necessary:** In analog simulation input constants and initial conditions are incorporated by setting up voltages and potentiometers then the various elements like amplifiers, voltmeters, etc., have to be connected on a patch board. Such setting up is not needed in a digital simulation.

### 7.3.2 Advantages of Analog Simulation

1. Analog computer is faster than a digital computer, because it can be solving many equations in a truly simultaneous manner.

**2.** The analog representation of a system is often more natural as it directly reflect the structure of the system and display of computed results.

### 7.3.3 Difference between Analog Simulation and Digital Simulation

| Analog Simulation | Digital Simulation |
|---|---|
| 1. An analog computer must usually be dedicated to one application at a time. | 1. A digital computer can be easily used for many different problems. |
| 2. The accuracy of analog simulation depends on the accuracy of the component being used. | 2. The result from digital simulation is more accurate. |
| 3. This limited accuracy cannot meet the need in simulating systems like missiles and space vehicles. | 3. For real time system, we need more accuracy, *i.e.*, result of digital simulation. |
| 4. Magnitude scaling is needed in analog simulation. | 4. Magnitude scaling is not needed in digital simulation due to floating point arithmetic. |
| 5. Analog simulation also needed time scaling. | 5. In digital simulation time scaling is generally not needed. |
| 6. Hardware set up is necessary in analog simulation. | 6. Hardware setting is not necessary in digital simulation. |
| 7. A simulation program on analog computer cannot be easily stored for reuse. | 7. A simulation program on digital computer can be easily stored for reuse. |
| 8. Calibration and accuracy test is needed for rapid switching from one simulation to another. | 8. Calibration and accuracy test is not needed for rapid switching from one simulation to another. |
| 9. Analog simulation provide higher speed of solution for certain application. | 9. Digital simulation provide slow speed compare to analog simulation. |
| 10. It can be solve many equations in a truly simultaneous manner (so it is faster). | 10. It can work only on one equation at a time. |

## 7.4 SIMULATION OF A WATER RESERVOIR SYSTEM [*UPTU 2006-07*]

Consider an example of constructing a dam across a river to create a reservoir. The reservoir is to be constructed at a specified site. The curve of the projected demand for the water from the reservoir has been determined.

Input to the reservoir is:

1. River inflow
2. Rainfall direct flow

Output of the reservoir is:

1. Seepage
2. Evaporation
3. The water supplied to meet the projected demand. Graphically we can represent this system as follows:

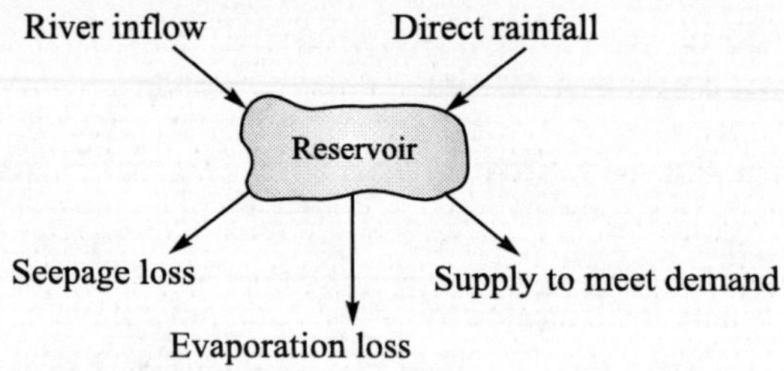


**Fig. 7.1.** A simple run-of-river storage demand system.

- The amount of seepage loss is not a constant quantity actually it depends on the volume of the water stored.
- The evaporation loss depends on the area of the exposed surface and the coefficient of evaporation (seasonal variation is also there).
- Supply that meets the demand.

We have to calculate these two losses, *i.e.,* seepage and evaporation loss for a given volume of water in the reservoir at time of the year.

In our real life, we cannot have any finite sized reservoir that can provide guarantee of meeting the demand 100% at every time because river inflow, the rainfall, the losses and the demand are all random variables. To build such a large dam which will always meet the demand throughout its entire life will generally be uneconomical.

But we can determines the reservoir size which will meet the demand with a specified risk of failure. For example, 4% failure means that once in 25 years reservoir would become empty before meeting the demand for water. Now our objective of the study is to determine the size of the reservoir with a specified risk of failure.

As the volume varies continuously with time, we are dealing with a continuous system.

We can take one month as the basic time interval for the purpose of simulation study.

**Example:** If want to simulate the system for 75 years; the simulation run length will be 900.

$$(12 \times 75 = 900)$$

The simulation will be repeated assuming several different capacities of the reservoir. The output will be in series shortages for each capacity.

**Process to Perform the Simulation**

It follows the number of steps:

**Step 1:** For the current month $C_m$ of the current year $C_y$, calculate total amount of river inflow ($R_{\text{flow}}$) and the total rainfall (Rain) directly over the reservoir. Let the sum of these two inputs is given by $T_{\text{in}}$.

$$\boxed{T_{\text{in}} = \text{Rain} + R_{\text{flow}}}$$

**Step 2:** Add the volume of total input $T_{\text{in}}$ to the volume left over in the reservoir at the end of the last month, $V(C_m - 1)$. So Gross Volume (GV) is given by

$$\boxed{GV = T_{\text{in}} + V(C_m - 1)}$$

**Step 3:** On the basis of the last month's volume $V(C_m - 1)$, we will calculate this month's seepage ($SPG$) and evaporation losses ($EVAP$) and add then as total loss ($T_{\text{loss}}$).

$$\boxed{T_{\text{loss}} = SPG + EVAP}$$

**Step 4:** From the demand data determine the demand of water for the current month, let it be *DM.*

**Step 5:** If $T_{loss} \geq GV$ then

(*i*) The reservoir becomes dry without supplying any water so shortage

$$\text{SHORT} = DM$$

(*ii*) The volume of water left at the end of the current month.

$$V(C_m) = 0$$

(*iii*) Spillage $\quad \text{Spill} = 0$

(*iv*) Goto **Step 8**

else if $\quad T_{loss} < GV$

Then the net water volume available to meet the demand is

$$V_{net} = GV - T_{loss}$$

**Step 6:**

$$\text{If } DM \geq V_{net} \text{ then}$$

(*i*) The reservoir runs dry

(*ii*) Shortage is given by

$$\text{Short} = DM - V_{net}$$

(*iii*) $\quad \text{Spill} = 0$

(*iv*) Go to step 8

else if $DM < V_{net}$, then the differentiation is given by diff (diff $= V_{net} - DM$) this amount of water is left over.

**Step 7:** If this difference (diff) left over water exceeds the capacity cap of the reservoir then there will be a spill over.

If $\quad$ diff > Cap then

(*i*) $\quad$ spill = diff – Cap

(*ii*) $\quad V(C_m) = \text{Cap}$

else if diff ≤ Cap then

(*i*) $\quad$ spill = 0

(*ii*) $\quad V(C_m) = \text{diff}$

**Step 8:** At the last, print out the spill and short for this month and move to the next month. If the period exceeds the intended simulation length then stop this process otherwise go to **step 1.**

The output will be series of monthly shortage and spills. The shortages could be combined into total annual shortages.

These annual shortages can be ranked according to the amount of shortage involved. We will compute series shortage for each capacity of reservoir to determine the capacity of reservoir or size of reservoir.

## 7.5 SIMULATION OF A SERVO SYSTEM

Simulation of servo system is very important application of continuous system simulation. It is most important and widely used forms of control system.

For simulation purpose let we study the behaviour of a second order nonlinear feedback system represented by the following differential equation.

$$\ddot{y} = M(1 - y^2)\ \dot{y} - Ny + e_{in} \qquad ...(1)$$

where $M$ and $N$ are positive constants

$e_{in}$ is value of input signal.

Block diagram of equation (1) is represented by Fig. 2.

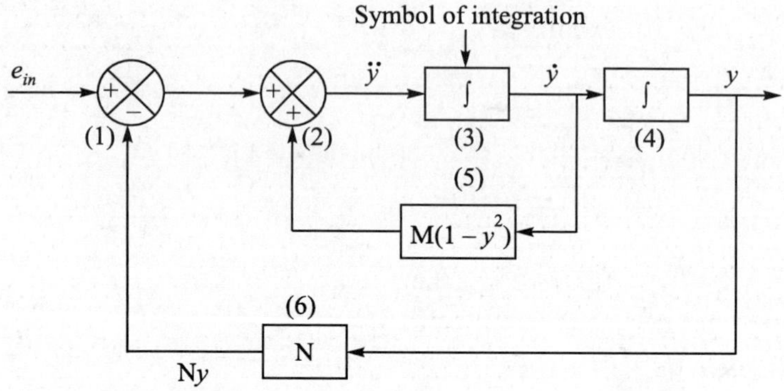


**Fig. 7.2.** A non-linear second order servo system.

This equation is described by the Van der Pol equation. Van der Pol introduced as a mathematical model of self-sustained oscillations of triode circuit with a cubic current-voltage characteristics. This equation serves as a basic model of self-sustained oscillations arising in systems of mechanical and electrical engineering, biology and biochemistry, and many other application.

Van der Pol equation:

$$\ddot{y} - \mu\,(1 - y^2)\ \dot{y} + y = 0 \quad \text{where } \mu > 0 \text{ is a constant.}$$

The above block diagram (Fig. 7.2) simply describe the equation.

Each blocks performs a mathematical operation. Blocks (1) and (2) provides mathematical operation, while (3) and (4) provide integration and block 5 provide $M(1 - y^2)$ and block (6) provide $N$ value of positive symbol. The interconnection of the blocks is self-explanatory.

Some example of man-made servo systems are:

- ✓ Beating of the heart
- ✓ Periodic opening and closing of flowers in response to the sunlight.
- ✓ A neon-lamp oscillator.
- ✓ Squeaking of door with rusty hings.

In equation (7.1), if the case of zero input signal then equation modified as

$$\ddot{y} = M(1 - y^2)\ \dot{y} - Ny \qquad ...(2)$$

Let us simulate this system. The second order differential equation can be written as a set of two simultaneous equations of first order.

Let take $\qquad y = y_1$

Then $\qquad \dot{y}_1 = y_2$

$$\dot{y}_2 = M(1 - y_1^2)\, y_2 - Ny_1$$

Let take some specific value of constant $M = 0.1$, $N = 1.0$ and let the initial condition be

$$y_1(0) = 1.0$$
$$y_2(0) = 0$$

Our equations become $\dot{y}_1 = y_2$

and $$\dot{y}_2 = 0.1(1 - y_1^2)\, y_2 - y_1 \quad ...(3)$$

An instantaneous description of the state of the system is given by the output of the two integrators, *i.e.,* by variables $y_1$ and $y_2$. We will use, the fourth-order Runge-Kutta method to obtain the values of $y_1$ and $y_2$ as a function of time.

We will choose the step-size $\Delta t = H = 0.001$ second and simulate the system for 5 seconds. So we can have $n = 5{,}000$ number of steps. This is very large number of outputs to be plotted or examined. We will therefore, print out the values of $y_1$ and $y_2$ only once every 100 integration steps.

## 7.6 SIMULATION OF AN AUTOPILOT SYSTEM [*UPTU 2007-08*]

For simulation of an autopilot system first we have to consider a mathematical model of the aircraft system. The error signal $\varepsilon_1$ has been defined as the difference between $\theta_i$ and $\theta_o$.

$$\varepsilon = \theta_i - \theta_o \quad ...(4)$$

$\theta_i$ = desired heading, or input, $\theta_i$

$\theta_p$ = Actual heading, or output, $\theta_o$.

Let assume the rudder is turned to an angle proportional to the error signal, so that the force changing the aircraft heading is proportional to the error signal.

Instead of moving the aircraft sideways, the force applies a torque which will turn the aircraft. The torque ($T$) acting on the aircraft can, therefore, be represented by the following equation

$$\text{Torque } (T) = K\varepsilon - R\dot{\theta}_o \quad ...(5)$$

$K$ and $R$ are constant

$T$ is torque produced by the rudder

$(K\varepsilon - R\dot{\theta}_o)$ is the viscous drag

The fundamental law of mechanics is that the acceleration of a body is proportional to the applied force. That was related to linear motion. The same law, applies to a turning motion: the angular acceleration of a body is proportional to the applied torque. Further, the coefficient of proportionality is the inertia of the body, denoted by $I$. Since the angular acceleration of the aircraft is the second derivative of its heading, the equation of motion is

$$I\ddot{\theta}_o = \text{torque} \quad ...(6)$$

Equations (4) and (5) and transposing terms, the resultant equation is

$$I\ddot{\theta}_o + R\dot{\theta}_o + K\theta_o = K\theta_i \quad ...(7)$$

Now divide both sides of equation (4) by $I$, then

$$\ddot{\theta}_o + \frac{R}{I}\dot{\theta}_o + \frac{K}{I}\theta_o = \frac{K}{I}\theta_i$$

Let $$\frac{R}{I} = 2\zeta\, w,\ \frac{K}{I} = \omega^2$$

The equation of motion relating output to input then takes the following form:

$$\ddot{\theta}_o + 2\zeta\omega\dot{\theta}_o + \omega^2\theta_o = \omega^2\theta_i \qquad ...(8)$$

This is a second order differential equation, it is in exactly the same form as our previous equation

$$M\ddot{x} + D\ddot{x} + Kx = KF(t)$$

It describes the relationship between output and input for the automobile wheel suspension system.

Let aircraft is initially flying a steady course which, we take to be the zero heading as per definition. If it is asked to change to a new heading at time zero, this corresponds to a unit step change of input results show that the aircraft will settle to the new heading, and the response will be oscillatory if $\zeta$ is less than 1.

Means $\boxed{\zeta < 1}$ for oscillatory

To simulate autopilot system we have to use a model in the form of the three individual equations (4), (5) and (6).

Important factor is error signal and the applied torque error signal appears as an electrical signal in the autopilot. It can easily be modified by using electrical circuits. The motion of the aircraft can also be measured by instruments giving electrical signals that can be added to the autopilot output. The modifications affect the aircraft response by changing the applied force. This can be done is to add to the error signal a component that is proportional to the integral of the error signal.

The applied force is than

$$F(t) = K\varepsilon + K_1 . \int \varepsilon \, dt$$

By this modification control theory shows that for the system to remain stable, we must have

$$K_1 < \frac{KR}{I}$$

In every case, the value of $K_1$ was chosen so that

$$\frac{K_1}{K} = \frac{0.2R}{I}$$

This result is called integral control. It will eliminate the lag when the aircraft is performing a steady run.

## 7.7 DISCRETE-EVENT-SIMULATION

Discrete-event-simulation concerns the modeling of a system as it evolves over time by a representation in which the state variables change instantaneously at separate points in time. These points in time are the ones at which an event occurs, where an event is defined as an instantaneous occurrence that may change the state of the system. OR we can define "system in which the state changes abruptly at **discrete points** in time are called discrete system".

Although discrete-event simulation could conceptually be done by hand calculations, the amount of data that must be stored and manipulated for most real-world systems dictates that discrete-event simulations be done on a digital computer.

**Example:**

1. Sale of an item from the stock or arrival of an order.
2. Arrival of a car at an intersection is an event if we are simulating street traffic.

## 7.8 TIME ADVANCE MECHANISM

As we know that the dynamic nature of discrete event simulation models; so we must need a mechanism to advance simulated time from one value to another. We call the variable in a simulation model that gives the current value of simulated time the simulation clock.

**Note:** Simulation clock is the current value of simulated time.

Two main approaches are suggested for advancing the simulation clock.

1. Next-event time advance or event to event.
2. Fixed-increment time advance or fixed time event.

The first approach is used by all major simulation software and by most people coding their model in a general-purpose language, and since the second is a special case of the first, we shall use the next-event time advance approach for all discrete-event simulation models.

**Note:** The system in which changes are discontinuous, is called as Discrete System. Each changes in the state of the system is called an event.

### 7.8.1 Event to Event Model

1. In event-to-event simulation model computer advances time to the occurrence of the next event.

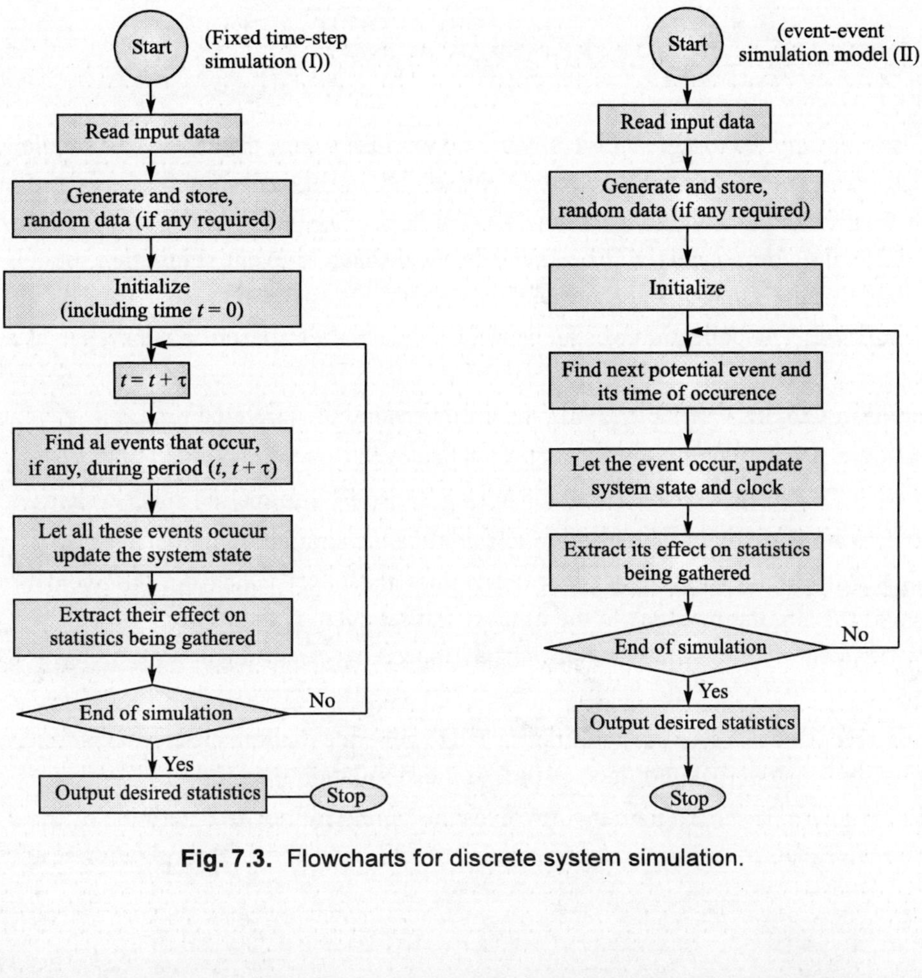


**Fig. 7.3.** Flowcharts for discrete system simulation.

2. It shifts from event to event.
3. The system state does not change in between. Only those points in time are kept track of when something of interest happens to the system.
4. In general, the event-to-event model is preferred because we do not waste any computer time in scanning those points in time when nothing takes place.
5. The only drawback of the event-to-event model is that usually its implementation (programming for it) turns out to be more complicated than the fixed time step model.

Flowcharts for these two methods of simulating a discrete system in their most general forms are shown in Fig. 7.3.

### 7.8.2 Fixed Time-Step Model

Following characteristics of fixed time-step model:

1. In a fixed time-step model a "timer" or "clock" is simulated by the computer.
2. This clock is updated by a fixed time interval $\tau$ and system is examined to see if any event has taken place during this time interval $\tau$.

   *OR*

   System is examined to see if any event has taken place during this time interval (min, hr, days, whatever).

## 7.9 COMPONENT AND ORGANIZATION OF A DISCRETE-EVENT SIMULATION

Simulation has been applied to a great diversity of real world systems, discrete-event simulation models all share a number of common components and there is a logical organization for these components that promotes the programming, debugging, and future changing of a simulation model's computer program.

The following components will be found in most discrete-event simulation models using the next-event time-advance approach programmed in a general-purpose language.

1. **System State:** The collection of state variables necessary to describe the system at a particular time.
2. **Simulation Clock:** A variable giving the current value of simulated time.
3. **Event List:** A list containing the next time when each type of event will occur.
4. **Statistical Counters:** Variables used for storing statistical information about system performance.
5. **Initialization Routine:** A subprogram to initialize the simulation model at time 0.
6. **Timing Routine:** A subprogram that determines the next event from the event list and then advances the simulation clock to the time when that event is to occur.
7. **Event Routine:** A subprogram that updates the system state when a particular type of event occurs.
8. **Library Routines:** A set of subprograms used to generate random observations from probability distributions that were determined as part of the simulation model.
9. **Main Program:** A subprogram that involves the timing routine to determine the next event and then transfers control to the corresponding event routine to update the system state appropriately.

The main program may also check for termination and invoke the report generator when the simulation is over.

10. **Report Generator:** A subprogram that compute estimates of the desired measures of performance and produces a report when the simulation ends.

## 7.10 EVENT GRAPHS

Event graphs were introduced by Schruben (1983) to develop simulations of discrete-event systems. The system dynamics are characterized by *events* that change the state of the system and the *logical and temporal relationships* among these events. Their utility and many examples are discussed in Schruben (1995). An event-graph model consists of several components. The state of the system is defined by the values of a set of state variables *S*. In the graph, the set of vertices *V* represents the events, and the set of directed edges *E* represents the temporal and logical relationships between pairs of events. Associated with each vertex *v* is a function *fv* that describes the state changes caused by the event. Each directed edge *eod* = (*vo*, *vd*) indicates that if the edge condition *ce* holds when the origin event *vo* occurs, the destination event *vd* will occur after some time delay *te.* It is possible for an edge to be unconditional or for *vd* to be scheduled to occur without any simulated delay. The basic building block of the event graph is as follows in Fig. 7.4.

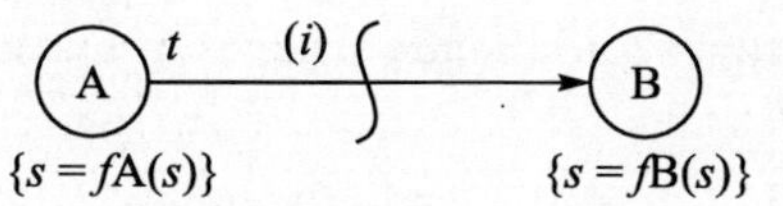


**Fig. 7.4**

This construct indicates that *"whenever event A occurs, the system state s changes to fA*(*s*)*. Then, if condition* (*i*) *is true, event B will he scheduled to occur after a delay of te.*" The model can be interpreted by reading each component in a similar way. Corresponding labels are omitted if the time delay is zero or if the scheduling is unconditional. The simulation maintains a *clock time* $\tau$ and an *events list* $\Lambda$ that is a record of scheduled (pending) events. Each event is listed along with the scheduled time of execution so the events list is a set of pairs ($\tau$ *i*, *vi)*, where $\tau$ *i* is the scheduled execution time and *vi* is the scheduled event. $\Lambda$ is a priority queue, prioritized according to increasing values of $\tau i$. At each step, the simulation executes the highest priority event in the set, which is the one with the earliest scheduled execution time. An event is executed by setting the simulation clock to the scheduled time of the pending event ($\tau \leftarrow \tau\ i$), then evaluating the function *fi* and the conditions {*ce*} on the outgoing edges. All destination events of scheduling edges with satisfied conditions are placed on the events list with $d \leftarrow \tau + te$.

The simulation is initialized by placing a set of events on the events list (with $\tau\ i \leftarrow 0$, usually) and continues until at least one of the following conditions is satisfied:

(*i*). The event being executed is designated as a termination or End event (*e.g.*, the closing of a bank).

(*ii*). $\tau > \tau$ end, where $\tau$ end is some predetermined termination time (*e.g.*, for a simulation of an eight-hour work day).

(*iii*). $\Lambda$ is empty (that is, there are no further events to be processed).

## SUMMARY


- Systems in which the state changes smoothly or continuously with time are called continuous systems.
- Accuracy of analog simulation depends on the accuracy of the components being used hence results from a digital simulation is more accurate than that from analog.
- Simulation of a water reservoir system is proposal of constructing a dam across a river to create a reservoir, objective is to determine the size of the reservoir with a specified risk of failure.
- Simulation of a servo system concerns the study of a second order nonlinear feedback system.
- Simulation of an autopilot system is concerns mathematical model of the aircraft system. It has discussed two important factor error signal (ε) and torque (τ).
- In discrete system simulation, there are two basic approaches for advancing the simulation clock:

  (*i*) Next event time or event-to-event model.

  (*ii*) Fixed increment time advance or fixedtime step.


## SOLVED QUESTIONS

**Q. 1. What are the main advantages and disadvantages of analog simulation?**

**Ans. Advantages:**

1. Analog simulation provides higher speed of solution.
2. There is direct accessing and immediate display of the computed results.

**Disadvantages:**

1. Inadequate accuracy
2. Need of scaling
3. Hardware set up is necessary.

**Q. 2. Discuss continuous dynamic system?**

**Ans.** Continuous dynamic systems are those systems in which the state or variables vary continuously with time. It can be generally described by differential equations.

**Q. 3. How can you prove that pure pursuit problem is an example of continuous system?**

**Ans.** The system of pure pursuit, is an example of continuous system because the position of the two aircraft can vary continuously with time.

**Q. 4. Compare numerical integration and a continuous system simulation.**

**Ans.**

| System simulation | Numerical integration |
|---|---|
| 1. System simulation always keep track of the state of the system explicitly. | 1. In numerical integration no such corresponding is preserved. |
| 2. There is one-to-one correspondence between what the computer does and what takes place in the system. | 2. It destroy the one-to-one correspondence between the computer steps and the original system. |
| 3. In simulation we look upon the equations as one of the steps in the entire process. | 3. In numerical calculation, we only see the given set of differential equations as a mathematical object and proceed to integrate them. |

## EXERCISES

1. Define following term :
   (*a*) Digital Analog Simulators (*b*) Hybrid Computers.
2. Draw the block diagram for the automobile suspension problem.
3. What is the objective of real-time simulation?
4. Explain in brief, the simulation of servo system.
5. Differentiate following:
   (*i*) Analog simulation *vs*. Digital simulation
   (*ii*) Numerical integration *vs.* Continuous system.
6. What do you mean by the simulation of autopilot system?
7. How simulation of water reservoir system is carried out?

# 8

# System Dynamics

INSIDE THIS CHAPTER



## 8.1 CONCEPTS OF SYSTEM DYNAMICS

System dynamics is the rigorous study of problems in system behaviour using the principles of feedback, dynamics and simulation. In more words system dynamics is characterized by:

- Searching of useful solutions to real problems, especially in social systems (business, schools, governments, ...) and the environment.
- Using methods and insights from feedback control engineering and other scientific disciplines to assess and improve the quality of models.
- System dynamics approach looks at systems at a very high level so it is more suited to strategic analysis.
- System dynamics analysis is continuous in time and it uses mostly deterministic analysis.

**System dynamics modeling approach are useful in following areas:**

- Short-term and long-term forecasting of agricultural produce with special reference to field crops and perennial fruits such as grapes, which have significant processing sectors of different proportions of total output where both demand and supply side perspectives are being considered.
- Long-term relationship between the financial statements of balance sheet, income statement and cash flow statement balanced against scenarios of the stock market's need to seek a stable/growing share price combined with a satisfactory divident and related return on shareholder funds policy.
- Managerial applications include the development and evaluation of short-term and long-term strategic plans, budget analysis and assessment, business audits and benchmarking.

System dynamic has come to used following different areas of student.

1. Initial studies of industrial systems, are called **Industrial Dynamics Studies.**
2. Dramatic changes in urban centers, are called **Urban Dynamics Studies.**

3. Many social and environmental problems, which can be seen world-wide, study of these problems are called **World Dynamics.**

There is no major differences in the techniques used in these different areas, the general name System Dynamics has come to be used.

**The Output of System Dynamics are:**

- It suggest some reorganization, or change in policy.
- It can solve an existing problem.
- It can guide developments away from potentially dangerous directions.
- Models most sometimes reduce qualitative factor as personal references, or social tension, or quantitive form.

## 8.2 EXPONENTIAL GROWTH MODEL [*UPTU 2007-08*]

Growth means a rate of change. Mathematically if we went to describe growth it means involvement of differential equation. There are so many example of exponential growth, some are:

1. **Human Population:** If the number of births and deaths per person per year were to remain at current levels.
2. **Nuclear Chain Reaction:** Each uranium nucleus that undergoes fission produce mulitiple neutrons, each of which can be absorbed by adjacent uranium atoms, causing them to fission in turn.
3. Growth of a capital fund that is earning compound interest.

If the growth rate coefficient is $r$, then the rate at which the population, fund or nuclear chain reaction grows is $r$ times the current size of fund or population, etc.

- Let $x$ is the current size of the fund then mathematically

$$dx/dt = rx \qquad dx/dt = \dot{x}$$

$$\dot{x} = rx \qquad ...(1)$$

$$x = x_0 \quad \text{at} \quad t = 0$$

This is first order differential equation whose solution is the exponential function. The solution is

$$x = x_0 e^{rt} \qquad e = 2.72 \qquad ...(2)$$

If we plot a graph for $x$ with various values of $r$, then following conclusions. We can have:

- Population or fund grows indefinitely, for all positive value of *n.*
- It grows faster with greater values of $r$.

In Fig. 8.1, we have seen that slope *X*, at $r = 0.2$ and $x = 2$ has certain value, *i.e.*, 3.46 that becomes double in slope *Y*, at $r = 0.2$ and $x = 4$, having certain values 6.93.

| Slope | $r$ | $x$ | Time ($t$) or certain value |
|---|---|---|---|
| *X* | 0.2 | 2 | 3.46 |
| *Y* | 0.2 | 4 | 6.93 |

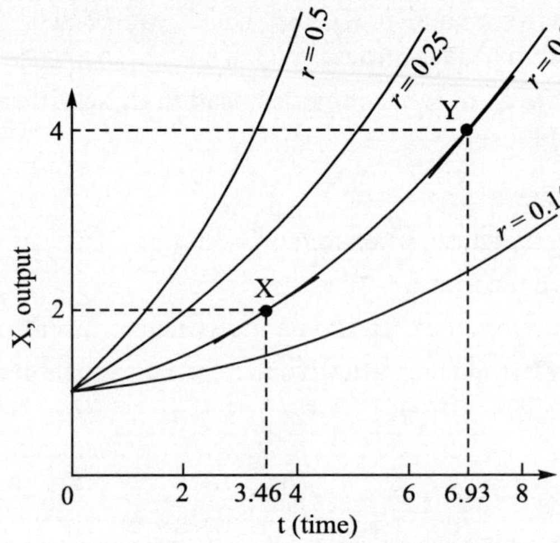


**Fig. 8.1.** Exponential growth curves.

Let $r = 0.2$, $x = 2$ and $x_0 = 1$ put this value in equation (2).

$$2 = 1 \times e^{.2t}$$

Taking log on both side

$$\log 2 = .2t$$

$$.2t = 0.693$$

$$t = \frac{0.693}{0.2} \approx 3.46$$

If $r = 0.2$, and $x = 4$ and $x_0 = 1$ then we have

$$4 = e^{.2t}$$

Taking log on both sides

$$\log 4 = .2t$$

$$2 \log 2 = .2 = .2t$$

$$2 \log 2 = .2t \Rightarrow 2 \times .693 = .2t$$

$$t = \frac{\not{2} \times .693}{.\not{2}} = 6.93$$

Slope *Y* has become twice because slope is measured as the first-order differential coefficient, this fact is simply the reflection of the law defining exponential growth:

"The growth rate is directly proportional to the current level."

There is another way of describing the exponential function that is—"logarithm of the variable increases linearly with time."

To represent exponential growth, the logarithms of the data should plotted against time.

If data appear to fall on a straight line the growth is **exponential**, and the slope of the straight line will be greater for larger growth rate coefficient.

Alternatively, the data can be plotted on semi logarithmic graph paper where horizontal lines are placed at logarithmic intervals.

**Example:** Figure 8.2 shows the growth of gross national products.

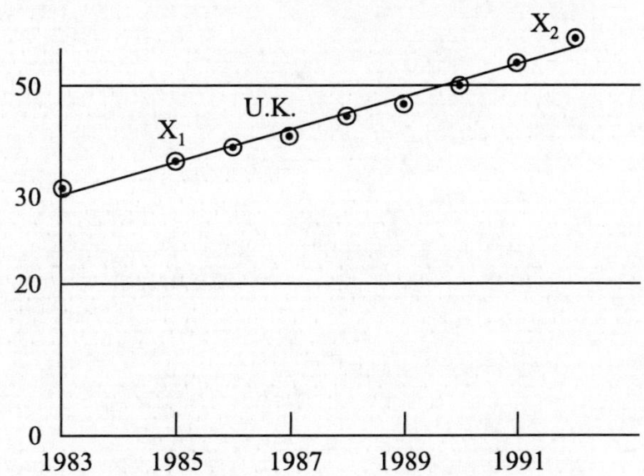


**Fig. 8.2.** Growth of gross national products.

### 8.2.1 How Growth Rate is Computed?

The growth rate co-efficient ($r$) can be computed by selecting two points of the straight line that best fitted on straight line, and take logarithm of the ratio of the values.

Let points are $x_1$ at $t_1$ and $x_2$ at $t_2$ $(t_2 > t_1)$

$$x_1 = x_0\, e^{t_1 r} \quad (x_0 = 1) \qquad ...(i)$$

$$x_2 = x_0\, e^{t_2 r} \qquad ...(ii)$$

or $$x_1 = e^{t_1 r} \quad \text{and} \quad x_2 = e^{t_2 r}$$

or $$\frac{x_2}{x_1} = \frac{e^{t_2 r}}{e^{t_1 r}} \quad \text{or} \quad e^r(t_2 - t_1)$$

Taking log

$$\ln\left(\frac{x_2}{x_1}\right) = r(t_2 - t_1) \text{ (log at base } e)$$

Let take log at base 10 then it becomes

$$\boxed{\log\left(\frac{x_2}{x_1}\right) = 0.434\,(t_2 - t_1)r} \qquad ...(iii)$$

Looking at the figure for UK in Table 8.1

$$t_1 = 1985 \quad t_2 = 1991$$
$$x_1 = 35.3 \quad x_2 = 55.6$$

We have selected above two points where data is best fitted on straight line

then ratio $$\frac{x_2}{x_1} = \frac{55.6}{35.3} = 1.575$$

By the formula (*iii*): $\log(1.575) = 0.435\,(1985 - 1991)$

$$0.1973 = 0.434 \times 6r$$

$$r = \frac{0.1973}{2.6}$$

$$r = 0.075$$

or $$r = 7\frac{1}{2}\%$$

**Table 8.1**

| Year | UK |
|---|---|
| 1983 | 30.2 |
| 1985 | 35.3 |
| 1986 | 37.6 |
| 1987 | 39.8 |
| 1988 | 42.9 |
| 1989 | 45.7 |
| 1990 | 49.9 |
| 1991 | 55.6 |

Sometime the coefficient $r$ is represented in the form

$$r = \frac{1}{T}$$

The solution for the exponential growth model then it takes following form

$$x = x_0\ e^{t/T}$$

## 8.2.2 The Constant *T*

$T$ is said to be a time constant since it provides a measure of how rapidly the variable $x$ grows.

- When $t$ equals to $T$, the variable is exactly $e$ times of its initial value $x_0$.

$$t = T$$
$$x = x_0\ e^{T/T}$$
$$x = x_0\ e$$
$x = ex_0$ : that is $e$ times of its initial value $x_0$

- The inverse relationship between $r$ & $T$ means that a large coefficient is associated with a small time constant so, a more rapid rate of increasing.

**Example:**
- If growth rate $r = 0.05$ or 5%, population will be double in 14 years.
- If growth $r$ = 10%, then population will be double at = 07 years.
- If growth $r$ = 20% then population will be double at = 3½ years.

We can say 14, 7, $3\frac{1}{2}$ ... etc., are the time constants of the three cases.

## 8.3 EXPONENTIAL DECAY MODEL *[UPTU 2006-07]*

We have another model that is closely related to the exponential growth model, in this model, variable decays from some initial value, $x_0$, at a rate proportional to the current value. This model can be represented as a negative growth model.

The equation for the model is:

$$\frac{dx}{dt} \propto -x$$

or $$\frac{dx}{dt} = -rx$$

$$x = x_0 \quad \text{at} \quad t = 0$$

Solution is $$x = x_0 \, e^{-rt}$$

Figure 8.3 is plotted for various values of $r$.

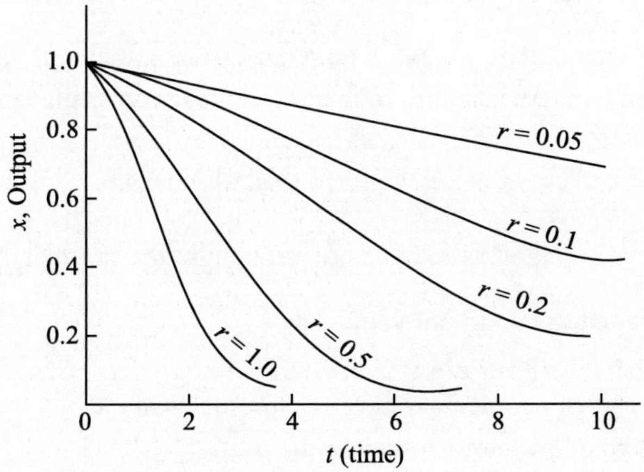


**Fig. 8.3.** Exponential decay curves.

**Characteristics of Exponential Decay Model:**

1. The constant $r$ is sometimes expressed in the form $\frac{1}{T}$.

*i.e.,* $$r = \frac{1}{T}$$

2. Level $x$, is divided by a constant factor for a given interval of time

$$x = x_0 \, e^{-rt} \quad r = \frac{1}{T} \text{ then}$$

$$x = x_0 \, e^{-t/T}$$

At $t = T$ $$x = x_0 \, e^{-1}$$

or $$x = \frac{x_0}{e}$$

In the interval of $T$ time units, the level is divided by $e$

since $e = 2.72$ and $x_0 = 1$

Then $x = 1/2.72 \approx 0.37$

(level is reduced by a factor .37)

3. Each successive interval of T reduces the level by the same factor.

**Example:** • Radioactive material decay.

• Decay model is an aging population, which is being diminished by deaths or breakdowns.

## 8.4 MODIFIED EXPONENTIAL GROWTH MODELS

The exponential growth model states that the rate of growth is directly proportional to the existing level and it implies indefinitely growth

$$\frac{dx}{dt} \propto x$$

So we cannot apply this model on market model. Market model describe how many items of a product will be sold. Generally, there is a limit to market, if we ignore replacement sales. We can make following assumption for market model.

- Growth rate is proportional to the no. of people who have not yet bought the product.

$$\frac{dx}{dt} \propto \text{people who have not yet bought the product} \quad ...(i)$$

- Let market is limited to maximum value max
  max → No. of people who might buy the product.
- Let $x$ be the number of people who have bought the product.
- Then the people who have not yet bought the product = (max – $x$).

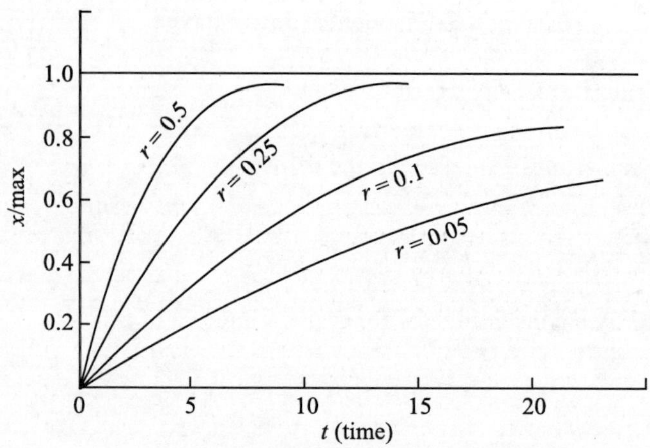


**Fig. 8.4.** Modified exponential curves.

- If $r$ is coefficient of proportionality then the equation for the model is

$$\frac{dx}{dt} \quad \text{or} \quad \dot{x} \propto (\max - x) \qquad \text{...(ii)}$$

$$\dot{x} = r(\max - x)$$

$$x = 0 \quad \text{at} \quad t = 0$$

Solution is $$x = \max(1 - e^{-rt})$$

Figure 8.4 plots the solution for the number of values of $r$ this type of curve is sometimes referred to as a modified exponential curve.

Maximum slope occurs at the origin and slope from book decreases as per as time increases. As a result, the curve approaches the limit more slowly the closer it gets, and never actually reaches the limit.

The constant $r$ plays the same role as the growth rate coefficient in exponential growth model. As $r$ increases, the sales grow more rapidly.

$$r = \frac{1}{T} \text{ as in exponential growth model.}$$

## 8.5 LOGISTIC CURVES

We have seen that modified exponential curve; considered as a market model, shows its maximum slope at the beginning

Generally, the sales rate, which is represented by the slope, started with a low value and initially increases as occurs in exponential growth model and eventually the slope begins to decline as curve of market growth like the modified exponential curve. It makes a $S$-shaped curve as shown in Fig. 8.5.

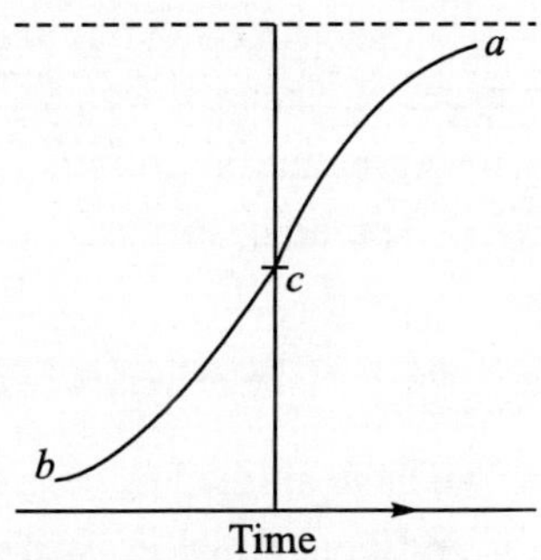


**Fig. 8.5.** $S$-shaped growth curve.

In Fig. 8.5 curve is divided into two part:

(*i*) $b$ to $c$ curve is like exponential growth model.

(*ii*) $c$ to $a$ curve is like modified exponential growth model.

**Example:** This example is related to some ideas and behaviour of people if they are going to buy some product. At the beginning when the product is new and unknown, knowledge of product is spread by advertising and word of mouth or the people who have really purchased this product. So sales rate will be proportional to the no. of sold which is the condition of driving the exponential growth model. As no. of products are sold, and the market begin to become saturated which is the condition of driving the modified exponential model. Finally, we can conclude that the logistic function is a combination exponential growth function and modified exponential growth function.

The differential equation defining the logistic function is

$$\frac{dx}{dt} \quad \text{or} \quad \dot{x} = rx(\max - x) \qquad \text{...(i)}$$

$x$ = No. of people who purchased the product

max = No. of people who might buy the product

**Consider Following Point**

1. If $x$ is very small than max (maximum market limit) the term (max $- x$) remains constant at the value max. The logistic function becomes

$$\dot{x} = r \max x$$

$$\max - x = \max \quad [\text{because } x << \max] \qquad ...(ii)$$

Let $r \max = R$ than equation (*ii*) becomes

$$\dot{x} = Rx$$

This equation is same as of exponential growth model with constant $r \max = R$.

2. When the market is almost saturated, the value of $x$ will be close to the value of max, *i.e.*, ($x \approx$ max), so the logistic function becomes

$$\dot{x} = r \max (\max - x) \qquad ...(iii)$$

Let $r \max = R_1$ then equation (*iii*) becomes

$$\dot{x} = R_1 (\max - x) \qquad ...(iv)$$

Equation (*iv*) is the same as of modified exponential growth model with a constant $r \max = R_1$.

3. The true differential equation for the logistic function is non-linear, because it contains the term $x^2$. Solution of non-linear equation is complicated.

$$\dot{x} = rx (\max - x)$$

$$\dot{x} = rx \max - x^2 \quad (x^2 \text{ component})$$

Logistic function has been described as a market model. However, it can be applicable for some other types of systems, *e.g.*,

- population growth
- spread of disease

## 8.6 GENERALIZATION OF GROWTH MODELS

So far we have studied three growth model.

- Exponential growth model
- Modified exponential growth model
- Logistic curves

Generally, these basic models depend upon certain coefficient which have been assumed to be constant. We can generalized these models by removing this assumption. Specially in the case of market where demand for a commodity cannot be likely to remain constant. However, some limit may exist, instead, it is similar to expand with population growth, or fluctuate with economic conditions, or respond to both these factors.

In logistic model equation is given by

$$\frac{dx}{dt} = rx (\max - x) \qquad ...(i)$$

Then (max $- x$) [no. of people who have not buy the product] is depend upon some economic factors such as

- No. of people or Target customer ($T_c$)
- Personal income ($P_i$)
- Local service revenue per item if applicable ($L_s$)

The (max – $x$) is directly proportional to these three factors and equation become

$$(\max - x) = cT_c^{a1} P_i^{a2} L_s^{a3} \quad \ldots(ii)$$

After finding the best values of the constants to fit the data to the equation of the model, the graph drawn is confined to the upper portion of a logistic curve.

## 8.7 SYSTEM DYNAMICS DIAGRAMS

In system dynamics, the system concentrates on the rates at which various quantities change and expresses the rates as continuous variables.

### Basic Structure of System Dynamics Diagrams

Basic structure of system dynamics diagrams consists of a number of reservoirs or levels, interconnected by flow paths. Following symbols are used to define system dynamics diagrams.

**1. Level**

Levels represent the accumulation of various entities in the system such as inventories of goods, unfilled orders, numbers of employees, etc. The current value of a level at any time represents the accumulated difference between the input and the output flow for that level.

**Symbol:** Level is represented by rectangular box, with name of level.

where $n = 1,2.... n.$

Symbol of level

**2. Rate of Flow**

Rates are used to define to represent the instantaneous flow to or from a level. They are also called decision functions or rate equations. It determines how the flow rates depend upon the levels.

**Symbol:** Rates are represented by following figure.

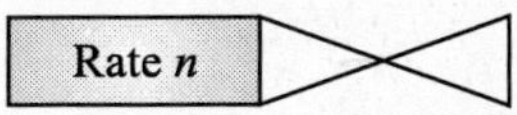


where $n = 1, 2, 3, ...... n$

Symbol of rate

**3. Constants**

The need for a constant in the model is necessary for completeness of the model.

**Symbol:** Constants are represented with a special symbol as shown in below figure.

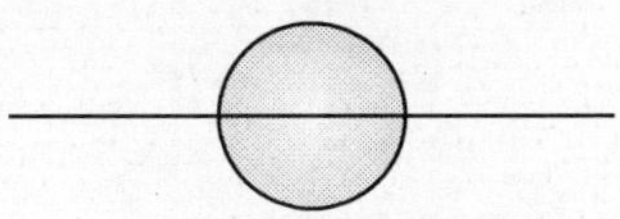

Symbol of constant

**4. Source/Sink**

Source and sink of entities are represented by following figure.

Symbol of source and sink

**5. Auxiliary Variable**

Auxiliary variable is a combination of other variables mathematically by some relationship other than the integration implied by a rate of equation and a level.

**Symbol:** Auxiliary variables are represented by circles.

**6. Solid Line**

Solid line in the diagram indicate the flow of tangible objects. Arrow represents direction.

**Symbol:**



Solid line

**7. Dotted Line**

Dotted lines indicate causal relationships.

**Symbol:**

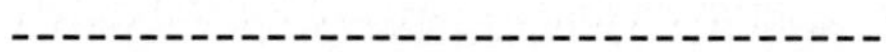

Dotted line

The simplest test for determining which quantities are to be regarded as levels is to imagine the system brought to rest. Any quantity that is a rate is then automatically zero, while any quantity that maintains a magnitude should be regarded as a level.

**Example:** The basic structure of a system dynamics model is shown in Fig. 8.6.

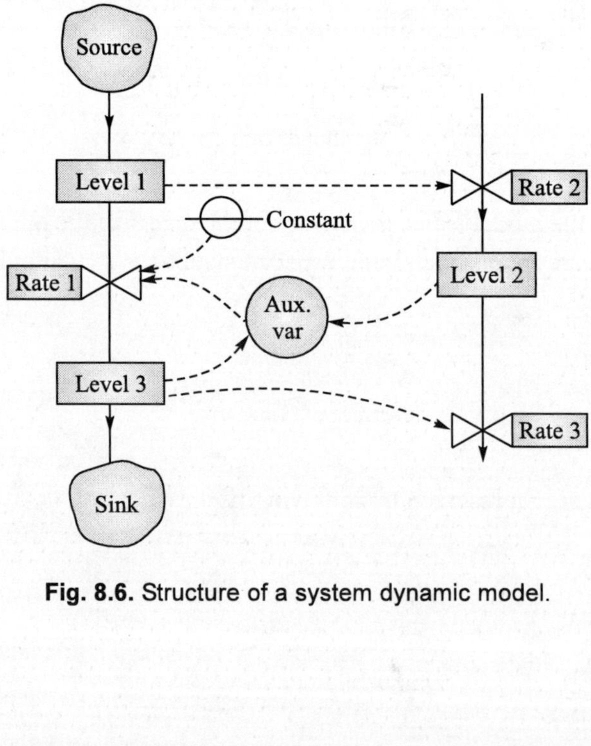


**Fig. 8.6.** Structure of a system dynamic model.

### 8.7.1 Simple System Dynamics Diagrams

The simple exponential growth model is concerned with a single level representing as population. The model is represented by the system dynamics (Fig. 8.7*a*).

Level $x$ is representing as population. It is increased by a single rate factor. The rate equation is given by

$$\dot{x} = r\,x$$

The dependence of the rate on the level, $x$ is indicated by the dashed line of Fig. 8.7 (*a*). The coefficient $r$ is shown as a constant.

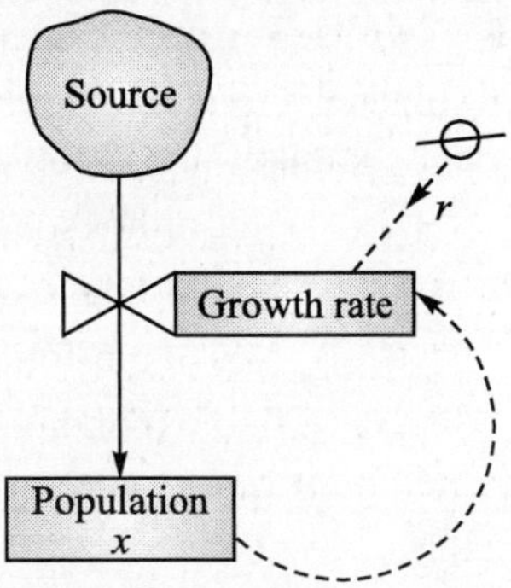


**Fig. 8.7.** (*a*) System dynamics diagrams of population growth.

Another way of modeling the population is to represent the births and deaths separately as shown in Fig. 8.7 (*b*)

- The level is now affected by two rate variables:

(*i*) One for births to increase the level

(*ii*) One for deaths to decrease the level

- Suppose both rates depend upon the current level of the population, the rate equation for births is the exponential growth model of following equation

$$\dot{x} = b_r x$$

$b_r$ is birth rate coefficient.

The rate equation for the deaths is the decay model. The equation is given by

$$\dot{x} = -d_r \cdot x$$

where $d_r$ is death rate coefficient.

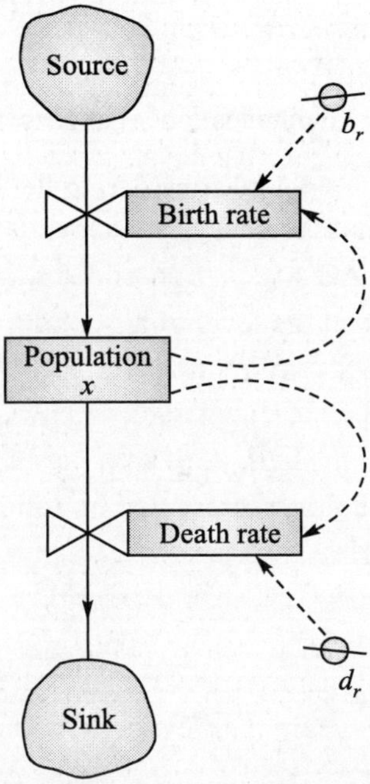


**Fig. 8.7.** (*b*) System dynamics diagram of population growth.

### 8.7.2 Multi-Segment Models

Let we consider a model of house air-conditioner sales to explain the general method of carrying out the numerical computations involved in a continuous simulation. The model consists:

HH No. of households

HS No. of houses sold

AC No. of air-conditioners installed

The modified exponential growth model showing how the number of houses increases, with a given fixed limit HH, is given by following equation

$$\dot{HS} = r_1 (HH - HS) \quad ...(i)$$

Similarly we have another equation of modified growth model for the number of air conditioners, but it has a limit of HS.

$$\dot{AC} = r_2 (HS - AC) \quad ...(ii)$$

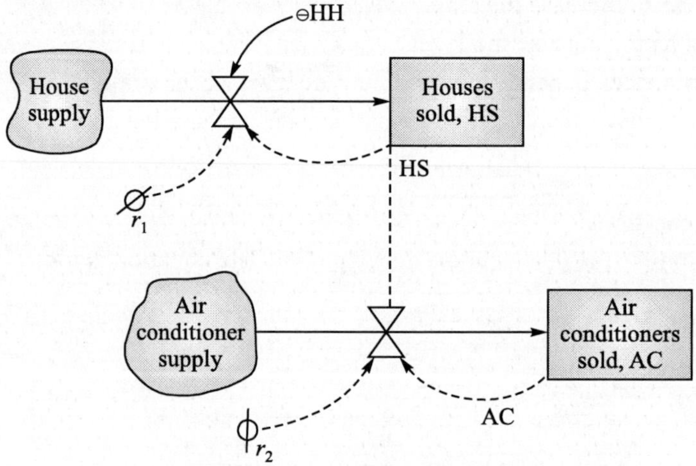


**Fig. 8.8.** System dynamics model of air-condition sales.

Suppose we want to extend this model by considering the breakdown of air-conditioner. Then a decay model, where we can assumed that the decay rate depends upon the current level means:

$$\dot{AC} = -r_3 AC \quad \text{(equation of decay model)} \quad ...(iii)$$

This needs to be added to diminish the level of air-conditioner

$$\dot{AC} = r_2 (HS - AC) + (-r_3 AC)$$

or

$$\dot{AC} = r_2(HS - AC) - r_3 AC \quad ...(iv)$$

We know that we cannot keep housing limit to be fixed so longer. So we assume an exponential growth model for the number of households, represented by following equation and Fig. 8.9.

Equation $\dot{HH} = r_4 HH$

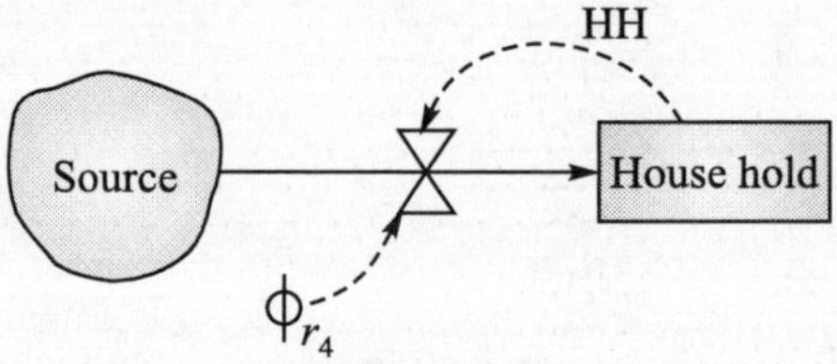


**Fig. 8.9**

The complete model consists equation (*i*), (*iv*) and (*v*).

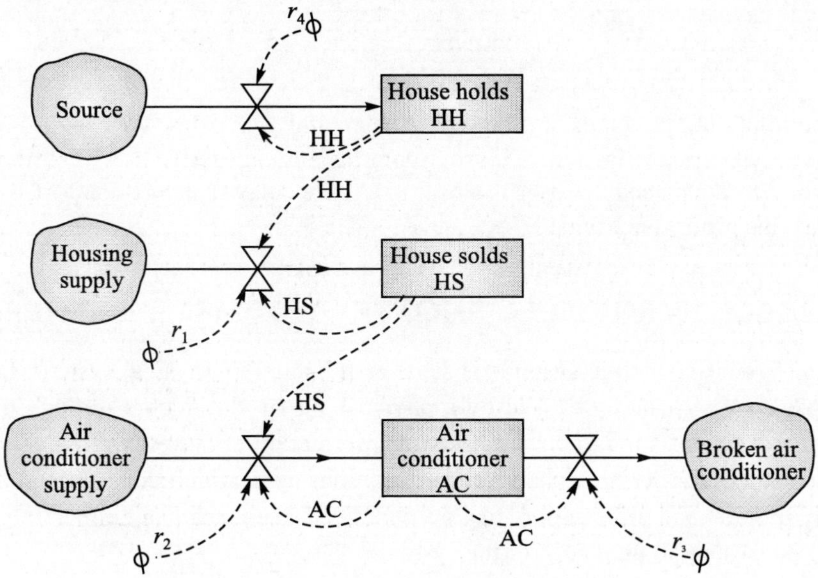


**Fig. 8.10.** Extended air-conditioner sales model.

## 8.8 REPRESENTATION OF TIME DELAYS

As we have studied in exponential growth model that coefficient involved in growth model ($r$), can be represented as time constant ($T$).

or $$r = \frac{1}{T}$$

So in air-conditioner model, the constants $r_1$, $r_2$ and $r_3$ can be represented as $\frac{1}{T_1}, \frac{1}{T_2}$ and $\frac{1}{T_3}$. It means model can be described as the fact that there is a time lag of $T_1$ time units before changes in the economy affect buying trends, that there is a lag of $T_2$ units before new houses get air-conditioner, or that the average of an AC is $T_3$ units.

If there is a need to represent a time delay, the simplest method is to introduce an appropriate time constant.

For example, suppose there is a level $x$ of outstanding orders, and it is known that it takes an average of $D$ time units to fill an order. According to system dynamics the outstanding order level is being depleted with a time constant of $D$ time units. It would include the decay equation

decay equation $\dot{x} = -rx$

Average time units $D$ then

$$r = \frac{1}{D}$$

or $$\dot{x} = -\frac{x}{D}$$

To represent delay, it is necessary to identify the level that is controlled by the delay and apply the rule

$$\text{Dealy} = \frac{\text{Level}}{\text{Rate}}$$

Due to the solution it produces, it is known as an exponential delay. We can do appropriate finite delay by using series of exponential delays if programming language only allows continuous equations. A pair of delays of size $T/2$ placed in series is closer to a finite delay than one delay of size $T$, and the fit becomes better as this process continued.

## 8.9 FEEDBACK IN SOCIO-ECONOMIC SYSTEMS [*UPTU 2007-08*]

Feedback occurs in socio-economic systems is same as it occurs in physical system. The feedback in socio-economic system not usually caused by a physical signal, but occurs through the feedback of information. Some decision is made on the basis of information about prevailing condition, and the action taken as a result of the decision changes the conditions upon which the information is based. The link between the decision and information may go unnoticed, feedback can deliberately be introduced in a socio-economic system to help control the system.

Let consider a example of inventory control. In which retailer will usually maintain:

- Inventory of goods.
- Keep a balance such that there should be balance between cost of holding goods and penality of losing sales if inventory become empty.
- There should not be loss of goodwill.

The retailer establishes a steady average order rate which balances the average rate of sales and maintains a constant inventory level. The changes represent changes with respect to the average order rate.

A simple model of inventory control system have required two levels and three rates, defined as follows:

I — Current inventory level
O — Outstanding level of orders placed with the supplier
} Two level

RO — Rate of ordering from supplier
RD — Rate of delivery from supplier
RS — Rate of sales
} Three rates

Two constants, we also need to defined

PI — Planned inventory level

DT — Average delivery time.

The model of system is shown in Fig. 8.11.

**Level (1):** It shows the level of outstanding orders being controlled by the rate at which orders are placed and the rate of delivery.

**Level (2):** The level of the inventory is controlled by the rate of delivery and the rate of sales.

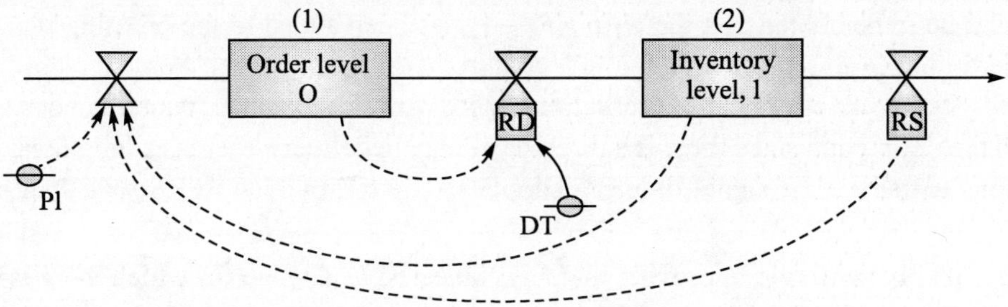


**Fig. 8.11.** Dynamics model of inventory control system.

To derive a mathematical model of the system, the equation can be written for the two levels are given below:

$$\dot{O} = RO - RD$$
$$\dot{I} = RD - RS$$

General formula for delays, the equation for the rate of delivery can be written as:

$$RD = \frac{O}{DT}$$

The rate of ordering is decision of management clearly there should be balance in steady sales rate, but simply to equates order rate to sales rate is not satisfactory.

When the sales rate increases, the inventory will become depleted because the delivery rate takes time to catch up with the new sales rate. Similarly, a drop in sales rate will result in the inventory level rising. The decision, is to add to the sales rate a term proportional to the difference between desired level and the actual level. When the inventory is low, this increases the order rate and, when the level is high, it decreases the rate. The order rate is therefore given by equation

$$RO = RS + r(PI - I)$$

where $r$ is a constant called as **ordering constant.**

Suppose average delivery delay DT be 3 days. Let the demand suddenly increases at time $t = 4$ to 6 items/day. The changes in inventory level for various value of $r$ are shown in Fig. 8.12. There can be seen oscillatory motion in Fig. 8.12.

The oscillation are more sophisticated when $r = 1$ (as in Fig. 8.12 clear oscillation). As per as $r$ decreases, oscillatory behaviour is disappearing as in $r = 0.0625$ or 0.075.

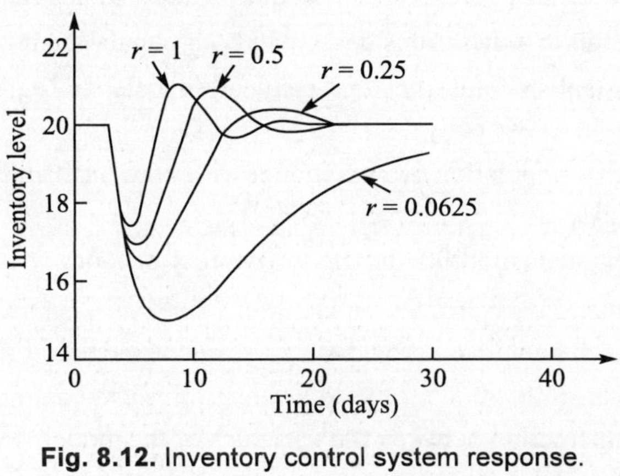


**Fig. 8.12.** Inventory control system response.

It should be remembered that the term $r$(PI – I) has been added to the ordering rate. It is not a one-time order intended to replace a deficit detected in the inventory level. Some insight can then be gained as to why large values of $r$ cause oscillations. With $r = 1$, for example, enough goods are shipped in a day to fill the deficit but, since there is a three-day delay in delivery, there is a long delay before the feedback begins to reduce the signal ordering supplies. The term added to RS will than overfill the original deficit.

It is reasonably good rule to provide that $r$ is chosen to be a value for which there is little or no oscillation. A suitable value for this example is

$r = \frac{1}{2}$ (approx). The value will, of course, change with the delay time DT. Some other inventory rules can also be used, but it needs a considerable amount of study to select a rule best suited for any particular system and type of demand.

The basic problem of adjusting rates so that a certain level is maintained appears repeatedly in socio-economic systems with the level representing less tangible items. The level might be money supply, with, $r$ representing the influence of interest rates.

## 8.10 WORD MODEL [*UPTU 2006-07*]

The most widely known example of a system dynamics model is the world model studied under the auspices of the club of Rome.

The club of Rome is an informal, international group of individuals concerns with the future of mankind. The club of Rome model have five aspects of global development.

- Population Growth
- Spread of industrialisation
- Pollution
- Depletion of natural resources
- Incidence of malnutrition

**The Main Conclusion Drawn are that**

- Present trends place a limit on growth, the limit is likely to be reached in the next 100 years.
- That limit will result in a disastrous decline in both population and, living condition.
- It is feasible to establish conditions that can lead to a state of equilibrium with satisfactory living conditions for every body.

This models also uses a population growth model with both birth and death rates is affected by many other factors like:

- The natural death rate is modified by the availability of food.
- Level of pollution.
- The degree to which people are crowded as a result of industrialization.

Conflicts between industralization and food production cause completion for capital and land, resulting in other forms of interaction between the variables of the model.

Critics of the model have questioned the assumptions and data of the model, while advocates have pointed out the persistence of the main conclusions, with relatively small changes in the time scale, for a wide range of input values.

Among the main topics of controversy has been the role of future technological changes, with the critics being unable to say what these will be, and the advocates finding it difficult to see feasible changes that would produce sufficient corrective action without changes in existing practices.

## SUMMARY

In this chapter the main focus of a system dynamics study is to understand the forces operating in a system in order to determine their influence on the stability or growth of the system.

- Growth implies a rate of change, the rate at which the fund grows is $r$ times the current size of the fund, called exponential growth model
$$(x = x_0 e^{-rt})$$
- The model in which a variable decays from some initial value, $x_0$, at a rate proportional to the current value called as exponential decay models
$$x = x_0\, e^{-rt}$$
- If growth rate is proportional to the number of people who have not yet bought the product *i.e.,* (max – $x$) then this model is called modified exponential growth models
$$x = \max\,(1 - e^{-rt})$$
Logistic function is a combination of exponential and modified exponential functions. We can define logistic function as
$$\dot{x} = rx(\max - x)$$
- We can do generalization of growth models by removing the dependency of basic models upon certain coefficients, which have been assumed to be constants.
- The model could be further developed by making more detailed models of its individual segments or by replicating segments of model to represent, different geographic regions, various products and a range of social group. This is called multisegment models.
- We can represent a time delay, by using appropriate time constant.
- Feedback in socio-economic systems is the same way as it occurs in physical systems but feedback is not usually caused by a physical signal, but occurs through the feedback of information.
- The most widely known example of a system dynamics model is the world model. World model investigates five aspects of global development like population growth, spread of industrialization, pollution, the depletion of natural resources and the incidence of malnutrition.

## SOLVED QUESTIONS

**Q. 1. How logistic function is combination of exponential and modified functions?**

**Ans.** Mathematically, logistic function is represented by
$$\dot{x} = rx\,(\max - x) \qquad ...(1)$$

- Initially, $x$ is very small than max or market limit

$$x << \max$$

so max $\quad -x \approx \max$

Equation (1) become $\dot{x} = rx \cdot \max$

or $\quad \dot{x} = r \max \cdot x$

Let $\quad r \cdot \max = R$ (any constant)

$$\dot{x} = Rx \qquad ...(2)$$

Equation (2), which is the equation for exponential growth model.

- When market is almost saturated then

$$x \approx \max$$

Then the equation (1) become

$$\dot{x} = r \cdot \max (\max - x) \text{ or}$$

Let $\quad r \cdot \max = R$

Then $\quad \dot{x} = R(\max - x) \qquad ...(3)$

Equation (3), *i.e.*, is the equation of modified function so we can say that logistic function is the combination of exponential and modified function.

**Q. 2. Define time constant *T*.**

**Ans.** The time constant $T$ is said to be a time constant since it provides a measure of how rapidly the variable $x$ grows.

**Q. 3. How feedback occurs in socio-economic system?**

**Ans.** Feedback occurs in socio-economic systems in such the same way as it occurs in physical system. The feedback is not usually caused by a physical signal, but occurs through the feedback of information. Some decision we have to make on the basis of information about prevailing conditions, and the action taken as a result of the decision changes the conditions upon which the information is based.

**Q. 4. During the 1980s the population of a certain city went from 100,000 to 205,000. Population by year listed in the table below:**

| Year | 1980 | 1981 | 1982 | 1983 | 1984 | 1985 | 1986 | 1987 | 1988 | 1989 |
|---|---|---|---|---|---|---|---|---|---|---|
| Population in thousands | 100 | 108 | 117 | 127 | 138 | 149 | 162 | 175 | 190 | 205 |

**This data is approximated well by the exponential growth model $X = 100\, e^{0.08t}$, where $t$ is the number of years since 1980. In other words the year 1980 corresponds to $t = 0$, etc. $X$ is population in thousands.**

**Answer the following questions:**

**(*a*) Draw exponential growth curve.**

**(*b*) Use the model to predict the population of city in 1991.**

**(*c*) According to our model when will the population reach 300.**

**Ans.** (*a*) **Exponential Growth Curve**

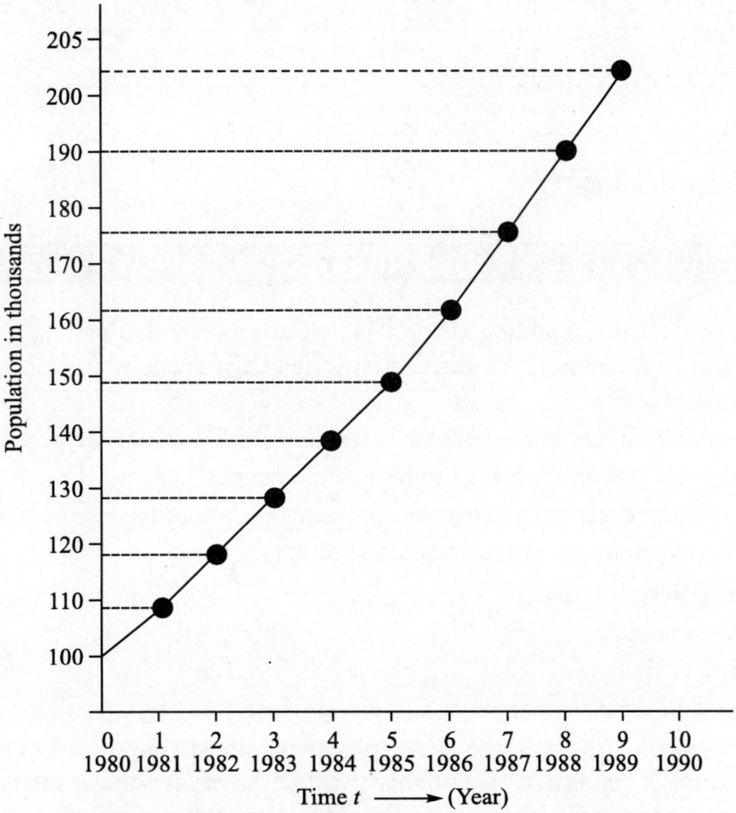


**Fig. 8.13.** Exponential growth curve.

(*b*) Model $X = 100\ e^{0.08t}$

Year at 1991, is correspond to $t = 11$

At $t = 11$, the population $X$ will be

$$x = 100 \times e^{0.08*11} = 241 \text{ thousand.}$$

(*c*) To solve this problem we set $100\ e^{0.08t}$ equal to 300 and solve for it.

$$100\ e^{0.08t} = 300$$

$$e^{0.08t} = 3$$

Take the natural logarithm of both sides

$$\ln e^{0.08t} = \ln 3$$

$$0.08t = \ln 3$$

$$t = \frac{(\ln 3)}{0.08} = 13.73 \text{ approximately}$$

Therefore, the population is expected to reach 300 thousand year about three fourths of the way through the year 1993.

**Q. 5. What are five aspects of global development in world models?**

**Ans.** There are following five aspects:

(*i*) Population growth

(*ii*) The spread of industrialization

(*iii*) Pollution

(*iv*) The depletion of natural resources

(*v*) The incidence of malnutrition

## EXERCISES

1. The birth rate of a country is adding 100,000 people a year to an initial population of 5,000,000. The average life expectancy is 65 years It is estimated that one ton of coal is consumed per annum for each individual.

   Draw a diagram showing how the country's resources of 500,000,000 tons are being depleted.
2. What is exponential growth models? Explain with example.
3. Show there is inverse relationship between *r* (the growth rate coefficient) and *T* (the time constant).
4. Discuss generalization of growth models along with an example.
5. Define following term:

   (*a*) Multisegment models

   (*b*) Time delays.
6. How is feedback in socio-economic system carried out? *[UPTU 2005-06]*
7. Two competing companies invest funds in capital equipment to improve their positions. The rate at which each invests funds decreases linearly as their own investment increases but increases linearly as their competitor's investment increases. Draw a diagram from which to simulate the competition and determine under what conditions the investments will stabilize.
8. What type of a model is the world model? Explain it in detail.
9. Explain exponential decay models. Also describe logistic curves.
10. Discuss, in brief, logistic curves?

9

# Random Number Generation

INSIDE THIS CHAPTER



A simulation of any system or process in which there are inherently random components requires a method of generating or obtaining numbers that are random, in some sense for example queuing theory, inventory models, etc.

## 9.1 RANDOM VARIABLE

An experiment is a process whose outcome is not known with certainty. The set of all possible outcomes of an experiment is called the sample space and is denoted by $S$. The outcomes themselves are called the sample points in the sample space. For example, if the experiment consists of flipping a coin, then

$$S = \{H, T\}$$

where $H$ and $T$ mean that the outcome is a head and a tail, respectively.

A random variable is a function that assigns a real number to each outcome of the experiment or each point of sample space $S$.

A *random variable* is a function defined on a sample space.

**Example:** Consider the experiment of rolling a pair of dice. Then

$$S = \{(1, 1), (1, 2), (1, 3), ...,(6, 6)\}$$

If $X$ is the random variable corresponding to the sum of the two dice, the $X$ assigns the value 4 to the outcome (1, 3).

**Example:** I select at random a student from the class and measure his or her height in centimetres. Here, the sample space is the set of students; the random variable is 'height', which is a function from the set of students to the real numbers: $h(St)$ is the height of student $St$ in centimetres.

**Note:** Random variable is denoted by capital letters such as $X$, $Y$, $Z$ and the values of the random variables take on by lower case letters such as $x$, $y$, $z$.

## 9.2 TYPES OF RANDOM VARIABLES

The distinction basically depends on the range of values that the random variable can take. Generally, there are two types of random variables.

- Discrete random variables.
- Continuous random variables.

### 9.2.1 Discrete Random Variable

A discrete random variable is one that can take on a countable number of values as in example-1 where the values are the integers from 2 to 12. "Countable" means that the set of possible values can be put in a one-one correspondence with the set of positive integers. Thus, a random variable that takes on only a finite number of values $x_1$, $x_2$,...$x_n$ is discrete. The probability that the discrete random variable X takes on the value $x_i$ is given by:

$$p(x_i) = P(X = x_i) \quad \text{for} \quad i = 1, 2...$$

and we must have

$$\sum_{i=1}^{\infty} p(x_i) = 1$$

where the summation means add together $p(x_i)$, $p(x_2)$,... all probability statements about $X$ can be computed from $p(x)$, which is called the probability mass function (pmf) for the discrete random variable $X$. if $I = [a, b]$, where $a$ and $b$ are real numbers such as $a \leq b$, then

$$p(X \varepsilon I) = \sum_{a \leq x_i \leq b} p(x_i)$$

Where symbol ε means "contained in" and summation means add together $p(x_i)$ of all $x_i$. Such that $a \leq x_i \leq b$. The distribution function $F(x)$ for the discrete random variable $X$ is given by:

$$F(x) = \sum_{x_i \leq x} p(x_i) \quad \text{for all} \quad -\infty < x < \infty$$

**Example** I toss a fair coin three times. The random variable $X$ gives the number of heads recorded. The possible values of $X$ are 0, 1, 2, 3, and its p.m.f. is

| A | 0 | 1 | 2 | 3 |
|---|---|---|---|---|
| $P(X = a)$ | 1/8 | 3/8 | 3/8 | 1/8 |

For the sample space is {*HHH, HHT, HTH, HTT, THH, THT, TTH, TTT*}, and each outcome is equally likely. The event $X = 1$, for example, when written as a set of outcomes, is equal to {*HTT, THT, TTH*}, and has probability 3/8. Two random variables $X$ and $Y$ are said to have *the same distribution* if the values they take and their probability mass functions are equal. We write $X - Y$ in this case.

In the above example, if $Y$ is the number of tails recorded during the experiment, then $X$ and $Y$ have the same distribution, even though their actual values are different (indeed, $Y = 3 - X$).

**Example:** For a discrete Random variable, the probability distribution is a table of all the events and their related probabilities. For example, in the roll of a die:

| Value | Probability | Value | Probability |
|---|---|---|---|
| 1 | 1/6 | 4 | 1/6 |
| 2 | 1/6 | 5 | 1/6 |
| 3 | 1/6 | 6 | 1/6 |

A probability distribution will contain all the outcomes and their related probabilities, and the probabilities will sum to 1.

From the distribution, we can find:

$$X = 3$$

$$Pr\ (X = 3) = 1{=}6$$

$$X = \text{even number}$$

$$Pr\ (X = 2 \text{ or } X = 4 \text{ or } X = 6) = 3 = 6 = 1 = 2$$

More interesting, we can also find:

$$Pr\ (X > 3) = 3 = 6 = 1 = 2$$

This is called a cumulative probability.

### 9.2.2 Continuous Random Variable

A continuous random variable is one that can take on an **uncountable number of values** as in example-2. A random variable whose values range over an interval of real numbers, or even over all real numbers, is continuous.

A random variable $X$ is said to be continuous if there exists a non-negative function $f(x)$ such that for any set of real numbers $R$ (*e.g.*, $R$ could be all real numbers between 1 and 2),

$$P(X \,\varepsilon\, R) = \int_R f(x)\,dx \quad \text{and} \quad \int_{-\infty}^{\infty} f(x)\,dx = 1$$

Thus, the total area under $f(x)$ is 1. If $X$ is a nonnegative random variable, the second range of integration is from 0 to $\infty$. All probability statements about $X$ can be computed from $f(x)$, which is called the probability density function (pdf) for the continuous random variable $X$. The pdf enables us to obtain the cdf by integrating under the pdf.

## 9.3 RANDOM NUMBER GENERATOR

Random variables play two important roles in simulation models. First, when we assume that within our models some delays will not have deterministic values, but instead will be represented by random variables; and second when a choice must be made within the behaviour of an entity and want the decision to be made probabilistically. Both cases will involve sampling a probability distribution to extract a value each time this part of the entity's behaviour is reached. Both cases rely on the random number generator.

To obtain a sequence of uniform random numbers, it is sufficient to be able to generate a sequence $X_i$ of integers in the range $[0, M-1]$ since the sequence $X_i/(M-1)$ will then be approximately uniformly distributed over (0, 1). In 1951, D.H. Lehmer discovered that the residues of successive powers of a number have good randomness properties.

**A random number:** Simply a particular value taken on by a random variable. A random variable is a function that assigns a number to each outcome of the experiment.

In this chapter, we discuss how random values can be conveniently and efficiently generated from a desired probability distribution for use in executing simulation models. Here we will avoid to speaking 'generating random variable' because random variable is defined in mathematical probability theory as a function satisfying certain conditions, hence we will adopt more precise terminology and speaking of 'generating random variate'.

Random variates generated from the $U(0, 1)$ distribution will be called random numbers.

There are three types of random numbers:

1. **Truly random number** (from physical generators)
2. **Pseudo-random number** (from mathematical generators)
3. **Quasi-random number** (special correlated sequences of numbers, used only for integration – give faster convergence than the standard Monte-Carlo integration).

### 9.3.1 Truly Random Number (from physical generators)

For example, tossing a coin, a roulette, radioactive decay, thermal noise in electronic devices (particularly "white noise"), cosmic ray arrival times, etc.

**Drawbacks of physical generators:**

- Too slow for typical calculational needs;
- Problems with stability – particularly generators based on physical processes, *e.g.*, small change in physical conditions of the source or its environment can cause major changes in probabilistic properties of produced random numbers ! additional testing and bias-correcting devices needed.

### 9.3.2 Pseudo-random Numbers

It is not possible to generating truly random numbers in a simulation. Only natural processes generate truly random numbers. If random numbers are generated using a known mathematical formula, then they can be replicated, so they are not truly random anymore but having the **appearance** of randomness, *i.e.*, their statistical properties are very close to the ones of the truly random numbers (someone who does not know the formula is not supposed to be able to tell that a formula was used rather than a physical process). So these numbers are called pseudo random numbers. "Pseudo", because generating numbers using a known method removes the potential for true randomness. Pseudo-random numbers are called simply random numbers, and mathematical algorithms for their generation are called random number generators (RNG).

The first mathematical generator is 'mid-square' generator proposed by John von Neumann and Metropolis in the 1940s.

**Midsquare Method**

One of the first pseudo-random number generators was the midsquare method. It starts with a fixed initial number, say of 4 digit integer, called the seed. This number is squared and the middle 4 digits of this square become the second number. The middle digits of this second number are then squared again to generate the third number and so on. Finally, we get realizations from the uniform (0, 1) distribution after placement of the decimal point (*i.e.*, after division by 10000). The choice of the seed is very important as the next examples will show.

**Example:**

- If we take as seed $Z_0 = 1234$, then we will get the sequence of numbers
0.1234, 0.5227, 0.3215; 0.3362, 0.3030; 0.1809; 0.2724, 0.4201;
0.6484; 0.0422; 0.1780, 0.1684, 0.8358, 0.8561; 0.2907
- If we take as seed $Z_0 = 2345$, then we get the sequence of numbers 0:2345; 0:4990; 0:9001, 0.0180, 0.0324; 0.1049; 0.1004, 0.0080; 0.0064; 0.0040;
.... Two successive zeros behind the decimal point will never disappear.
- If we choose $Z_0 = 2100$, then we get the sequence 0.2100; 0.4100; 0.8100, 0.6100; 0.2100, 0.4100 .... Only after four numbers the sequence starts to repeat itself.

[How? $Z_0 = 1234$, $Z_1 = 1234^2 = 1\underline{5227}56 = 5227$, $U_1 = 5227/1000 = 0.5227$]

It is not random at all, in the sense of being un predicatable. If we know the one number next is completely determined since the rule to obtain it is fixed.

**Properties of good random number generator:**

A good random number generator should satisfy the following properties:

1. The generated numbers should satisfy the property of uniformity.
2. The generated numbers should satisfy the property of independence.
3. The random numbers should be replicable.
4. It should take a long time before numbers start to repeat.
5. The routine should be fast.
6. The routine should not require a lot of storage.

### 9.3.3 Use of Random Numbers

Random number is very important in different areas of computer application. In inventory simulation (many other also) we have seen, demand of particular product can go **up, down,** or **stay at the same level** at any particular day. Let the probabilities of these three outcomes are 0.3, 0.5 and 0.2 respectively. Then we need a random selector which reproduces outcome, in the long run, with relative frequencies of 0.3, 0.5 and 0.2.

They can be obtained with a device generating 10 unit random numbers 0, 1, 2, ... 9 with equal probability and associating the sets of numbers {0, 1, 2} – with first outcome, *i.e.,* the demand going up, {3, 4, 5, 6, 7} – with second outcome, *i.e.,* the demand coming down {8, 9} – with third outcome, *i.e.,* the demand remaining unchange. So we have a device that simulates the stochastic phenomenon of the demand going up, down or remaining unchanged with the specified probabilities (of 0.3, 0.5 and 0.2 respectively). By using random number we can do our experiment. We are having number of techniques to generate random number that we will discuss one by one. These techniques are:

1. Linear Congruential Generators.
2. Mixed Generators.
3. Multiplicative Generators.
4. Combined Linear Congruential Generators.
5. Tausuortho and Related Generators.

## 9.4 TECHNIQUES FOR GENERATING RANDOM NUMBERS

### 9.4.1 Linear Congruential Generators (LCGs) *[UPTU 2007-08]*

Many random number generators in used today are linear congruential generator (LCG), introduced by Lehmer (1951). This produce following sequence of integers $Z_1, Z_2$ ... is defined by the recursive formula.

$$Z_i = (a\,Z_{i-1} + c)\ (\text{mod}\ m) \qquad \ldots(1)$$

where

$m$ — Modulus

$a$ — Multiplier or constant

$c$ — Increment

$Z_o$ — Seed or starting value

To obtain $Z_i = (a\,Z_{i-1} + c)/m$ {$Z_i$ is remainder of this division}. So it is $0 \le Z_i \le m - 1$.

To obtain random numbers $U_i$ (for $i$ = 1, 2...)

$$U_i = Z_i/m$$

In addition to non-negativity, the integers $m$, $a$, $c$ and $Z_o$ should satisfy $0 < m$, $a < m$, $c < m$ and $Z_o < m$. In equation (1) $c \ne 0$ resulted equation is called **Mixed Congruential Method.**

If $c = 0$ resulted form is called **Multiplicative Congruential Method.**

The selection of the values for $a$, $c$, $m$ and $Z_o$ drastically affects statistical properties and the cycle length.

**Two Objections could be Raised Against LCGs:**

1. It is common to all pseudo randoms number generators, that is defined by equation (1), are not really random at all. Every $Z_i$ is completely determined by $m$, $a$, $c$ and $Z_o$.
2. The second objection to LCGs might be that the $U_i$'s can take on only the rational values 0, $1/m$, $2/m$ ... $(m-1)/m$.

**Example:** Consider the LCG defined by $m = 16$, $a = 5$, $c = 3$, and $Z_o = 7$ table 9.1 gives $Z_i$ and $U_i$ (upto 3 decimal places) for $i$ = 1, 2, ... 19.

Note that $Z_{17} = Z_1 = 6$, $Z_{18} = Z_2 = 1$, and so on.

Means from $i$ = 17 through 32, we shall obtain exactly the same values of $Z_i$ (and hence $U_i$).

(So we do not seriously suggest that anyone use this generator since $m$ is so small)

**Table 9.1. The LCG = $(5Z_{i-1} + 3)$ (mod 16) with $Z_o = 7$**

| $i$ | $Z_i$ | $U_i$ | $i$ | $Z_i$ | $U_i$ | $i$ | $Z_i$ | $U_i$ | $i$ | $Z_i$ | $U_i$ |
|---|---|---|---|---|---|---|---|---|---|---|---|
| 0 | 7 | – | 5 | 10 | 0.625 | 10 | 9 | 0.563 | 15 | 4 | 0.250 |
| 1. | 6 | 0.375 | 6 | 5 | 0.313 | 11 | 0 | 0.000 | 16 | 7 | 0.430 |
| 2. | 1 | 0.063 | 7 | 12 | 0.750 | 12 | 3 | 0.188 | 17 | 6 | 0.375 |
| 3. | 8 | 0.500 | 8 | 15 | 0.938 | 13 | 2 | 0.125 | 18 | 1 | 0.063 |
| 4. | 11 | 0.698 | 9 | 14 | 0.875 | 14 | 13 | 0.813 | 19 | 8 | 0.500 |

So we have seen that same sequence of values is generated, and this cycle repeats itself endlessly.

The length of a cycle is called the period of a generator. It is clear that if period is $m$, the LCG is said to have full period clearly if a generator is full period, the cycle length could in fact depend on the particular value of $Z_0$ chosen or seed value.

**Note:** $Z_0, Z_1, Z_2 ... Z_i$ are random integer

$U_0, U_1, U_2 ... U_i$ are random number

**Theorem:** LCG is defined in equation (1), has full period if and only if the following three conditions hold:

(*a*) The only positive integer that (exactly) divides both $m$ and $c$ is 1.

(*b*) If $q$ is a prime number (divisible by only itself and 1) that divides $m$, then $q$ divides $a - 1$.

(*c*) If 4 divides $m$, then 4 divides $a - 1$.

(*d*) $c$ is relatively prime to $m$.

**Maximum Density**

Maximum density means that the values assumed by $U_i$, $i = 1, 2...n$ leave no large gaps on [0, 1]. To achieve maximum density, and to avoid cycling in practical applications, the generator should have the largest possible period. Maximum period can be achieved if we take correct choice of $a$, $c$, $m$ and $Z_0$.

### 9.4.2 Mixed Generators

To get maximum density and high period, we want $m$ to be large. Furthermore dividing by $m$ to obtain remainder in equation (1) is a relatively slow arithmetic operation, so avoid this division explicity. We can have following good choice for longest period and maximum density.

(*a*) $m = 2^b$, where $b$ is the no. of bits in a word on the computer being used that are available for actual data storage.

(*b*) $c \neq 0$, $c$ is relatively prime to $m$.

(*c*) $a = 1 + 4k$, where $k$ is integer.

**Integer Overflow**

Many computers and compilers have 32-bit words, left most bit being a sign bit, so $b = 31$. If $b \geq 31$, then $m \geq 2^{31} > 2.1$ billion.

Choosing $m = 2^b$ does allow us to avoid explicit division by $m$ on most computers by taking advantage integer overflow. The largest integer can be represented is $2^{b-1}$ and any attempt to store a larger integer $w$ (with, Say $h > b$ binary digits) will result in loss of the left (most significant) $h - b$ binary digits of this oversized integer. What remains in the retained $b$ bits is precisely $w$ (mod 2). The following example illustrates modulo division by overflow for $m = 2^b$.

**Example:** Computer with $b = 4$ bits per word available for data storage: $m = 16 = 2^b$, conveniently. How can overflow be used to obtain $Z_7 = 12$ from $Z_6 = 5$? Now generator is $Z_7 = 5Z_6 + 3$.

$$Z_7 = 5\,Z_6 + 3 \quad i.e. \quad 5 * 5 + 3 \;\Rightarrow\; 28$$

Binary representation of 28 is 11100.

Since our 4-bit computer can store only four binary digits the leftmost digit of the binary number 11100 is dropped, leaving the binary number $1100 \Rightarrow 8 + 4 + 0 + 0 \Rightarrow 12$ which is the binary representation of $12 = Z_7$.

### 9.4.3 Multiplicative Generators

Following two choice is the for modulus ($m$), increment ($c$) and multiplier ($a$).

**Choice 1:**

- In multiplicative LCG, we don't need $c$ as, $c = 0$.
- $m$ is choosen as $m = 2^{b-2}$, if $Z_0$ is odd.
- $a$ is $8k + 3$ or $8k + 5$ for some $k = 0, 1, ...$
- In this case, only one fourth of integers $[P = m/4 = 2^{b-2}]$ 0 through $m - 1$ can be obtained as values for the $Z_i$'s for longest possible period (P).

**Choice 2:**

- $m$ is prime number and $c = 0$.
- $P = m - 1$ is achieved by providing the multiplier $a$.
- $a$ has the property that the smallest integer $k$ is such that $a^k - 1$ is divisible by $m$, is $k = m - 1$.

**Example:** Let $m = 10^2 = 100$, $a = 19$, $c = 0$, and $Z_0 = 63$, generate a sequence of random integers using linear congruential method.

In this example, as $c = 0$, the form of equation (1) is known as the multiplicative congruential methods.

$$\begin{aligned} Z_0 &= 63 \\ Z_1 &= 19 * 63 \bmod 100 \\ &= 1197 \bmod 100 = 97 \\ Z_2 &= 19 * 97 \bmod 100 \\ &= 1843 \bmod 100 = 43 \\ Z_3 &= 19 * 43 \bmod 100 \\ &= 817 \bmod 100 = 17 \text{ and so on} \end{aligned}$$

**Example.** *Let*

$$\begin{aligned} a &= 7^5 = 16{,}807 \\ m &= 2^{31} - 1 = 2{,}147{,}483{,}647 \text{ (a prime no.)} \\ c &= 0 \\ Z_0 &= 123{,}457 \end{aligned}$$

*Generate Random no. using linear congruential method.*

**Solution:** These choice satisfy the condition that ensure a period of $P = m - 1$. The first few no. generated are as follows:

$$\begin{aligned} Z_1 &= 7^5(Z_0) \bmod (2^{31} - 1) \\ Z_1 &= 2{,}074{,}941{,}\ 799 \\ U_1 &= \frac{Z_1}{2^{31}} = 0.9662 \\ Z_2 &= 7^5(Z_1) \bmod (2^{31} - 1) \\ Z_2 &= 559{,}872{,}162 \\ U_2 &= Z_2/2^{31} = 0.2607 \\ Z_3 &= 7^5\ (559{,}872{,}160) \bmod (2^{31} - 1) \\ &= 1{,}645{,}525 \\ U_3 &= Z_3/2^{31} = 0.7662, \text{ and so on.} \end{aligned}$$

### 9.4.4 Combined Linear Congruential Generators

- As computing power has increased, the complexity of the system that we are able to simulate has also increased.
- A random number generator with period $2^{31} - 1 \simeq 2 \times 10^9$ is no longer adequate for all applications.
- So one approach is to combine two or more multiplicative congruential generators in such a way that the combined generator has good statistical properties and a longer period.

Special case of generator defined by

$$Z_i = g(Z_{t-1}, Z_{i-2}....) \pmod m \qquad ...(2)$$

For LCGs, the function $g$ is,

$$g(Z_{i-1}, Z_{i-2}, ...) = a_1 Z_{i-1} + c$$

or

$$g(Z_{i-1}, Z_{i-2}, ...) = a_1 Z_{i-1} + a_2 Z_{i-2} + .... a_q Z_{i-q} \qquad ...(3)$$

$a_1, a_2 ... a_q$ are constants.

The above equation – (2) is defined by Multiple Recursive Generators (MRGs).

The way to combine two generators is discussed and evaluated by L'Ecuyer (1988). Let $\{Z_{1t}\}$ and $\{Z_{2i}\}$ denote the integer sequences generated by two different LGGs with different moduli, then let,

$$Z_i = (Z_{1i} - Z_{2i}) \pmod m \text{ for some integer } m$$

finally

$$U_i = Z_i/m.$$

This idea could clearly be extended to more than two generators, and has several advantages:

(*a*) The period is very long (at least $10^{18}$).

(*b*) The multipliers in each of the component generators can be small and the resulting generating is quite fast.

(*c*) Statistical properties of these generators also appear to be very good.

This type of generators is obtained by combining Multiplicative Recursive Generators (MRGs) in a particular way, and was developed and studied by L'Ecuyer. In general, $K$ different MRGs are implemented simultaneously to form the sequence $\{Z_1, i)$, $\{Z_2, i\}$ ..., $\{Z_K, i\}$. Letting $m_1$ be the modulus in from equation ($i$) used in the first of these MRGs, and letting $\delta_1, \delta_2, ... \delta_k$ be specified constants, define $Y_i + (\delta_1 Z_{1,i} + \delta_2, Z_{2,i} + ... + \delta_k Z_{k,i}) \pmod{m_1}$ and finally return $U_i = Y_i/m_1$ as the $i$th random number use in the simulation. $K$ is number of different MRGs.

Parameter for these kinds of generators must be chosen very carefully, extremely long periods and exceptionally good statistical properties are possible.

L'Ecuyer and Cote (1991) describe a portable software package to implement these type of generators.

### 9.4.5 Tausworthe and Related Generators

Several interesting generators have been developed on the basis of a paper by Tausworthe (1965) by equation (4)

$$b_i = (b_{i-r} + b_{i-q}) \pmod 2 \qquad ...(4)$$

$r$ and $q$ is integers $[0 < r < q]$

Execution of equation is equivalent to the exclusive or instruction on bits, that is equation (4) can be represented as

$$b_i = \begin{cases} 0 \text{ if } b_{i-r} = b_{i-q} \\ 1 \text{ if } b_{i-r} \neq b_{i-q} \end{cases}$$

The maximum period here is $2^q - 1$.

**Example:** Let $r = 3$ and $q = 5$ in equation (4), and let $b_1 = b_2 = b_3 = b_4 = b_5 = 1$. Thus, for $i \geq 6$, $b_i$ is the "exclusive or" of $b_{i-3}$ with $b_{i-5}$. The first 45 bit are then, 11111 000 11 0 111 0 1 0 1 0000

repeated

100 1 0 11 00 11111 000110 111

*How?*
$$b_i = \begin{cases} 0 \text{ if } b_{i-r} = b_{i-q} \\ 1 \text{ if } b_{i-r} \neq b_{i-q} \end{cases}$$

| | | |
|---|---|---|
| $i = 6$ | then | $b_{6-3} = b_{6-5} \Rightarrow b_3 = b_1 = 1$ so $b_6 = 0$ |
| | because | $b_i = (1 + 1) \pmod 2 = 0$ |
| $i = 7$ | then | $b_{7-3} = b_{7-5} \Rightarrow b_4 = b_2 = 1$ so $b_7 = 0$ |
| $i = 10$ | | $b_{10-3} = b_{10-5} \Rightarrow b_7 = b_5 = 0 \neq 1$ so $b_{10} = 1$ |
| $i = 11$ | | $b_{11-3} = b_{11-5} \Rightarrow b_8 = b_6, 0 = 0$ so $b_{11} = 0$ |

and so on.

Note Maximum period here is : $2^q - 1$

It means $2^5 - 1 \Rightarrow 31$

With this sequence of $\{bi\}$ so defined, the question arises: How random number is generated. One natural possibility is to combine together on consecutive $bi$'s to form an $m$-bit binary integer between 0 to $2^m - 1$, which is then divided by $2^m$. In this example choosing $m = 4$ results in the sequence of random numbers.

$$\underbrace{1111}_{4}\,\underbrace{1000}\,\underbrace{1101}\,\underbrace{11\ldots}$$

$$\frac{1\times 2^3 + 1\times 2^2 + 1\times 2^1 \times 1\times 2^0}{2^4}, \frac{1\times 2^3 + 0\times 2^2 + 0\times 2^1 \times 0\times 2^0}{2^4}, \ldots$$

$$\left\{\frac{15}{16}, \frac{8}{16}, \frac{13}{16}, \frac{13}{16}, \frac{4}{16} \ldots\right\}$$

However, $m$ would have to be no more than the word size of the computer.

There is disagreement over how well this works, so another possibility is to use that $m$ consecutive bits for the first integer, then skip some number of the following $bi$'s then use the next $m$ bits for the second integer etc.

- A good bit-selection algorithm was given by Lewis and Payne (1975) but $r$ and $q$ are specificed as in Tausworthe generators.
- Advantage of Tauswontha is – size $(2^{521} - 1 > 10^{156}$ more)

## 9.5 TESTING OF RANDOM NUMBERS

It is clear that the sequence of numbers generated by any arithmetic method is not truly 'random' in the sense that the entire sequence is predetermined and predictable. If the process is started with the same $Z_0$, exactly the same sequence will occur. Since $m$ is finite, one of the numbers generated will eventually be exactly the same as a previously generated for these reasons the name pseudo-random is used for these generators. The generated sequence must have some of the important 'random like' characteristics. A variety of tests have been proposed for testing the randomness of a sequence.

The basic properties that we have needed to test random numbers is:

(*i*) Uniformity (*ii*) Independence

Some consequences of the uniformity and independence properties are the following:

(*a*) If the interval (0, 1) is divided into *n* classes, or sub-intervals of equal length, the expected number of observations in each interval is $N/n$, where $N$ is the total no. of observations.

(*b*) The probability of observing value in particular interval is independent from the other values.

In testing for uniformity, the hypothesis are as follows:

$$H_0 : Ui \sim U(0, 1)$$
$$H_1 : Ui \nsim U(0, 1)$$

The null hypothesis, $H_0$, reads that the numbers are distributed uniformly on the interval [0, 1]. Failure to reject the null hypothesis means that no evidence of non-uniformity has been detected on the basis of this test. This does not imply that further testing of the generator for uniformity is unnecessary.

In testing for independence, the hypothesis are as follows:

$$H_0 : Ui \sim \text{independently}$$
$$H_1 : Ui \nsim \text{independently}$$

This null hypothesis, $H_0$ reads that the numbers are independent. Failure to reject the null hypothesis means that no evidence of dependence has been detected on the basis of this test. This does not imply that further testing for the generator for independence is unnecessary.

It is always possible that the $H_0$ is true, but we rejected it because a sample landed in the $H_1$ region, leading us to reject $H_0$. This is known as *Type I* error. Similarily if $H_0$ is false, but we didn't reject it, this also results in an error, known as *Type II* error.

**Level of Significance:**

- A level of statistical significance has to be given. The level α is the probability of rejecting the *H* null while the *H* null is true (thus, Type I error).

$$\alpha = P(\text{reject } H_0/H_0 \text{ true})$$

We want the probability as little as possible. Generally, α is set to 0.01 or 0.05 (Hypothesis)

- Decreasing the probability of Type I error will increase the probability of Type II error. We should try to strike a balance.

| | Actually true | Actually False |
|---|---|---|
| Accept | 1 – α | β (Type II error) |
| Reject | α (Type I error) | 1 – β |

### 9.5.1 Uniformity Test

Uniformity test counts how often numbers in a given range occur in the sequence to ensure that the numbers are uniformly distributed. There should be no number that can occur more frequently.

Two methods are given to perform uniformity test.

(*i*) Kolmogorov-Smirnov test (*ii*) Chi-square test

**Kolmogorov-Smirnov Test**

In this test we compare the continuous cdf (continuous distributed function), $F(x)$, of the uniform distributed to the empirical cdf, $E_N(x)$, of the sample of $N$ observations.

**Note:** Uniformity test is also called frequency test

**Definition:** $F(x) = x, \quad 0 \le x \le 1$

If sample of random number generator is $U_1, U_2 \ldots U_n$, then the empirical cdf, $E_N(x)$, is defined by

$$E_N(x) = \frac{\text{Number of } U_1, U_2 \ldots U_n \text{ which are} \le x}{N}$$

$N$ becomes larger, $E_N(x)$ should become a better approximation to $F(x)$, provided that the null hypothesis is true.

Kolmogorov-Smirnov test is based on the largest absolute deviation between $F(x)$ and $E_N(x)$ over the range of the random variable. It is based on the statistics.

$$D = \max \mid F(x) - E_N(x) \mid$$

$D$ is known sample distribution. It is given as a function of $N$. For testing against a uniform cdf, the test procedure are follows:

(*i*) Divide the data from smallest to largest $U_{(i)}$ denote the *i*th smallest observation, so that

$$U_{(1)} \le U_{(2)} \le \ldots \le U_{(n)}$$

(*ii*) Calculate $$D^+ = \max_{1 \le i \le N} \left[ \frac{i}{N} - U_{(i)} \right]$$

$$D^- = \max_{1 \le i \le N} \left[ U_{(i)} - \frac{i-1}{N} \right]$$

(*iii*) Calculate $$D = \max (D^+, D^-)$$

(*iv*) Determine the critical value, $D\alpha$ from, table for the specified significance level $\alpha$ and given sample size $N$.

(*v*) If $D$ is greater than $D\alpha$ then null hypothesis that the data are a sample from a uniform distribution is rejected. If $D \le D\alpha$ then there is no difference between the true distribution of $\{U_1, U_2 \ldots U_n\}$ and the uniform distribution.

**Chi-Square Test**

It is very important and useful statistical test to determine how often certain observed data fit the theoretically expected data.

**Formula**

$$\text{Chi}^2 = \sum_{i=1}^{n} \frac{(O_i - E_i)^2}{E_i}$$

$n$ — No. of classes

$O_i$ — No. of times the **observed data** falls in each class $i$ for $i = 1, 2 \ldots n$

$E_i$ — **Expected No.** of occurrences, in each class $i$.

- For the uniform distribution, $E_i$, the expected number in each *class* is given by

  $E_i = N/n$

  $N$ = total no. of observation

- Now we have to measure how far the observed frequency deviates from the expected value of the above formula.

### Degree of Freedom

To read a chi²-table we must have another parameter, called the degree of freedom. The degree of freedom $d_f$ is given by

$$d_f = n - 1$$

where $n$ is the no. of classes into which data is divided.

**Example:** If we have a sequence of 4,000, 3 digit number (from 000 to 999). Then we can take in the range (00 to 99) there are about 400 numbers, similarly in the range of 100 to 199 there is 400 amd so on.

- But we cannot expect that there are exactly 400 numbers in each of the 10 ranges, *i.e.*, 00 to 99, 100 to 199,..., 900 to 999.
- If we found that there are exactly 400 numbers in each range, we should suspect some non-randomness.
- If deviation from 400 is too much then we would suspect about non-uniformity.

So how much deviation is accepted in sequence to be uniformly distributed is given by **chi²-test.**

In our example $E_i = 400\left(E_i = \dfrac{4000}{10} = \dfrac{(N)}{(n)}\right)$ for $i = 1, 2, ... 10$

Then the chi² statistic can be calculated as follows:

| $i$ | Range | $O_i$ No. of observed occurrences | $E_i$ No. of expected occurrences | $(O_i - E_i)^2$ | $\dfrac{(O_i - E_i)^2}{E_i}$ |
|---|---|---|---|---|---|
| 1. | 00–99 | 425 | 400 | 625 | 1.5625 |
| 2. | 100–199 | 378 | 400 | 784 | 1.96 |
| 3. | 200–299 | 395 | 400 | 25 | 0.0625 |
| 4. | 300–399 | 415 | 400 | 225 | 0.5625 |
| 5. | 400–499 | 340 | 400 | 3600 | 9 |
| 6. | 500–599 | 370 | 400 | 900 | 2.25 |
| 7. | 600–699 | 410 | 400 | 100 | 0.25 |
| 8. | 700–799 | 382 | 400 | 324 | 0.81 |
| 9. | 800–899 | 365 | 400 | 1225 | 3.0625 |
| 10. | 900–999 | 394 | 400 | 36 | 0.09 |

Total chi² = 19.61

Now question is that how large or small a computed value of chi² can we accept before concluding that the discrepancy between the observed frequency. For this we have statistical tables which give critical values of chi² which can be exceeded by chance with various probabilities, such as .001, .01, .05, .1 etc. These are called chi²-tables. We need degree of freedom to read chi² table.

So $d_f = n - 1 \quad n = 0$

or $d_f = 10 - 1$

$d_f = 9$

Now read the chi² tables and find the following now of entries for 9.

| $d_f = 9$ | .995 | .99 | .95 | .90 | .75 | .50 | .25 | .10 | .05 | .01 | .005 |
|---|---|---|---|---|---|---|---|---|---|---|---|
| | 1.73 | 2.09 | 3.33 | 4.17 | 5.90 | 8.34 | 11.4 | 14.7 | 16.9 | 21.7 | 23.6 |

It means 99.5% probability of chi$^2$ exceeding – 1.73
99% probability of chi$^2$ exceeding – 2.09
1% probability of chi$^2$ exceeding – 21.7
.5% probability of chi$^2$ exceeding – 23.6

Thus probability of the chi$^2$ statistic (for $d_f = 9$) falling below 1.73 and above 23.6 is only 1.25%. That is 988 out of 1,000 sequences from a perfectly uniform random number generator would have given

$$1.73 \leq \text{chi}^2 \leq 23.6$$

If we take 1 as the cut off point, we would reject all sequences with chi$^2$ below 2.09 and above 21.7. In particular, the random number sequence, whose chi$^2$ we have just calculated and found to be **19.61.** It is barely acceptable.

### 9.5.2 Independence Test

Sometimes the generators pass the Kolmogorov-Smirnov and chi-square tests for uniformity, but the numbers generated are not independent. Hence there is the need of independence test. There are following test to check Independence of generators:

1. Auto correlation test
2. Gap test
3. Runs test
4. Poker test

**Auto-correlation test:** We know that uniformity and independence are two most important properties of random number. If generated random number sequence has pass uniformity test then it's not necessary that it will be pass independence test.

**Note:** Independence: No one value in the sequence can be inferred from the others.

Auto correlation is a statistical test that determines whether a random number generator is producing independent random numbers in a sequence. The tests for autocorrelation are concerned with the dependence between numbers in a sequence. The test computes the auto correlation between every *m* numbers (*m* is also known as the lag) starting with the *i*th number (*i* is also known as the index).

**Consider the following steps for auto correlation test:**

1. Computation of the auto-correlation between every *m* (the lag) numbers starting with $i^{th}$ number.

   Thus auto correlation $\rho_{im}$ between the following number: *Ri*, *Ri* + *m*, *Ri* + 2*m*, *Ri* + (*M* +1) *m*, where *M* is the largest integer such that $i + (M + 1)\, m \leq N$, where *N* is the total number of values in the sequence.

   For example, $N = 17$, $i = 3$, $m = 4$, then the above sequence would be 3, 7, 11, 15 ($M = 2$). The reason we require $M + 1$ instead of *M* is that we need to have at least two numbers to test ($M = 0$) the auto correlation.

2. Since a non-zero auto correlation implies a lack of independence, the following test is appropriate

$$H_0 : \rho_{im} = 0$$
$$H_1 : \rho_{im} \neq 0$$

3. For large values of *M*, the distribution of the estimator $\rho_{im}$, denoted as $\hat{\rho}_{im}$ is approximately normal if the values $R_i$, $R_{i+m}$, $R_{i+2m}$, ... $R_{i+(M+1)m}$ are uncorrelated.

4. Form the test statistic

$$Z_0 = \frac{\rho_{im}}{\sigma_{\hat{\rho}_{im}}}$$

which is distributed normally with a mean of zero and a variance of one.

If $\rho_{im} > 0$, the subsequence has positive auto correlation. High random numbers tend to be followed by high ones, and vice versa.

If $\rho_{im} < 0$, the subsequence has negative autocorrelation. Low random numbers tend to be followed by high ones, and vice versa.

5. The actual formula for $\hat{\rho}_{im}$ and the standard deviation is

$$\hat{\rho}_{im} = \frac{1}{M+1}\left[\sum_{k=0}^{M} R_{i+km} R_{i+(k+1)m}\right] - 0.25$$

and $$\sigma_{\hat{\rho}_{im}} = \frac{\sqrt{13M+7}}{12(M+1)}$$ [by Schmidt and Taylor – 1970]

6. After computing $Z_0$, do not reject the null hypothesis of independence if

$$-z_{\alpha/2} \le Z_0 \le z_{\alpha/2}$$

where $\alpha$ is the level of significance.

**How do you find out the value of *M*?**

The value of $M$ can find out by using some simple algebra. For example, the equation: $i + (M+1)\, m \le N$ must be solved using the given values $i$ the index. $N$ the number of elements in the sequence, and $m$ the lag.

**For example:** $i = 1$, $m = 5$, $N = 20$ and $M = ?$

$$i + (M+1)m \le N$$
$$i + (M+1)5 \le 20$$
$$1 + (5M+5) \le 20$$
$$5M + 6 \le 20$$
$$5M \le 14$$
$$M \le 2.8$$

Since the value $M$ must be an integer, the 0.8 is truncated and the 2 is saved as the value of $M$. Thus $M$ is equal to 2.

**Disadvantages of autocorrelation test:**

1. The test is not very sensitive for small values of $M$, particularly when the numbers being tests are on the low side.
2. Problem when "fishing" for auto correlation by performing numerous tests:
   - If $\alpha = 0.05$, there is a probability of 0.05 of rejecting a true hypothesis.

     If 10 independence sequences are examined, The probability of finding no significant auto correlation, by chance alone, is $(0.95)^{10} = 0.60$. Hence, the probability of detecting significant auto correlation when it does not exist = 40%
   - If $\alpha$ is 0.10 and 10 tests are conducted, there is a 65% chance of finding auto correlation by chance alone.

### Gap Test

- The gap test is used to determine the significance of the interval between recurrence of the same digit.
- A gap of length $x$ occurs between the recurrence of some digit.

The Gap *Test* measures the number of digits between successive occurrences of the same digit. The following example illustrates the length of gaps associated with the digit 3.

4, 1, 3, 5, 1, 7, 2, 8, 2, 0, 7, 9, 1, 3, 5, 2, 7, 9, 4, 1, 6, 3
3, 9, 6, 3, 4, 8, 2, 3, 1, 9, 4, 4, 6, 8, 4, 1, 3, 8, 9, 5, 5, 7
3, 9, 5, 9, 8, 5, 3, 2, 2, 3, 7, 4, 7, 0, 3, 6, 3, 5, 9, 9, 5, 5
5, 0, 4, 6, 8, 0, 4, 7, 0, 3, 3, 0, 9, 5, 7, 9, 5, 1, 6, 6. 3, 8
8, 8, 9, 2, 9, 1, 8, 5, 4, 4, 5, 0, 2, 3, 9, 7, 1, 2, 0, 3, 6, 3

There are a total of eighteen 3's in the list. Thus only 17 gaps can occur and the first gap is of length 10 (count digit after first 3 digit to second 3 digit), the second gap is of length 7, and so on.

The probability of a particular gap length can be determined by a Bernoulli trail.

$$P(\text{gap of } 10) = P(x \neq 3)\, P(x \neq 3) \ldots P(x \neq 3)\, P(x = 3) = (0.9)^{10}\,(0.1)$$

Since the probability that any digit is not a 3 is 0.9, and that any digit is a 3 is 0.1. In general. If we are only concerned with digits between 0 and 9, then

$$P(\text{gap of } n) = 0.9^n\, 0.1 \quad \text{where } n = 0.1, 2\ldots$$

The theoretical frequency distribution for randomly ordered digit is given by

$$F(x) = 0.1 \sum_{n=0}^{x} (0.9)^n = 1 - 0.9^{(x+1)} \qquad \ldots(5)$$

**Note:** Observed frequencies for all digits are compared to the theoretical frequency using the Kolmogorov-Smirnov test

**Note:** Probability of occurrence for certain digit is 0.1.

**There are following step for gap test:**

1. Specify the cdf for the theoretical frequency distribution given by Equation (5) based on the selected class interval width (See Table 9.6 for an example).
2. Arrange the observed sample of gaps in a cumulative distribution with these same classes.
3. Find $D$, the maximum deviation between $F(x)$ and $S_N(x)$, as in Equation $D = \max\,[F(x) - S_N(x)]$.
4. Determine the critical value, $D_\alpha$, from standard table for the specified value of $a$ and the sample size $N$.
5. If the calculated value of $D$ is greater than the tabulated value of $D_\alpha$, the null hypothesis of independence is rejected.

**Example:** Based on the frequency with which gaps occur, analyze the 110 digits above to test whether they are independent. Use $\alpha = 0.05$. The number of gaps is given by the number of digits minus 10, or 100. The number of gaps associated with the various digits are as follows:

| **Digit** | 0 | 1 | 2 | 3 | 4 | 5 | 6 | 7 | 8 | 9 |
|---|---|---|---|---|---|---|---|---|---|---|
| **No. of gaps** | 7 | 8 | 8 | 17 | 10 | 13 | 7 | 8 | 9 | 13 |

**Gap Test Example**

**Table 9.2**

| Gap length | Frequency | Relative frequency | Cum. Relative frequency | $F(x)$ | $\lvert F(x) - SN(x) \rvert$ |
|---|---|---|---|---|---|
| 0–3 | 35 | 0.35 | 0.35 | 0.3439 | 0.0061 |
| 4–7 | 22 | 0.22 | 0.57 | 0.5695 | 0.0005 |
| 8–11 | 17 | 0.17 | 0.74 | 0.7176 | 0.0224 |
| 12–15 | 9 | 0.09 | 0.83 | 0.8147 | 0.0153 |
| 16–19 | 5 | 0.05 | 0.88 | 0.8784 | 0.0016 |
| 20–23 | 6 | 0.06 | 0.94 | 0.9202 | 0.0198 |
| 24–27 | 3 | 0.03 | 0.97 | 0.9497 | 0.0223 |
| 28–31 | 0 | 0.00 | 0.97 | 0.9657 | 0.0043 |
| 32–35 | 0 | 0.00 | 0.97 | 0.9775 | 0.0075 |
| 36–39 | 2 | 0.02 | 0.99 | 0.9852 | 0.0043 |
| 40–43 | 0 | 0.00 | 0.99 | 0.9903 | 0.0003 |
| 44–47 | 1 | 0.01 | 1.00 | 0.9936 | 0.0064 |

The gap test is presented in table 9.6.

The critical value of $D$ is given by

$$D_{0.05} = 1.36/\sqrt{100} = 0.136$$

Since $D = \max \lvert F(x) - S_N(x) \rvert = 0.0224$ is less than $D_{0.05}$, do not reject the hypothesis of independence on the basis of this test.

**Runs Tests**

Tests the runs up and down or the runs above and below the mean by comparing the actual values to expected values. The statistic for comparison is the chi-square.

1. **Runs up and down:** The runs test examines the arrangement of numbers in a sequence to test the hypothesis of independence. Consider the 40 numbers; both the Kolmogorov-Smirnov and Chi-square would indicate that the numbers are uniformly distributed. But, not quite independent.

   0.08 0.09 0.23 0.29 0.42 0.55 0.58 0.72 0.89 0.91
   0.11 0.16 0.18 0.31 0.41 0.53 0.71 0.73 0.74 0.84
   0.02 0.09 0.30 0.32 0.45 0.47 0.69 0.74 0.91 0.95
   0.12 0.13 0.29 0.36 0.38 0.54 0.68 0.86 0.88 0.91

- A run is defined as a succession of similar events proceeded and followed by a different event. For example, in a sequence of tosses of a coin, we may have

  H T T H H T T T H T

  Six runs: length 1, 2, 2, 3, 1, 1
- Two characteristics: number of runs and the length of run
  - ❑ An up run is a sequence of numbers each of which is succeeded by a larger number;
  - ❑ A down run is a squence of numbers each of which is succeeded by a smaller number
- If a sequence of numbers have too few runs, it is unlikely a real random sequence, e.g., 0.08, 0.18, 0.23, 0.36, 0.42, 0.55, 0.63, 0.72, 0.89, 0.91, the sequence has one run, an up run. It is not likely a random sequence.

- If a sequence of numbers have too many runs, it is unlikely a real random sequence, e.g., 0.08, 0.93, 0.15, 0.96, 0.26, 0.84, 0.28, 0.79, 0.36, 0.57. It has nine runs, five up and four down. It is not likely a random sequence.

For sequence of random numbers, we can define up runs and down runs (successive numbers are increasing or decreasing)

–0.87, + 0.15, + 0.23, + 0.45, –0.69, –0.32, –0.30, + 0.19, 0.24.

+ + + – – – +

–0 87, + 0.15, + 0.23, + 0.05, –0.69, + 0.32, –0.40, + 0.19, 0.24.

+ + + – + – +

**Note:**
1. If first no. has (–) sign it means second no. is less than first and if first no. has (+) sign it means second no. is greater than the first no. and so on.
2. If $N$ is the number of numbers in a sequence, the maximum number of runs is $N - 1$, and the minimum number of runs is one.

- If $a$ is the total number of runs in a truly random sequence, the mean and variance of $a$ is given by

$$\mu_a = \frac{2N - 1}{3} \quad \text{...(6)}$$

and

$$\sigma^2_a = \frac{16N - 29}{90} \quad \text{...(7)}$$

**Example:** In this sequence – + + + – + – + what is the value of $a$?

**Answer:**

| | |
|---|---|
| – | (1) |
| + + + | (2) |
| – | (3) |
| + | (4) |
| – | (5) |
| + | (6) |

Hence value of $a = 6$.

- For $N > 20$, the distribution of $a$ is reasonably approximated by a normal distribution, $N(\mu_a, \sigma_a^2)$. Converting it to a standardized normal distribution by

$$Z_0 = \frac{a - \mu_a}{\sigma_a}$$

Substituting equation (6) for $\mu_a$ and the square root of equation (7) for $\sigma_a$ yields:

$$Z_0 = \frac{a - [(2N - 1)/3]}{\sqrt{(16N - 29)/90}} \quad \text{where } Z_0 \sim N(0, 1).$$

- Failure to reject the hypothesis of independence occurs when $-z_{\alpha/2} \leq Z_0 \leq z_{\alpha/2}$, where the $\alpha$ is the leve of significance.

Acceptance region for hypothesis of independence $- z_{\alpha/2} \leq Z_0 \leq z_{\alpha/2}$

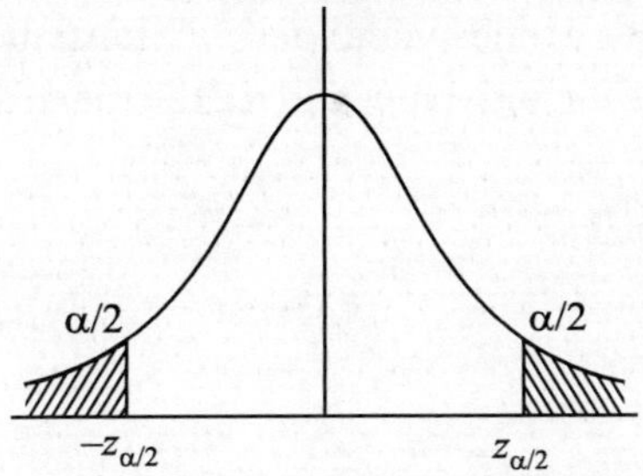


Fig. 9.1

**Example:** Based on runs up and runs down, determine whether the following sequence of 40 numbers is such that the hypothesis of independence can be rejected where $\alpha = 0.05$.

0.41 0.68 0.89 0.94 0.74 0.91 0.55 0.62 0.36 0.27

0.19 0.72 0.76 0.08 0.54 0.02 0.01 0.36 0.16 0.28

0.18 0.01 0.95 0.69 0.18 0.47 0.23 0.32 0.82 0.53

0.31 0.42 0.73 0.04 0.83 0.45 0.13 0.57 0.63 0.29

The sequence of runs up and down is as follows:

+ + + – + – + – – – + + – + – – + – + – – + – – + – + + – – + + – + – – + + –

There are 26 runs in this sequence. With $N = 40$ and $a = 26$,

$$\mu_a = \{2(40) - 1\} / 3 = 26.33 \text{ and}$$

$$\sigma_a^2 = 16(40) - 29\} / 90 - 6.79$$

Then,

$$Z_0 = (26 - 26.33) / \sqrt{(6.79)} = -0.13$$

Now, the critical value is $Z_{0.025} = 1.96$, so the independence of the numbers cannot be rejected on the basis of this test.

**2. Runs above and below the mean:**

- The previous tests for up runs and down runs are important. But they are not adequate to assure that the sequence is random. Consider the following 40 numbers :

  0.63 0.72 0.79 0.81 0.52 0.94 0.83 0.93 0.87 0.67

  0.54 0.83 0.89 0.55 0.88 0.77 0.74 0.95 0.82 0.86

  0.43 0.32 0.36 0.18 0.08 0.19 0.18 0.27 0.36 0.34

  0.31 0.45 0.49 0.43 0.46 0.35 0.25 0.39 0.47 0.41

- Check the sequence of numbers where they pass the runs up and down test. But it display the phenomenon that the first 20 numbers are above the mean, while the last 20 are below the mean.

- Let $n_1$ and $n_2$ be the number of individual observations above and below the mean, let $b$ the total number of runs. Then for a given $n_1$ and $n_2$, the mean and variance of $b$ can be expressed as

$$\mu_b = \frac{2n_1n_2}{N} + \frac{1}{2}$$

and
$$\sigma^2_{\ b} = \frac{2n_1n_2\,(2n_1\,n_2 - N)}{N^2(N-1)}$$

- For either $n_1$ or $n_2$ greater than 20, $b$ is approximately normally distributed

$$Z_0 = \frac{b-(2n_1n_2/N)-1/2}{\left[\dfrac{2n_1n_2\,(2n_1n_2 - N)}{N^2(N-1)}\right]^{1/2}}$$

- Failure to reject the hypothesis of independence occurs when $-z_{\alpha/2} \le Z_0 \le z_{\alpha/2}$, where $\alpha$ is the level of significance.

**Poker Test**

The poker test for independence is based on the frequency with which certain digits are repeated in a series of numbers. Suppose we are given three independent decimal digits in three-digit numbers there are only three possibilities, as follows:

- The individual numbers can all be different
- The individual numbers can all be the same
- There can be one pair of like digits.

The probability associated with each of these possibilities is given by the following:

$P$ (three different digits) = $P$ (second different from the first) × $P$ (third different from the first and second)

= (0.9) (0.8) = 0.72

$P$ (three like digits) = $P$ (second digit same as the first) × $P$ (third digit same as the first)

= (0.1) (0.1) = 0.01

$P$ (exactly one pair) = 1 – 0.72 – 0.01 = 0.27

To measure the amount of deviation between the expected and actual distribution we once again use the chi-square test with degree of freedom.

**Example 9.2.** A sequence of 1000 three-digit numbers has been generated and an analysis indicates that 680 have three different digits, 289 contain exactly one pair of like digits, and 31 contain three like digits. Based on poker test, are these independent? Let $\alpha = 0.05$.

The test is summarized in following table:

| *Combination i* | *Observed frequency,* $O_i$ | *Expected frequency,* $E_i$ | $\frac{(O_i - E_i)^2}{E_i}$ |
|---|---|---|---|
| Three different digits | 680 | 720 | 2.22 |
| Three like digits | 31 | 10 | 44.10 |
| Exactly one pair | 289 | 270 | 1.33 |
| | 1000 | 1000 | 47.65 |

No. of class interval $n = 3$

So degree of freedom $v = n - 1 = 2$

Since $47.65 > \chi^2_{00.5,2} = 5.99$, the independence of the numbers is rejected on the basis of numbers.

## 9.6 RANDOM VARIATE GENERATION

Random values can be conveniently and efficiently generated from desired probability distribution, such as an exponential or normal distribution, etc., for use in executing simulation models. So as to avoid speaking of 'generating random variables', the most suitable term is random variate generation. This entire chapter is devoted to methods of generating random variates from the uniform distribution on the interval [0, 1] or *U*[0, 1].

- A sequence of random variates generated from *U*[0, 1] distribution will be called **random number.**
- The sequence is transformed to produce a sequence of random values which satisfy the desired distribution, *e.g.,* exponential, normal, gamma or poisson. This step is called random variate generation.

There are three way to generate random variate.

1. General approaches to generating random variates
2. Generating continuous random variates
3. Generating discrete random variates

**General Approaches to Generating Random Variates:**

There are three general approaches to generate random variates. All these three approaches are used to generate continuous and discrete random variate by applying particular distribution either continuous or discrete.

1. The *inverse transform* method.
2. The *acceptance-rejection* method.
3. The *composition* method.

### 9.6.1 The Inverse Transformation Method

Let we have probability distribution function $F(Z)$, which is continuous, and we want to generate $n$ random samples $Z_1, Z_2 \ldots Z_n$ from this distribution. Since $F(Z)$ is continuous so it takes all values between 0 and 1; therefore for any number $U$, $0 \le U \le 1$ there exist a unique $Z_u$ such that

$$F(Z_u) = U$$
$$Z_u = F^{-1}(U) \text{ symbolically}$$

This value of $Z$ may be represented by $F^{-1}(U)$, hence it is also called inverse transformation method as in Fig. 9.2.

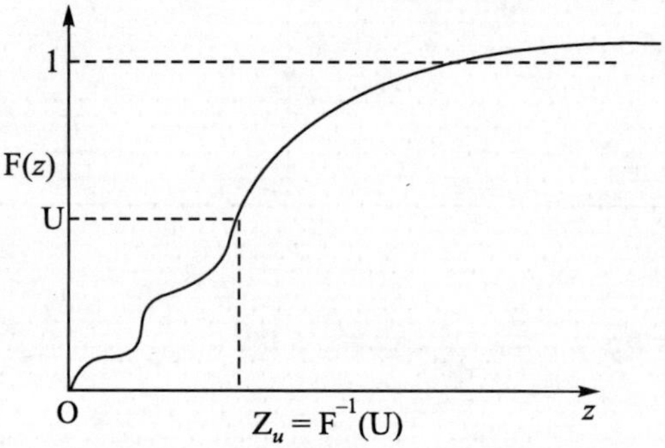


**Fig. 9.2**

If we have $n$ uniformly distributed random numbers $U_1$, $U_2$ ... $U_n$ in interval (0, 1), then their inverse transforms are $F^{-1}(U_1)$, $F^{-1}(U_2)$ ... $F^{-1}(U_n)$.

In general $Z_i = F^{-1}(U_i)$ for $i = 1, 2 .... n$ can be seen as follows.

Consider any two number $p$ and $q$ in the range (0, 1) such that $p \leq q$ from the definition – probability of uniform random number in the range (0, 1) lies between $p$ and $q$ is

$$(q - p) \; \forall \; 0 \leq p \leq q \leq 1$$

Let $F^{-1}(p) = Z_p$ and $F^{-1}(q) = Z_q$

The inverse transform method would hold of the actual probability of the random variable $Z$ lying between $Z_p$ and $Z_q$ equals the generated probability $(p - q)$. Means, actual probability of $Z$ lying between $Z_p$ and $Z_q$ should be $F(Z_p) - F(Z_q)$, which indeed is the case because

$$q - p = F(Z_q) - F(Z_p)$$

**How the Inverse Transform $F^{-1}(U_i)$ Works?**

Let generate a random sample from an exponential distribution function.

**Exponential Distribution Function**

In probability theory and statistics, the exponential distributions are characterized as:

- Probability density function (pdf)
- Cumulative distribution function (cdf)
- Alternate parameterization next line Probability density function (pdf) is given by

$$f(x; \lambda) = \begin{cases} \lambda^{-\lambda x}, & x \geq 0 \\ 0, & x \leq 0 \end{cases}$$

where $\lambda > 0$ is a parameter of the distribution, often called the rate parameter.

Cumulative distribution function is given by

$$F(x; \lambda) = \begin{cases} 1 - e^{-\lambda x}, & x \geq 0 \\ 0, & x < 0 \end{cases}$$

- The mean or expected value of an exponentially distributed random variable $X$ with rate parameter $\lambda$ is given by

$$E[X] = \frac{1}{\lambda} \quad X = x_1, x_2 \ldots x_i$$

- Variance is given by

$$\text{Var}[X] = \frac{1}{\lambda^2}$$

A step-by step method for the inverse transform technique, illustrated by the exponential distribution, is as follows :

**Step 1:** Calculate the cdf of the desired random variable $Z$. For the exponential distribution, the cdf is $F(x) = 1 - e^{-\lambda x}$, $x \geq 0$

**Step 2:** Set $\quad F(Z) = U$ on the range of $Z$.

then $\quad 1 - e^{-\lambda z} = U$ on the range of $x \geq 0$

Because $Z$ is a random variable then $1 - e^{-\lambda z}$ is also a random variable, say $U$.

$U$ has a uniform distribution over interval (0, 1), the solution is as follows:

**Step 3:** Solution $F(Z) = U$ for $Z$ in terms of $U$

$$1 - e^{-\lambda z} = U$$

$$1 - U = e^{-\lambda z}$$

$$\ln(1 - U) = -\lambda z$$

$$Z = -\frac{1}{\lambda}\ln(1 - U) \qquad \ldots(1)$$

This equation is called a random-variate generator for the exponential distribution. Equation (1) can be written as

$$Z = F^{-1}(U)$$

We can generate a sequence of values by using step 4.

**Step 4:** Generate uniform random numbers $U_1, U_2, \ldots U_n$ and calculate the desired random variates by

$$Z_i = F^{-1}(U_i)$$

From equation (1) $\quad Z_i = -\frac{1}{\lambda}\ln(1 - U_i)$

For $i = 1, 2, 3, \ldots$ Let for simplification put $1 - U_i = U_i$

Then $\quad Z_i = -\frac{1}{\lambda}\ln U_i$

Actually $U_i$ and $(1 - U_i)$ both are uniformly distributed on (0, 1).

Sometime it is very difficult to apply the inverse transformation method for many probability distributions. Since it is not easy to find out mathematical expression for the inverse function $F^{-1}$.

Then we use rejection method. In rejection method we don't need to use explicit expression for $F^{-1}$.

**Advantages of Inverse Transform Method**

1. Monotonocity
   — We have already seen how this can be useful

2. Variance reduction techniques
   — inducing correlation
   — 1 to 1′, *i.e.,* one $U(0; 1)$ variable produces one $Z$ variable

**Limitations of Inverse Transform Method**

1. It is usually required that a cumulative distribution function is known analytically and can inverted analytically.
2. Only a small number of functions satisfy these conditions.
3. In principle, one can use numerical integration of pdf or tabulated cdf (histogram) and, instead of analytical, perform numerical inversion of cdf.
4. This is usually slower and less accurate, therefore not so often realized in practice.

### 9.6.2 The Rejection Method

The rejection method is use to obtain samples from a given non-uniform distribution basically works by generating uniform random numbers repeatedly and accepting only those that meet a certain requirement. This requirement of acceptance is so designed that the accepted numbers appear to be drawn from the given distribution. For the rejection method to be applicable, the probability density function $f(t)$ of the distribution must be non-zero only over a finite interval, say $(X, Y)$. Let $f(t)$ be bounded by some upper limit $Z$ as in Fig. 9.3.

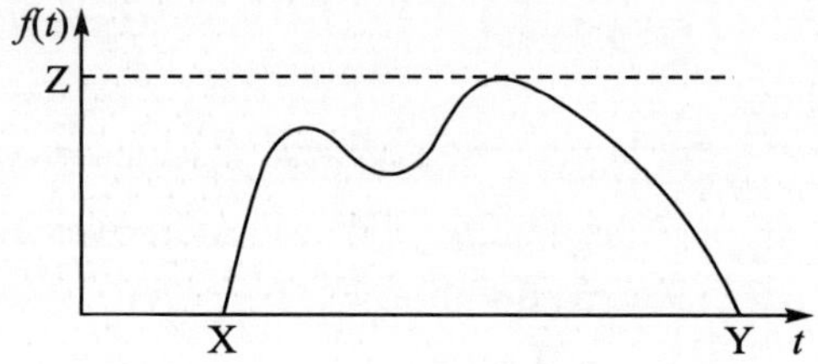


**Fig. 9.3** Probability density function.

**The procedure of rejection method consists following steps:**

**Step 1:** Generate a pair of uniform random number $U_1$, $U_2$ in the interval (0, 1).

**Step 2:** Using $U_1$ locate a point $m$ on the horizontal axis as

$$m = X + (Y - X) \cdot U_1$$

**Step 3:** Using $U_2$, locate a point $n$ on the vertical axis as

$$n = p \cdot U_2$$ (**Note:** $Z$ is upper limit)

**Step 4:** If $n > f(m)$ reject the pair and go to the Step 1, otherwise, accept $m$ as the value of a sample from the desired distribution. This procedure is repeated until the required number of samples have been generated.

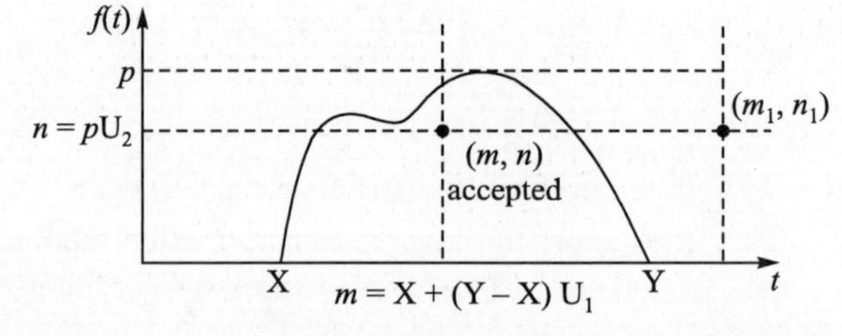


**Fig. 9.4** Rejection method.

The above figure shows how the reject method can be used to generate samples from probability density function.

**Some Conclusions:**

1. All the points above the curve $f(t)$ in the interval $(X, Y)$ are rejected, for example point $(m_1, n_1)$.
2. The points that are accepted, are within the boundary of the curve and therefore are distributed according density function $f(t)$ *e.g.* point $(m, n)$.
3. The restriction is that, this method will work only for a finite interval.

### 9.6.3 Composition Method

This applies when the distribution function $F$ from which the random number is desired can be expressed as a convex combination of other distribution functions $F_1$, $F_2$, .... Suppose we can find other cdfs $F_1$, $F_2$, ... (finite or infinite list) and weights $p_1, p_2, \ldots$ ($p_j \geq 0$ and $p_1 + p_2 + \ldots = 1$) such that for all $x$,

$$F(x) = p_1F_1(x) + p_2F_2(x) + \ldots$$

(Equivalently, can decompose density $f(x)$ or mass function $p(x)$ into convex combination of other density or mass functions). We can often write as:

$F(x) = \sum_{j=1}^{\infty} p_j F_j(x)$, where $p_j \geq 0$ for all $j$ and $\sum_{j=1}^{\infty} p_j = 1$, and each $F_J$ is distribution function.

This also works if : $f(x) = \sum_{j=1}^{\infty} p_j f(x)$, where $p_j \geq 0$, $\sum_{j=1}^{\infty} p_j = 1$

**Algorithm:**

1. Generate a positive random integer $J$ such that $P(J = j) = p_j$ ... for $j = 1, 2, \ldots$
2. Return $X$ with cdf $F_J$ (given $J = j$, $X$ is generated independent of $J$)

**Proof:** For fixed $x$,

$$P(X \leq x) = \sum_{j=1}^{\infty} P(X \leq x \mid J = j)\, P(J = j) \quad \text{(condition on } J = j)$$

$$= \sum_{j=1}^{\infty} P(X \leq x \mid J = j)\, P_j \quad \text{(distribution of } J)$$

$$= \sum_{j} F_j(x)\, p_j \quad \text{(given } J = j, X \sim Fj)$$

$$= F(x) \quad \text{(decomposition of } F)$$

The trick is to find $Fj$'s from which generation is easy and fast. Sometimes can use geometry of distribution to suggest a decomposition.

## 9.7 GENERATING CONTINUOUS RANDOM VARIATES

Here we will discuss some particular algorithm for generating random variates from commonly occurring continuous distribution. Following are common occurring continuous distribution.

1. Uniform
2. Exponential
3. Gamma
4. Weibull
5. Normal
6. Beta
7. Lognormal
8. m-Erlang

**Uniform:**

The distribution function of a $U(a, b)$ random variable is easily inverted by solving $u = F(x)$ for $x$ to obtain, for $0 \leq u \leq 1$,

$$x = F^{-1}(u) = a + (b - a)\,u$$

Thus, we can use the inverse-transform method to generate $X$:

- Generate $U \sim U(0, 1)$
- Return $X = a + (b - a)\,U$.

If many $X$ values are to generated, the constant $b - a$ sholud, of course, be computed and stored for use in the algorithm.

## 9.8 GENERATING DISCRETE RANDOM VARIATES

We can generate discrete random variates by applying some particular algorithm from commonly occurring discrete distribution. Following are common occurring distcrete distribution.

1. Bernoulli
2. Discrete Uniform
3. Binomial
4. Geometric
5. Negative
6. Posisson.

The inverse-transform method may not be the most efficient way to generate a random variate from given distribution. One other general method can be used for generating any discrete random variate having a finite range of values. This method is called **alias method.**

## 9.9 NON-UNIFORM RANDOM NUMBER GENERATION

Random numbers of distributions other than uniform are usually obtained from uniformly distributed random numbers by applying some transformation methods.

Many simulation methods need random samples from non-uniform distribution such as normal, exponential, Beta, gamma and many more other distribution.

## SUMMARY

- Systems in which the state changes are on discrete points in time are called discrete systems, *e.g.,* inventory system.
- Deterministically generated numbers which appear to be random are called pesudorandom numbers.
- Number of method or generator of producing random numbers, such as
  - ➢ Linear congruential generator (LCG)
  - ➢ Mixed generators
  - ➢ Multiplicative generater
  - ➢ Combined LCG
  - ➢ Tausuortho and Related generator
- A continuous random variable is one that can take on an uncontable number of values while a discrete random variable: is one that can take on a countable number of Values.

- The sequance is transformed to produce a sequence of random values which satisfy the desired distribution *e.g.*, exponential, normal, gamma or poisson. This step is called random variate generation.

To ensure that desirable properties of random numbers *i.e.*, uniformity and independence, are achieved by some test like frequency test, chi-square test and poker test.

- There is also non-uniformly distributed random numbers. They are generated by inverse transformation and rejection method.

## SOLVED QUESTIONS

**Q. 1. Suppose a random variable $X$ takes on the values –3, –1, 2 and 5 with respect probabilities**

$$\frac{2k-3}{10}, \frac{k+1}{10}, \frac{k-1}{10}, \frac{k-2}{10}$$

**Determine the distribution of $X$.**

**Ans.** Set the sum of probabilities equal to 1, and find $k$.

$$\frac{2k-3}{10} + \frac{k+1}{10} + \frac{k-1}{10} + \frac{k-2}{10} = 1$$

$$5k - 5 = 10$$

$$5k = 15$$

$$k = 3$$

Then put $k = 3$ into the above probabilites, yielding

0.3, 0.4, 0.2, 0.1

Thus the distribution of $X$ follows:

| $x$ | –3 | –1 | 2 | 5 |
|---|---|---|---|---|
| $P(X = x)$ | 0.3 | 0.4 | 0.2 | 0.1 |

**Q. 2. What do you mean by expectation of a finite random variable?**

**Ans.** Suppose $X$ is a random variable whose distribution $f$ is as follows:

| $x$ | $x_1$ | $x_2$ | $x_3$ | ... | $x_n$ |
|---|---|---|---|---|---|
| $f(x)$ | $f(x_1)$ | $f(x_2)$ | $f(x_3)$ | ... | $f(x_n)$ |

Then the mathematical expectation or expected value or, simply, the expectation of $X$, denoted by $E(X)$ or simply, $E$ is defined by

$$E = E(X) = x_1 f(x_1) + x_2 f(x_2) + \ldots + x_n f(x_n) = \sum_i x_i f(x_i)$$

or, when notation $(x_i, p_i)$ is used insted of $[x, f(x)]$,

$$E = E(X) = x_1 p_1 + x_2 p_2 + \ldots x_n p_n = \sum_i x_i p_i$$

or we can say if $x_i$ are numerical outcomes of an experiment, then $E$ is the expected value of the experiment.

**Q. 3. Explain normal distribution or Gaussian distribution.**

**Ans.** Normal or Gaussian distribution is the most familiar and frequently encountered probability distribution. Its density function is a bell-shaped curve given by the expression

$$f(x) = \frac{1}{\sigma\sqrt{2\pi}} e^{\frac{-1}{2}\left(\frac{x-\mu}{\sigma}\right)^2} \quad (pdf)$$

for $-\infty < x < \infty$ and variance $\sigma^2 > 0$.

The parameter μ is the expected value and σ is the standard deviation.

If parameters of the normal distribution have values $\mu = 0$ and $\sigma = 1$, we call it the standardized normal distribution. It is expressed by

$$f(s) = \frac{1}{\sqrt{2\pi}} e^{-S^2/2}$$

The density function and its integral *i.e.*, the cumulative distribution function are shown in Fig. 9.6.

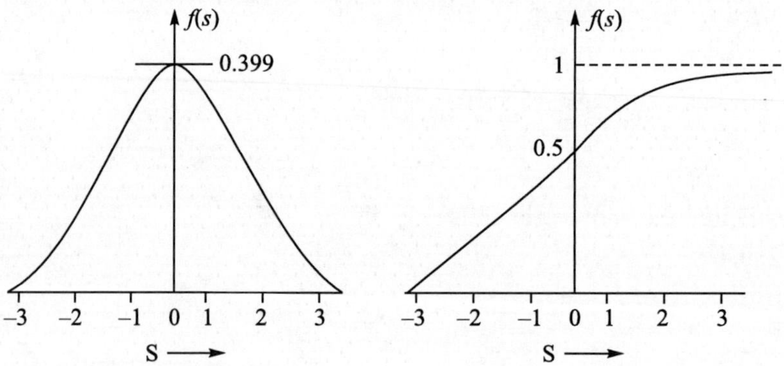


(*i*) Density function (*ii*) Distribution function

**Fig. 9.6** Standardized normal distribution.

**Note:** There does not exist an explicit expression for the cumulative distribution function *F*(*s*), although very extensive tables are available in handbooks of statistics.

**Q. 4. Multiplicative congruential generator is given by $x_{i+1} = x_i \cdot a$ (modulo *m*). If seed $x_0 = 3$ and parameter $a = 7$ and $m = 17$. Then generate successive random integers.**

**Ans.** Given generator $x_{i+1} = x_i \cdot a$ (modulo *m*)

$x = 3 \quad a = 7 \quad m = 17$

at $i = 0 \quad x_1 = x_0 \times a$ (modulo *m*)

or $x_1 = 3 \times 7$ (modulo 17) $\equiv 4$

at $i = 1 \quad x_2 = 4 \times 7$ (modulo 17) $\equiv 11$

$i = 2 \quad x_3 = 11 \times 7$ (modulo 17) $\equiv 9$

$x_4 = 11 \times 9$ (modulo 17) $\equiv 14$

and so on, all between 0 and 16.

**Q. 5. Use the linear congruential method of generate a sequence of random number where $Z_0 = 27$, $a = 13$, $c = 37$ and $m = 100$.**

**Ans.** $Z_{i+1} = (a\,Z_i + c) \bmod m \quad i = 0, 1, 2....$

The sequence of $Z_0 = 27$ and subsequent $Z_i$ values is computed as follows:

$$Z_1 = (13 \times 27 + 37) \bmod 100 = 388 \bmod 100 = 88$$

$$U_1 = \frac{88}{100} = 0.88$$

$$Z_2 = (13 \times 88 + 37) \bmod 100 = 1181 \bmod 100 = 81$$

$$U_2 = \frac{88}{100} = 0.81$$

$$Z_3 = (13 \times 81 + 37) \bmod 100 = 1090 \bmod 100 = 90$$

$$U_3 = \frac{90}{100} = 0.90$$

$$Z_4 = (13 \times 90 + 37) \bmod 100 = 1207 \bmod 100 = 7, \text{ and so on.}$$

$$U_4 = 0.07 \text{...so on.}$$

These random integers should appear to be uniformly distributed on the integers zero to 99. Random numbers between zero and one can be generated $U_1\ U_2 ....U_i$ are random numbers.

**Q. 6. Write the algorithm to generate random numbers for 32–bit computers $L^1$ Ecuyer suggests combining $J = 2$ generators with $m_1 = 2147483563$, $a_1 = 40014$, $m_2 = 2147483399$, and $a_2 = 40692$.**

**Ans.** Following algorithm, is used to generate random number:

1. Select seed $Z_{1,\,0}$ in the range (1, 2147483562) for the first generator, and seed $Z_{2,\,0}$ in the range (1, 2147483398). Set $K = 0$.
2. Evaluate each individual generator

$$Z_{1,\,K+1} = 40014\ Z_{1,\,K} \bmod 2147483563$$
$$Z_{2,\,K+1} = 40692\ Z_{2,\,K} \bmod 2147483399$$

3. Set

$$Z_{K+1} = (Z_{1,\,K+1} - Z_{2,\,K+1}) \bmod 21474383562$$

4. Return

$$U_{K+1} = \begin{cases} \dfrac{Z_{K+1}}{2147483563}, & Z_{K+1} > 0 \\ \dfrac{2147483562}{2147483563}, & Z_{K+1} = 0 \end{cases}$$

5. Set $K = K + 1$ and goto step 2.

This combined generator has period $(m_1 - 1)\,(m_2 - 1)/_2 \approx 2 \times 10^{18}$. However, such a long period may not be adequate for all applications.

**Q. 7. What problems or errors are occur during the generation of pseudorandom numbers.**

**Ans.** Some errors or problems are following:

1. The generated numbers may not be uniformly distributed.
2. The mean of the generated numbers may be too high or too low.
3. The variance of the generated numbers may be too high or too low.
4. There may be dependence.

**Q. 8. Discuss various disadvantage of LCGs.**

**Ans.** There are following disadvantage of LCGs:

- LCGs should not be used for applications where high quality randomness is critical.
- It is not suitable for Monte-Carlo simulation because of the serial correlation.
- They should also not be used for cryptographic applications.

**Q. 9. Why we preferred next-event model?**

**Ans.** We preferred next-event model because we don't want to waste any computer time in scanning those points in time when nothing takes place. This waste is bound to occur if we pick a reasonably small value for $\tau$. On the other hand, if $\tau$ is so large that one or more events must take place during each interval then our model become unrealistic and don't give meaningful results.

**Q. 10. Consider following 100 two-digits value in Table 9.3. Use the chi-squares test with $\alpha = 0.05$ to test whether the data shown below are uniformly distributed, where $\chi_{0.05,\,9} = 16.9$.**

**Table 9.3**

| | | | | | | | | | |
|---|---|---|---|---|---|---|---|---|---|
| 0.34 | 0.90 | 0.25 | 0.89 | 0.87 | 0.44 | 0.12 | 0.21 | 0.46 | 0.67 |
| 0.83 | 0.76 | 0.79 | 0.64 | 0.70 | 0.81 | 0.94 | 0.74 | 0.22 | 0.74 |
| 0.96 | 0.99 | 0.77 | 0.67 | 0.56 | 0.41 | 0.52 | 0.73 | 0.99 | 0.02 |
| 0.97 | 0.30 | 0.17 | 0.82 | 0.56 | 0.05 | 0.45 | 0.31 | 0.78 | 0.05 |
| 0.79 | 0.71 | 0.23 | 0.19 | 0.82 | 0.93 | 0.65 | 0.37 | 0.39 | 0.42 |
| 0.99 | 0.17 | 0.99 | 0.46 | 0.05 | 0.66 | 0.10 | 0.42 | 0.18 | 0.49 |
| 0.37 | 0.51 | 0.54 | 0.01 | 0.81 | 0.28 | 0.69 | 0.34 | 0.75 | 0.49 |
| 0.72 | 0.43 | 0.56 | 0.97 | 0.30 | 0.94 | 0.96 | 0.58 | 0.73 | 0.05 |
| 0.06 | 0.39 | 0.84 | 0.24 | 0.40 | 0.64 | 0.40 | 0.19 | 0.79 | 0.62 |
| 0.18 | 0.26 | 0.97 | 0.88 | 0.64 | 0.47 | 0.60 | 0.11 | 0.29 | 0.78 |

**Ans.** The test uses $n = 10$ intervals of equal length, say [0, 0.1], [0.1, 0.2], ... [0.9, 1.0] from Table 9.2.

Suppose $n = 1$ [0,0.10]

$n = 2$ [0.10, 0.20]

$n = 3$ [0.20, 0.30] ....

⋮ ⋮

$n = 10$ [0.90, 1.0]

We have observed from table that frequency of occurrence of 0.0 to 0.10 digit is 8, and 0.10 to 0.20 range digit is 8, and so on.

**Table 9.4.** Contains the Necessary Computations

| Interval | $O_i$ | $E_i$ | $O_i - E_i$ | $(O_i - E_i)^2$ | $(O_i - E_i)^2/E_i$ |
|---|---|---|---|---|---|
| 1 | 8 | 10 | –2 | 4 | 0.4 |
| 2 | 8 | 10 | –2 | 4 | 0.4 |
| 3 | 10 | 10 | 0 | 0 | 0.0 |
| 4 | 9 | 10 | –1 | 1 | 0.1 |
| 5 | 12 | 10 | 2 | 4 | 0.4 |
| 6 | 8 | 10 | –2 | 4 | 0.4 |
| 7 | 10 | 10 | 0 | 0 | 0.0 |
| 8 | 14 | 10 | 4 | 16 | 1.6 |
| 9 | 10 | 10 | 0 | 0 | 0.6 |
| 10 | 11 | 10 | 1 | 1 | 0.1 |
| | 100 | 100 | 0 | | 3.4 |

The value of $\chi^2$ is 3.4. This is compared with given critical value $\chi^2_{0.05,9}$, the null hypothesis of a uniform distribution is not rejected or it can be accepted.

**Q. 11. The sequence of numbers 0.54, 0.73,0.98, 0.11, and 0.68 has been generated. Use the kolmogorov -Smirnow test with a=0.05 and $D_{0.05}$ = 0.565 to determine if the hypothesis that the numbers are uniformly distributed on the interval [0, 1] can be rejected.?**

| Ri | 0.11 | 0.54 | 0.68 | 0.73 | 0.98 |
|---|---|---|---|---|---|
| $i$/N | 0.20 | 0.40 | 0.60 | 0.80 | 1 |
| $i/N–R_i$ | 0.09 | – | – | – | 0.02 |
| $R_i – (i\text{-}1)/N$ | 0.11 | 0.34 | 0.28 | 0.13 | 0.18 |

**Ans.** $D^+ = \text{Max}\ (i/N - R_i) = 0.09$

$D^- = \text{Max}\ (R_i - (i–1)/R_i) = 0.34$

The critical value $D_\alpha$ is given D0.05 = 0.565.D< D0.05. The hypothesis that there is no difference between the true distribution of {R1, R2... R5} and the uniform distribution on [0,1] can not be rejected on the basis of this test.

**Q. 12. Consider the 50 two-digit values.**

| | | | | |
|---|---|---|---|---|
| 0.34 | 0.9 | 0.25 | 0.89 | 0.87 |
| 0.83 | 0.76 | 0.79 | 0.64 | 0.7 |
| 0.96 | 0.99 | 0.77 | 0.67 | 0.56 |
| 0.47 | 0.3 | 0.17 | 0.82 | 0.56 |
| 0.79 | 0.71 | 0.23 | 0.19 | 0.82 |
| 0.99 | 0.17 | 0.99 | 0.46 | 0.05 |
| 0.37 | 0.51 | 0.54 | 0.01 | 0.81 |
| 0.72 | 0.43 | 0.56 | 0.97 | 0.3 |
| 0.06 | 0.39 | 0.84 | 0.24 | 0.4 |
| 0.18 | 0.26 | 0.97 | 0.88 | 0.64 |

Based on runs up and runs down , determine whether the hypothesis of independence can be rejected, where $\alpha = 0.05$.

**Ans.** The numbers are given a "+" or a "-" depending on whether they are followed by a larger or smaller number and assign to different number. The last number is followed by "no event" and hence will get neither a + nor a -.:

$$+ - + - - - + + + + - + - + + + - - + + + - - - -$$

$$+ + + - + - + - - + - + - + - - - + + - - + +$$

There are $a$ = 27 runs in this sequence.

For $N = 50$,

$$\mu_a = (2N\text{- }1)/3 = 33, \text{ and}$$

$$\sigma_a^2 = \{16N - 29)/90 = 8.5667$$

$$Zo = (a - \mu_a)/\sigma_a = -2.05$$

$$Z_{\alpha/2} = z._{025} = 1.96$$

Since $Z_0 < -z._{025}$, the null hypothesis of independence can be rejected.

**Q. 13.** A sequence of 1000 four-digit numbers has been generated and an analysis indicates the following combinations and frequencies.

| Combination i | Observed Frequency, Oi |
|---|---|
| Four different digits | 565 |
| One pair | 392 |
| Two pairs | 17 |
| Three like digits | 24 |
| Four like digits | 2 |
| Total | 1000 |

**Ans.**

| Combination (i) | Observed ($O_i$) | Expected ($E_i$) | $(O_i - E_i)^2/E_i$ |
|---|---|---|---|
| 4 different digits | 565 | 504 | 7.3829 |
| 1 pair | 392 | 432 | 3.7037 |
| 2 pairs | 17 | 27 | 3.7037 |
| 3 like digits | 24 | 36 | |
| 4 like digits | 2 | 1 | 3.2703 |
| Totals | 1000 | 1000 | 18.0606 |

$$\chi^2_{0.5,3} = 7.81 < \chi^2 = 18.0806$$

Reject the null hypothesis of independence based on this test. Notice that we grouped "3 like digits" and "4 like digits" into a single cell with $Oi = 26$ and $E_i = 37$.

## EXERCISES

1. Differentiate next-event time advance approach and fixed-event time advance approach.
2. What do you mean by discrete-event simulation? Explain with example.
3. What are different techniques for generating random number? Explain them. [*UPTU 2005-06*]
4. Write a program to generate and print 5,000 uniformly distributed numbers between 0.000 and .999 (3 significant digits). Test the numbers for uniformity by dividing them into 10 ranges and using chi-square test.

5. Discuss the time advance mechanism in discrete event system simulation.
6. Define following term:
   (*i*) Normal distribution (*ii*) Inverse transformation method (*iii*) Combined LCG.
7. How to test independence, explain with example?
8. Explain with example, how we can test random number.
9. Differentiate between Monte-Carlo computation and stochastic simulation.
10. What is the method of testing random number generation of non-uniformly distributed random numbers? [*UPTU 2006-07*]
11. The probability distribution function of the Weibull distribution is defined by
$$F(y) = 1 - e^{-\lambda y^c}, y \geq 0, \;\; c > 1, \lambda \geq 0$$
   Using the inverse transformation method, write a function subprogram WE1 *B* to generate *N* random samples from the Weibull distribution.
12. Write a programe to generate uniformly distributed random numbers between 0 and 1 (with as many significant digits as your computer word can hold). Study and comment on the behaviour of the least significant digit of these numbers generated.
13. Develop the poker test for:
   (*a*) Four-digit numbers
   (*b*) Five-digit numbers.
14. Consider the first 50 two-digits values in solve problem 8 based on runs up and runs down, determine whether the hypothesis of independence can be rejected, where $\alpha = 0.05$.
15. Consider the following sequence of 120 digits:

| 1 | 3 | 7 | 4 | 8 | 6 | 2 | 5 | 1 | 6 | 4 | 4 | 3 | 3 | 4 | 2 | 1 | 5 | 8 | 7 |
|---|---|---|---|---|---|---|---|---|---|---|---|---|---|---|---|---|---|---|---|
| 0 | 7 | 6 | 2 | 6 | 0 | 5 | 7 | 8 | 0 | 1 | 1 | 2 | 6 | 7 | 6 | 3 | 7 | 5 | 9 |
| 0 | 8 | 8 | 2 | 6 | 7 | 8 | 1 | 3 | 5 | 3 | 8 | 4 | 0 | 9 | 0 | 3 | 0 |  | 2 |
| 2 | 3 | 6 | 5 | 6 | 0 | 0 | 1 | 3 | 4 | 4 | 6 | 9 | 9 | 8 | 5 | 6 | 0 | 1 | 7 |
| 5 | 6 | 7 | 9 | 4 | 9 | 3 | 1 | 8 | 3 | 3 | 6 | 6 | 7 | 8 | 2 | 3 | 5 | 9 | 6 |
| 6 | 7 | 0 | 3 | 1 | 0 | 2 | 4 | 2 | 0 | 6 | 4 | 0 | 3 | 3 | 3 | 6 | 8 | 1 | 5 |

   Test whether these digits can be assumed to be independent based on the frequency with which gaps occur. Use $\alpha = 0.05$.
16. Use the LCG method to generate a sequence of three two-digit random integers. Let $x_0 = 27$, $a = 8$, $c = 47$, and $m = 100$.
17. Use the multiplicative congruential method to generate a sequence of four three digit random integers. Let $x_0 = 107$, $a = 41$, and $m = 1000$.
18. The sequence of numbers 0.54, 0.73, 0.98, 0.11, and 0.6 has been generated. Use the Kolmogorov-Smirnov test with2 = 0.05 to determine if the hypothesis that the numbers are uniformly distributed on the interval [0, 1] can be rejected.
19. Use the mixed congruential method to generate a sequence of three two-digit random numbers with $z_0 = 37$, $a = 7$, $c = 29$, and $m = 100$.
20. Use the mixed congruential method to generate a sequence of three two-digit random integers between 0 and 24 with $z_0 = 13$, $a = 9$, and $c = 35$.

**21.** Use the principles described in this chapter to develope your own linear congruential random number generator.

**22.** Consider the 50 two-digit values.

| | | | | |
|---|---|---|---|---|
| 0.44 | 0.12 | 0.21 | 0.46 | 0.67 |
| 0.81 | 0.94 | 0.74 | 0.22 | 0.74 |
| 0.41 | 0.52 | 0.73 | 0.99 | 0.02 |
| 0.05 | 0.45 | 0.31 | 0.78 | 0.05 |
| 0.93 | 0.65 | 0.37 | 0.39 | 0.42 |
| 0.66 | 0.1 | 0.42 | 0.18 | 0.49 |
| 0.28 | 0.69 | 0.34 | 0.75 | 0.49 |
| 0.94 | 0.96 | 0.58 | 0.73 | 0.05 |
| 0.64 | 0.4 | 0.69 | 0.79 | 0.62 |
| 0.47 | 0.6 | 0.11 | 0.29 | 0.78 |

Based on runs up and runs down, determine whether the hypothesis of independence can be rejected, where a=0.05.

**23.** Consider the following sequence of 120 digits:

| | | | | | | | | | | | | | | | | | | | |
|---|---|---|---|---|---|---|---|---|---|---|---|---|---|---|---|---|---|---|---|
| 1 | 3 | 7 | 4 | 8 | 6 | 2 | 5 | 1 | 6 | 4 | 4 | 3 | 3 | 4 | 2 | 1 | 5 | 8 | 7 |
| 0 | 7 | 6 | 2 | 6 | 0 | 5 | 7 | 8 | 0 | 1 | 1 | 2 | 6 | 7 | 6 | 3 | 7 | 5 | 9 |
| 0 | 8 | 8 | 2 | 6 | 7 | 8 | 1 | 3 | 5 | 3 | 8 | 4 | 0 | 9 | 0 | 3 | 0 | 9 | 2 |
| 2 | 3 | 6 | 5 | 6 | 0 | 0 | 1 | 3 | 4 | 4 | 6 | 9 | 9 | 8 | 5 | 6 | 0 | 1 | 7 |
| 5 | 6 | 7 | 9 | 4 | 9 | 3 | 1 | 8 | 3 | 3 | 6 | 6 | 7 | 8 | 2 | 3 | 5 | 9 | 6 |
| 6 | 7 | 0 | 3 | 1 | 0 | 2 | 4 | 2 | 0 | 6 | 4 | 0 | 3 | 9 | 3 | 6 | 8 | 1 | 5 |

Test whether these digits can be assumed to be independent based on the frequency with which gasps occur. Use a=0.05.

**24.** Develop the poker test for:

(*a*) Four-digit numbers

(*b*) Five-digit numbers.

10

# Simulation of Queuing System

**INSIDE THIS CHAPTER**



## 10.1 QUEUING SYSTEM

Queues are common in computer systems. Thus, there are queues of inquiries waiting to be processed by an interactive computer system, queue of data base requests, queues of I/O requests, etc. Typically a queue (or queuing system) has one service facility, although there may be more than one server in the service facility, and a waiting room (or buffer) or finite or infinite capacity. Customers from a population or source enter a queuing system to receive some service. Here the word customer is used in its generic sense, and thus maybe a packet in a communication network, a job or a program in a computer system, a request or an inquiry in a database system, etc.

Upon arrival a customer joins the waiting room if all servers in the service center are busy. When a customer has been served, he leaves the queuing system.

A special notation, called **Kendall's notation**, is used to describe a queuing system. The notation has the form A|B|c|K

where

- A: describes the interarrival time distribution
- B: represents the service time distribution
- c: represents the number of servers
- K represents system capacity

The symbols traditionally used for A and B are:

- M for exponential distribution (M stands for Markov or memory less)
- D for deterministic distribution
- $E_k$ (Erlang of order $k$)
- H for hyperexponential
- G for general distribution, and GI for general independent.

When the system capacity is infinite ($K$ = infinite) one can simply uses the symbol $A/B/c$. For instance, M/M/1, M/M/C, M/G/1 and G/M/1, and M/D/1 are very common queuing systems. The notation can be extended with an extra letter to cover other queuing models. For example, a system with exponential interarrival and service times, one server and a waiting room only for N customers (including the one in service) is abbreviated by the code M/M/l/N.

A queuing system thus consists of arriving units or using common terminology customers waiting for some kind of service facility. The use of these terminologies, not in the literal sense, greatly facilitates the study of queues. On analysis it is found that the following features characterize a queuing system which distinguish one system from another.

1. **System Capacity:** This represents the number of customer allowed in the system. The system capacity can be considered as finite or infinite.
2. **Number of Servers and System Configuration:** To enhance system capacity the number of servers is often more than one. Servers in many cases such as ticket counters, bank, etc., are configured in parallel.
3. **Arrival Process:** The arrival process is random and is best described by probability model. Based on simple hypothesis we can derive the probability distribution of the arrival process. One important hypothesis in the derivation is that the customers originate from a population of unlimited size (Infinite source).
4. **Service:** Except in special situations, the service time process in most cases is random, described by probability distribution.
5. **Queue Discipline:** This represents the order in which customers are selected from a queue for service. The most common discipline is "First come first serve" (FCFS). Other disciplines are "Last come first serve" (LCFS) and "Service in random order" (SIRO). Queue discipline may also be based on some priority, such as rush jobs at a shop processed ahead of regular jobs.

## 10.2 BIRTH-DEATH PROCESS

The birth-death process is a special case of CTMC means the Continuous Time Marko Chain. For a CTMC the Markov property can be stated as

$$P(X(t)) = x \mid X(t_n) = x_n, X(t_{n-1}) = x_{n-1}, \ldots, X(t_0) = x_0) = P(X(t) = x \mid X(t_n) = x_n)$$

The states represent the current size of a population and where the transitions are limited to births and deaths. Birth-death processes have many applications in queuing theory, performance engineering, or in biology.

When a birth occurs, the process goes from state $n$ to $n + 1$. When a death occurs, the process goes from state $n$ to state $n - 1$. The process is specified by birth rates $\{\lambda_i\}_i = 0 \cdots \infty$ and death rates $\{\mu_i\}_{i=1 \ldots \infty}$.

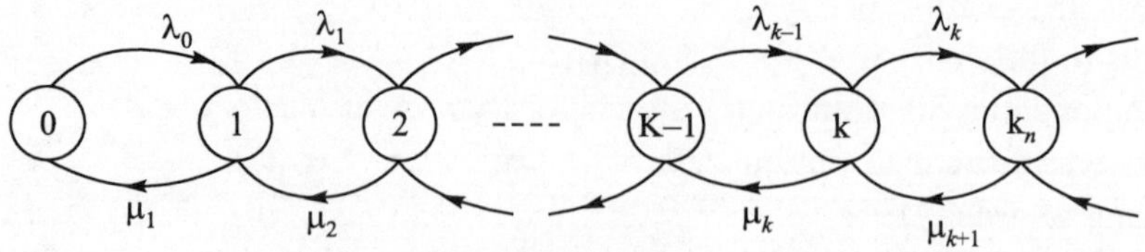


Fig. 10.1

**Examples 10.1:** A pure birth process is a birth-death process where

$$\mu_i = 0 \ \forall \ i \geq 0.$$

A pure death process is a birth-death process where

$$\lambda_i = 0 \ \forall \ i \geq 0.$$

A (homogenous) **Poisson process** is a pure birth process where

$$\lambda_i = \lambda \ \forall \ i \geq 0.$$

M/M/1 model and M/M/c model, both used in queuing theory, are birth-death processes used to describe customers in an infinite queue.

## 10.3 ARRIVAL PROCESS AND SERVICE DISTRIBUTION

Here we will proceed to determine the probability distribution of the 2 processes.

***(a)* Service Time Distribution:** At a server the service job is taken up one after another at random points of time, which is called event [arrival/departure event]. The process of succession of events is stationary by which it is meant that the probability of an event occurring in a time interval $(t, t + h)$ depends only on the length $h$. Beginning at an event, during service time $t$, let the probability of no further event be denoted by $p_0(t)$, then at service time $(t + h)$.

$$p_0(t + h) = \text{Probability of no event at time } t \times \text{Probability of no event in time } h$$

$$= p_0(t) \times p_0(h) \qquad \ldots(10.1)$$

where $p_0(0) = 1$: probability of no event in 0 time.

The solution of equation (10.1) is the exponential function. By Taylor series expansion we can write

$$p_0(h) = p_0(0) + hp'_0(0) + O(h^2) \quad \text{[put in equation (10.1)]}$$

$$p_0(t + h) = p_0(t) \times [p_0(0) + hp'_0(0) + O(h^2)]$$

$$p_0(t + h) = p_0(t) \times [1 + h(-\mu) + O(h^2)] \quad [\text{put } p_0(0) = 1] \text{ and } [\text{put } p'_0(0) = \mu]$$

$$p_0(t + h) = p_0(t) \times [1 - h\mu + O(h^2)]$$

$$p'_0(t) = \lim_{h \to 0} \frac{p_0(t+h) - p_0(t)}{h} = \lim_{h \to 0} [-\mu + O(h)] = -\mu$$

Hence by integration, $p_0(t) = e^{-\mu t}$ satisfying the initial condition $p_0(0) = 1$. Since $0 \leq p_0(t) \leq 1$ we must have $\mu > 0$. The function $p_0(t)$ decreases from 1 to 0. On the other hand the cumulative distribution function $F(t)$ of the random variable always increases from 0 to 1. Hence we must have

$$F(t) = 1 - p_0(t) = 1 - e^{-\mu t}$$

So the probability density function (pdf), $f(t)$ of the distribution is

$$f(t) = F'(t) = \mu e^{-\mu t},\ t \geq 0 \qquad \ldots(10.2)$$

Hence the service time has an exponential distribution, which rapidly tends to zero starting width $f(0) = \mu$.

Mean value of the service time distribution is

$$E\{f(t)\} = \int_0^\infty t\mu e^{-\mu t}\, dt = \frac{1}{\mu} \qquad \ldots(10.3)$$

So, $\frac{1}{\mu}$ is equal to mean service time.

***(b)* Distribution of Arrivals:** In the queue customers arrive randomly. Let the number of customers in the queue at time $t$ be $n$ with probability $p_n(t)$, then at time $(t + h)$ after infinitesimely small time $h$, the probability of $n$ number of customers in the queuing system is

$$p_n(t + h) = \text{Probability of } n \text{ customers at time } t \times \text{Probability of no customers in time } h + \text{Probability of } (n - 1) \text{ customers at time } t \times \text{Probability of one customers in time } h$$

$$= p_n(t) \times p_0(t) + p_{(n-1)}(t) \times p_1(t) \qquad \ldots(10.4)$$

Here we are writing λ, instead of μ (as in case of service time distribution)

$$p_0(h) = 1 - \lambda h + O(h^2),\ \lambda > 0 \qquad \text{...[10.5(a)]}$$

$$\begin{aligned} p_1(h) &= \text{Probability of 1 call in time } h \\ &= 1 - p_0(h) \\ &= \lambda h - O(h^2) \end{aligned} \qquad \text{...[10.5(b)]}$$

Since in small time $h$, 0 to 1 arrivals is a certainty. Substituting from equations (10.5) in (10.4), we get

$$p_n(t+h) = p_n(t)\,[1 - \lambda h + O(h^2)] + p_{n-1}(t)[\lambda h - O(h^2)]$$

Hence,

$$p_n'(t) = \lim_{h \to 0} \frac{p_n(t+h) - p_n(t)}{h} = -\lambda p_n(t) + \lambda p_{n-1}(t) \qquad \text{...(10.6)}$$

Considering equation (10.6), as a first order linear differential equation in $p_n(t)$, we get

$$p_n(t)\, e^{\lambda t} = \lambda \int_0^t p_{n-1}(t)\, e^{\lambda t}\, dt$$

Since $p_n(0) = 0$ for $n \geq 1$. Now as in $(a)$, $p_0(t) = e^{-\lambda t}$; hence

$$p_1(t) e^{\lambda t} = \lambda t \text{ or, } p_1(t) = \lambda t e^{-\lambda t}$$

$$p_2(t) e^{\lambda t} = \lambda \int_0^t \lambda t\, dt = \frac{\lambda^2 t^2}{2} \text{ or } p_2(t) = \frac{\lambda^2 t^2}{\underline{|2}} e^{-\lambda t}$$

Finally we get

$$p_n(t) = \frac{\lambda^n t^n}{\underline{|n}} e^{-\lambda t},\ n = 0, 1, 2, \ldots, \infty \qquad \text{...(10.7)}$$

which is a poisson distribution of discrete variable $n$.

The mean value of the distribution is

$$\begin{aligned} E\{p_n(t)\} &= \sum_{n=0}^{\infty} n \frac{(\lambda t)^n}{\underline{|n}} e^{-\lambda t} \\ &= \lambda t e^{-\lambda t} \sum_{n=1}^{\infty} \frac{(\lambda t)^{n-1}}{\underline{|n-1}} = \lambda t \end{aligned} \qquad \text{...(10.8)}$$

Since the exponential series equals $e^{\lambda t}$.

For $t = 1$, the mean value of λ. Hence λ equals mean arrival rate and so 1/λ equals **mean interarrival times.**

The interarrival process is exponential as in the case of service time distribution. In this respect Poisson arrivals in a queue are also called **Markovian.**

## 10.4 GENERALIZED POISSON QUEUING SYSTEM

Now we are assuming that mean arrival and service rates vary depending on the number of customers $n$ the system (queue + service). Let $\lambda_n$, $\mu_n$ be the mean arrival and service rates under this assumption. The probability $p_n$ of $n$ customers in the system is in general a function of time $t$, particularly during initial build up of the queue. If the system stays in operation for the sufficiently long time, under proper conditions, its behaviour generally speaking is expected to stabilize around some norm. This **steady-state** behaviour is of actual interest in queuing theory.

The steady-state queuing system consists of $n$ states where $n$ is the number of customers in the system with probability $p_n$. $n = 0, 1, 2, \ldots, \infty$ if the system capacity is assumed infinite. The states can be shown as following diagram.

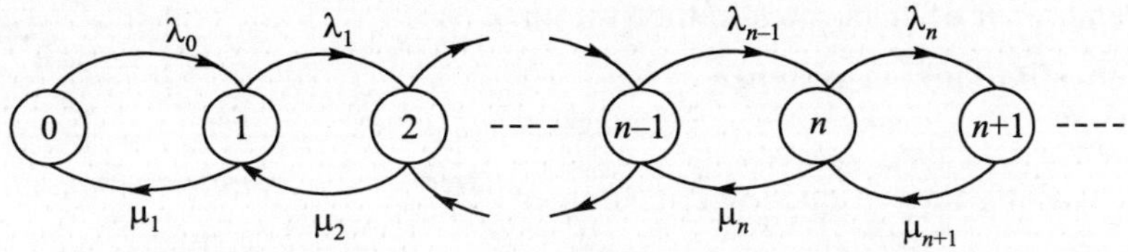


**Fig. 10.2**

In a Poisson queue there can be at most one arrival or one departure after service in a small time, the node reached when either one arrival takes place from node $(n - 1)$ with rate $\lambda_{n-1}$ or one departure takes place from node $(n + 1)$ with rate $\mu_{n+1}$. It is very clear that node 0 can be reached only when one departure takes place from node 1 at rate $\mu_1$ and node 1 can be reached by either one arrival from node 0 at rate $\lambda_0$ or by one departure from node 2 at rate $\mu_2$. $\mu_0$ is undefined because no departure can take place when queue is empty. At any node, the expected rate of inflow equals to expected rate of outflow. Hence we obtain the balance equation at node $n$ as

$$\lambda_{n-1}p_{n-1} + \mu_{n+1}p_{n+1} = (\lambda_n + \mu_n)p_n,\ n = 1, 2, \ldots, \infty \qquad \ldots[10.9(a)]$$

and at node 0,

$$\mu_1 p_1 = \lambda_0 p_0 \qquad \ldots[10.9(b)]$$

equations [10.9(*a*)] and [10.9(*b*)] are difference equations, which can be solved recursively for determining $p_n$ for the system from equation [10.9(*b*)] we have

$$p_1 = \frac{\lambda_0}{\mu_1} p_0$$

putting $n = 1$, in equation [10.9(*a*)] we have,

$$\lambda_0 p_0 + \mu_2 p_2 = (\lambda_1 + \mu_1)p_1$$

or

$$\mu_2 p_2 = -\lambda_0 p_0 + (\lambda_1 + \mu_1)p_1$$

or

$$\mu_2 p_2 = -\lambda_0 p_0 + \lambda_1 p_1 + \mu_1 p_1$$

$$\mu_2 p_2 = -\lambda_0 p_0 + \lambda_1 p_1 + \lambda_0 p_0$$

or

$$\mu_2 p_2 = \lambda_1 p_1$$

or

$$p_2 = \frac{\lambda_1 p_1}{\mu_2} \text{ or } \frac{\lambda_1 \lambda_0}{\mu_0 \mu_1} p_0 \qquad \text{[from 10.9(b)]}$$

putting $n = 2$, similarly,

$$\mu_3 p_3 = -\lambda_1 p_1 + (\lambda_2 + \mu_2)p_2 = \lambda_2 p_2, \text{ or } p_3 = \frac{\lambda_2 \lambda_1 \lambda_0}{\mu_3 \mu_2 \mu_1} p_0$$

Similarly we get,

$$p_n = \frac{\lambda_{n-1}\lambda_{n-2}\ldots\lambda_0}{\mu_n \mu_{n-1}\ldots\mu_1} p_0 \qquad \ldots(10.10)$$

$p_0$ is given by $\sum_{n=0}^{\infty} p_n = 1$ or, by

$$p_0 \sum_{n=0}^{\infty} \frac{\lambda_{n-1}\lambda_{n-2}\ldots\lambda_0}{\mu_n \mu_{n-1}\ldots\mu_1} = 1$$

If the $\lambda_n$, $\mu_n$ are such that the infinite series fails to converge we conclude that the queuing system cannot be in steady-state.

## 10.5 PERFORMANCE MEASURES

Relevant performance measures in the analysis of queuing models are:

- Average number of customers in the system.
- Average wating time in system.
- Busy time of servers.
- Average waiting time in queue per customer.
- Server idle ratio.

In particular, we are interested in mean performance measures, such as the mean waiting time, the mean sojourn time and the mean queue length.

## 10.6 LITTLE'S LAW

In queuing theory, Little's result, theorem, or law are three names for the same law. In simple English, it says: "The average number of customers in a stable system (over some time interval) is equal to their average arrival rate, multiplied by their average time in the system.

$$N = \lambda T$$

where $\lambda$ = Mean job arrival rate

$N$ = Number of jobs in the system

$T$ = Mean response time

Little law relates mean queue length to arrival rate and mean response time and applies to any "black box" queue under the following assumptions:

— System is work conserving

— Number of jobs entering is same as number leaving (system is stable).

We will first demonstrate this law using the M/M/l queue. This queue can be translated as this: has a memoryless arrival process–memoryless departure process–uses 1 server, it is a Markovian process. Using Little's law, if the mean calculated is equal to the variance, then both results are the same. To change this we varied arrival rate and measure the delay. We then obtained $N$, repeat the process several times and we then computed the confidence interval.

## 10.7 THE M/M/1 QUEUE

An M/M/1 queue is a Stochastic process whose state space is the set {0,1,2,3, ...} where the value corresponds to the number of customers in the system, including any currently in service.

- Arrivals occur at rate $\lambda$ according to a poisson process and move the process from state $n$ to $n + 1$.
- Service times have an exponential distribution with parameter $\mu$ in the M/M/1 queue.
- A single server serves customers one at a time from the front of the queue, according to a first-come, first-served discipline. When the service is complete the customer leaves the queue and the number of customers in the system reduces by one.

- The buffer is of infinite size, so there is no limit on the number of customers it can obtain.

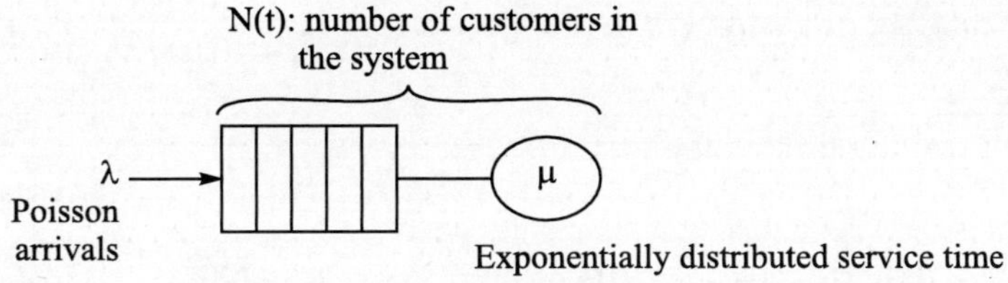


**Fig. 10.3** M/M/1

In steady-state the probability of $n$ customers in the queue is denoted by $p_n$. The balance equations for the system like (10.9) are:

$$\lambda p_{n-1} + \mu p_{n+1} = (\lambda + \mu)p_n,\ n \geq 1 \qquad \text{...[10.11(a)]}$$

$$\mu p_1 = \lambda p_0 \qquad \text{...[10.11(b)]}$$

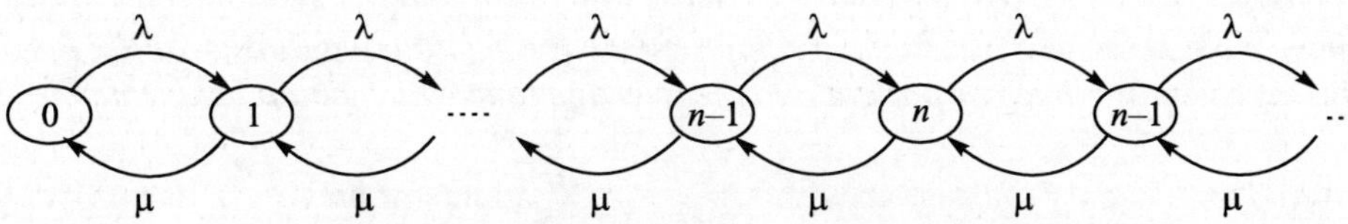


**Fig. 10.4** Markov chain of the M/M/1 queue

The solution of the differential equations (10.11) as in (10.10) is

$$p_n = \frac{\lambda^n}{\mu_n} p_0 = \rho^n p_0 \qquad \text{...(10.12)}$$

where $\rho = \dfrac{\lambda}{\mu}$,

$$= \frac{\text{Arrival rate}}{\text{Service rate}} = \text{traffic intensity of the system.}$$

$$\text{Balance equations} \begin{cases} \text{state 0:} & p_1\mu = p_0\lambda \\ \text{state } 0-1: & p_2\mu = p_0\lambda \\ \vdots & \\ \text{state } 0-1-\ldots-n: & p_{n+1}\mu = p_n\lambda \\ \vdots & \end{cases}$$

**Normalization equations:** $\sum_{n=0}^{\infty} p_n = 1$, $p_0(1 + \rho + \rho^2 + \ldots) = 1$ if $\rho < 1$ that is $\lambda < \mu$, the geometric series can be summed and we can obtain and from (10.11),

$$p_0 = 1 - \rho,\ p_n = \rho^n(1 - \rho) \qquad \text{...(10.13)}$$

**Performance Measures of M/M/1 Queue**

$L_s$ = Average number of customers in the system = $\dfrac{\rho}{(1-\rho)} = \dfrac{\lambda}{(\mu-\lambda)}$

$W_s$ = Expected waiting time in the system = $\dfrac{1}{(1-\rho)\mu} = \dfrac{1}{(\mu-\lambda)}$

$L_q$ = Average no. of customers in queue = $\dfrac{\lambda^2}{(\mu-\lambda)\mu} = L_s - \rho$

$W_q$ = average waiting time in the queue per customer = $\lambda/(\mu-\lambda)\mu = W_s - 1/\mu$

$TH$ = departure rate = $\lambda$

Server utilization ratio = $\rho$

Server idle ratio = $P_0 = 1-\rho$

$P\{n > k\}$ = Probability of more than $k$ customers = $\rho^{k+1}$

**Example 10.1:** *The emergency of a hospital receives on an average 72 patients in 24 hour. On an average a patient can be given 12 minutes of active attention and the section can handle only one patient at a time. What is the average number of patients in the section and those waiting for attention? What is the average waiting time per patient and the average time which could be devoted to a patient?*

**Solution:** There is only one server with $\lambda = \dfrac{72}{24}$ = 3 patients/hr and $\dfrac{1}{\mu}$ = 10 min per patient or

$\mu = \dfrac{1}{12}$ patient/min = 6 patients/hr.

Hence, $L_s$ = Average number of patients in the section = $\dfrac{\lambda}{(\mu-\lambda)}$

$\Rightarrow \quad \dfrac{3}{6-3} \Rightarrow 1$ patients

$L_q$ = Average number of patients in waiting = $\dfrac{\lambda^2}{\mu(\mu-\lambda)}$

$= \dfrac{3\times 3}{6(6-3)} \Rightarrow \dfrac{9}{18} = \dfrac{1}{2}$ patients

$W_q$ = Average waiting time in queue = $\dfrac{\lambda}{\mu(\mu-\lambda)}$

$= \dfrac{3}{6\times 3} = \dfrac{\not{3}}{\not{18}} = \dfrac{1}{6}$ hr or 10 min.

$\rho$ = Average number of patients under service in 10 min = $\dfrac{\lambda}{\mu}$

$= \dfrac{3}{6} = \dfrac{1}{2}$ patients

Hence, one patient can be given $\dfrac{2}{1}\times 10$ = 20 min.

## 10.8 SINGLE SERVER M/M/1/N MODEL

In this model the system capacity is restricted to $N$ customers and no additional customers are allowed in. Thus in this model there is a chance of loosing customers if the queue is full to its capacity $N$. The system states are $n = 0, 1, 2, \ldots, N$.

If as before $p_n$ is the steady-state probability for the $n^{th}$ state, the balance equations with arrival rate $\lambda$ and service rate $\mu$ are given by equation (10.11)

$$\lambda p_{n-1} + \mu p_{n+1} = (\lambda + \mu)p_n \qquad n \geq 1 \text{ and} \qquad \ldots[10.11(a)]$$

$$\mu p_1 = \lambda p_0$$

but with the restriction at the last $N^{th}$ node equation [10.11(*b*)] becomes $\mu p_N = \lambda p_{N-1}$. The solution of the difference equations is evidently given by

$$p_n = \rho^n p_0,\ n = 0, 1, 2, \ldots, N \qquad \ldots(10.14)$$

where $\rho = \dfrac{\lambda}{\mu}$. Since $\sum_{n=0}^{N} p_n = 1$, we must have

$$p_0(1 + \rho + \rho^2 + \ldots + \rho^N) = 1$$

or,

$$\frac{p_0(1-\rho^{N+1})}{(1-\rho)} = 1 \text{ thus,}$$

$$p_0 = \frac{1-\rho}{1-\rho^{N+1}} \qquad \ldots(10.15)$$

The right side of (10.15) always lies between 0 and 1 for all $\rho$. Thus it is not necessary that $\rho$ be less than 1 ($\lambda < \mu$) as in the M/M/1 model. If $\lambda > \mu$ and the queue is full, new arrivals are simply denied entry in to system.

The probability of denial of entry into the queue is $p_N$ when the system is full to its capacity. So the probability of entry is $1 - p_N$ and therefore the effective arrival rate is

$$\lambda_{\text{eff}} = \lambda(1 - p_N) = \lambda\frac{1-\rho^N}{1-\rho^{N+1}} \qquad \ldots(10.16)$$

**The performance measures of M/M/1/N:**

1. $L_s = \begin{cases} \dfrac{\rho}{1-\rho} - \dfrac{N+1}{1-\rho^{N+1}}\rho^{N+1} & \rho \neq 1 \\ \dfrac{N}{2} & \rho = 1 \end{cases}$

2. Average number of customers in queue $(L_q) = L_s - \rho$ or $L_s - \left(\dfrac{\lambda}{\mu}\right)$

3. Expected waiting time in system $W_s = \dfrac{L_s}{\lambda}$

4. Expected waiting time in queue $W_q = \dfrac{L_q}{\lambda}$

5. Utilization $= \rho$

6. Blocking probability $(P_B) = \begin{cases} \dfrac{(1-\rho)\rho^N}{1-\rho^{N+1}} & \rho \neq 1 \\ \dfrac{1}{N+1} & \rho = 1 \end{cases}$

7. Throughput $= \rho(1 - P_B)$

## 10.9 MULTIPLE SERVER M/M/c QUEUING MODEL

In this model, $c(\geq 1)$ number of servers arranged in parallel and an arriving customer can wait for any one of the servers. The system capacity is thus assumed unlimited having states $n = 0, 1, 2, \ldots\infty$.

- Arrival occur at rate $\lambda$ according to a mean Poisson process.
- Service times have an exponential distribution with parameter $\mu$ in M/M/c queue.
- If $n \leq c$ all the customers may be served simultaneously with $(c - n)$ number of servers remaining idle and the mean rate of service becomes $\mu_n = n\mu$, $n = 0, 1, 2, \ldots, c$.
- If $n \geq c$, all the servers will be busy and $\mu_n = c\mu$, $n = c, c + 1, \ldots, \infty$.

In steady-state the balance equations of the system as in (9) are

$$\lambda p_{n-1} + (n+1)\mu p_{n+1} = (\lambda + n\mu)p_n,\ n = 1, 2, \ldots, c \qquad \ldots[10.17(a)]$$

$$\lambda p_{n-1} + c\mu p_{n+1} = (\lambda + c\mu)p_n,\ n = c+1, c+2, \ldots, \infty \qquad \ldots[10.17(b)]$$

$$\mu p_1 = \lambda p_0 \qquad \ldots[10.17(c)]$$

The solution of these equations following (10) is

$$p_n = \frac{\lambda^n}{\mu \cdot 2\mu \cdot 3\mu \ldots n\mu} p_0 = \frac{1}{\lfloor n}\left(\frac{\lambda}{\mu}\right)^n p_0\ \forall\, n \leq c$$

and

$$p_n = \frac{\lambda^n}{\mu \cdot 2\mu \cdot 3\mu \ldots c\mu(c\mu)^{n-c}} p_0 = \frac{1}{\lfloor c\, c^{n-c}}\left(\frac{\lambda}{\mu}\right)^n p_0\ \forall\, n \geq c$$

Setting $\rho = \dfrac{\lambda}{c\mu}$ = traffic intensity of the system the above expressions can be written as

$$p_n = \frac{(c\rho)^n}{\lfloor n} p_0 \quad \text{for } n \leq c \qquad \ldots[10.18(a)]$$

$$= \frac{c^n \rho^n}{\lfloor c} p_0 \quad \text{for } n \geq c \qquad \ldots[10.18(b)]$$

Since $\sum_{n=0}^{\infty} p_n = 1$, we obtain for $p_0$

$$p_0\left[\sum_{n=0}^{c} \frac{(c\rho)^n}{\lfloor n} + \frac{c^c}{\lfloor c} \sum_{n=c+1}^{\infty} \rho^n\right] = 1$$

So that

$$p_0 = \left[\sum_{n=0}^{c} \frac{(c\rho)^n}{\lfloor n} + \frac{c^c}{\lfloor c} \frac{\rho^{c+1}}{(1-\rho)}\right]^{-1} \qquad \ldots(10.19)$$

Provided that $\rho < 1$. For $\rho \geq 1$, the queue will grow and steady-state will never be reached.

From equation [10.18(*a*)], $p_c = \left[\frac{(c\rho)^c}{\lfloor c}\right] p_0$,

**System performance of M/M/c queue model**

1. Average number of customers in the system $(L_s) = \frac{\lambda}{\mu} + \left(\frac{\left(\frac{\lambda}{\mu}\right)^c \rho}{\lfloor c\,(1-\rho)^2}\right) p_0$

2. Average number of customers in the queue $(L_q) = \left(\frac{\rho p_0}{(1-\rho^2)}\right) \Rightarrow \frac{\rho p_c}{(1-\rho)^2}$

3. Expected waiting time in system $(W_s) = \frac{L_s}{\lambda}$

$$W_q = \frac{L_q}{\lambda}$$

4. Expected number of busy server = traffic intensity × number of busy servers

$$= \rho \times c \text{ or } \frac{\lambda}{\not{c}\mu} \times \not{c} \Rightarrow \frac{\lambda}{\mu}$$

## 10.10 MULTIPLE SERVER M/M/c/c MODEL

In the M/M/c model, we assume the system capacity to be limited to *c*. This model was given by A.K. Erlang in which the servers constituted a telephone exchange and calls represented the customers.

The system states are therefore $n = 0, 1, 2, \ldots, c$ and customers are lost if $n > c$. If $\lambda$ is is the average arrival rate and $\mu$ the service rate of a server, so that $\mu_n = n\mu$ is the service rate of all the servers, we get from equation [10.18(*a*)] of the previous section

$$p_n = \frac{(c\rho)^n}{\lfloor n} p_0, \forall\, n \le c$$

where $\rho = \frac{\lambda}{c\mu}$ is the traffic intensity of the system. From $\sum_{n=0}^{c} p_n = 1$ we get as in equation (10.19)

$$p_0 = \left[\sum_{n=0}^{c} \frac{(c\rho)^n}{\lfloor n}\right]^{-1}$$

Unlike the unlimited capacity case $\rho$ may not be less than 1.

The probability that the system may not be full to its capacity is $1 - p_c$, where $p_c = \left[\frac{(c\rho)^c}{\lfloor c}\right] p_0$ is the probability of blocking arrival. So the effective arrival rate is

$$\lambda_{\text{eff}} = (1 - p_c)\lambda$$

**System Performance:**

(*i*) $L_q = 0$

(*ii*) $L_s$ = Number of busy server = $\rho c = \dfrac{\lambda}{N}$

(*iii*) $W_s = \dfrac{\rho c}{\lambda_{\text{eff}}} = \dfrac{\cancel{\lambda} \cdot \cancel{c}}{\cancel{c}\mu \cdot (1 - p_c)\cancel{\lambda}} = \dfrac{1}{\mu \cdot (1 - p_c)}$

(*iv*) $W_q = 0$

**Note:** The model M/M/c/N can be similarly developed, where system capacity limited to $c$ and system capacity is restricted to N customers and no additional customers are allowed in.

## 10.11 APPLICATIONS OF QUEUING THEORY

Queuing theory was born in the early 1900s with the work of A. K. Erlang of the Copenhagen Telephone Company, who derived several important formulas for telegraphic engineering that today bear his name. The range of applications has following area:

(*i*) Telecommunications and computer science (For Scheduling)

(*ii*) Manufacturing

(*iii*) Air traffic control (Airports–runway layout, luggage collection, shops, passport control, etc.)

(*iv*) Military logistics

(*v*) Banking system

Other areas:

1. Supermarkets
2. Restaurants
3. Manufacturing processes
4. Bus scheduling
5. Hospital appointment bookings
6. Printer queues
7. Minimizing page faults in computing.

Queuing theory is considered to be one of the standard methodologies (together with linear programming, simulation, etc.) of operations research and management science, and is standard fare in academic programs in industrial engineering, manufacturing engineering, etc.

In computer science, queuing is a necessary consideration in contention for processing resources. Given that the highest-cost component in advanced computation is processing power, systems are moving increasingly toward network solutions in which processing power is distributed. The result of this trend toward greater distribution is that systems win contend for access to a network and to diverse processors and files.

### 10.11.1 Application of Queuing Theory in Computer System

There are following example of computer system and network where Queuing theory is playing his important role:

1. N terminals request to use of a computer (server) to process transactions. The length of time that the terminal takes to generate a request for the computer is called "think" time. The

length of time from the instant a terminal generates a transaction until the computer completes the transaction (and instantaneously responds by communicating this fact to the user at the terminal) is called "response time". We would like to know, for example, the rate at which transactions are processed (which in steady-state equals the rate at which they are generated) is called "throughput", which is one the most important performance measures showing the system's processing power.

2. Trunk mobile systems, telephone lines are interfaced with the radio system at the repeaters which serve dispatch type mobile subscribers and telephone line users. Let us consider a system which serves two different types of communication traffic:
   (*i*) dispatch traffic has short average service time, and
   (*ii*) interconnect traffic of telephone line users. Both types of users are assumed to arrive from a finite population. The dispatch users are allowed to access all repeaters while interconnect users can occupy only a fixed number of repeaters. A sharing service algorithm to derive blocking probabilities of dispatch and interconnect users and average dispatch delay is the find.
3. Example 6 let us examine the dynamic behaviour of a local area network based on the non-persistent Carrier Sense Multiple Access with Collision Detection (CSMA/CD) protocol. In such a network a finite number, say N, of users (or active terminals) are connected by a single channel (bus). Under the specific protocol, if a terminal has a message ready for transmission, the terminal immediately senses the channel to see whether it is idle or busy. If the channel is sensed busy, it re-senses the channel after a random amount of time. On the other hand, if the channel sensed is idle, it starts transmitting the message immediately. Due to non-zero propagation delay, within a certain time interval after the terminal starts transmitting the message, other terminals (if any) with "messages ready for transmission may also sense the channel idle and transmit their messages. This phenomenon is referred to as a collision. Each terminal involved in a collision abandons its transmission and re-sense the channel at a later time as if it had sensed a busy channel. A collision usually lasts for a certain amount of time during which no terminals are allowed to transmit, that is, a recovery time is needed by the channel to be free again. This kind of system can be modeled as retrial queuing system with server's vacation.
4. Let us consider a memory system where N disk units share a disk controller (server) and transmit information when they find the controller idle. Unsatisfied requests are repeated after a disk's rotation which can be modeled as a constant repetition interval.

### 10.11.2 Application of Queuing System in Manufacturing

In manufacturing, queuing is a necessary element of flexible systems in which factors of production may be continually adjusted to handle periodic increases in demand for manufacturing capacity. Excess capacity in periods of low demand may be converted into other forms of working capital, rather than be forced to spend those periods as idle, non-productive assets.

The primary focus of queuing theory on flexible manufacturing remains centered on machine reliability and depreciation and processing and cycle times.

## SUMMARY

- Queuing Theory is mainly seen as a branch of applied probability theory. Its applications are in different fields, *e.g.*, communication networks, computer systems, machine plants and so forth.

- Queuing systems have many parameters for analysis such as distributions of the arrival time, interarrival time and service time, number of servers, the size of the waiting line (infinite or finite), the service discipline.
- Queuing discipline is typically specified using Kendall's notation (*A/B/c/K*), where letters correspond to five queue attributes:
- *A*: interarrival time distribution
- *B*: service time distribution
- *C*: number of servers
- *K*: the size of the system
- Service time is defined as time required by one server to serve one customer. The service time may be deterministic, or it may be random variable.
- The average number of customers in a stable system (over some time interval) is equal to their average arrival rate, multiplied by their average time in the system, this is known as little formula.

$$N = \lambda T$$

- There are different type of queuing model such as M/M/1, M/M/c, M/M/1/N, M/M/c/N and M/G/1.
- In M/M/1, first M = Exponentially distributed interarrival times, second M = Exponentially distributed service times, number of server = One server and other parameter is assumed as Infinite number of buffers, Infinite population size, First-come-first-serve service discipline.
- The M/M/c-Queue model has the same interarrival time and service time distributions as the M/M/1 queue, but differ in number of server. There are *c* servers in the system and the waiting line is infinitely long.
- The M/M/1/N-Queue has exponential interarrival time and service time distributions, each with the respective parameters. The customers are served in FIFO-Order, there is a single server but the system can only hold up to *N* customers.
- M/M/c/N Queue model is same as M/M/1/N model but differ in number of server. Number of server is more than 1 in M/M/c/N queuing model.
- In M/G/1 Queuing model, there is General service time distribution (*G*), Otherwise, similar to M/M/1 queues. It is the most complex, readily solvable single queue. Service history matters and denote service time already received by $X_0(t)$.

## SOLVED QUESTIONS

**Q.1. Discuss the various applications of Little Law.**

**Ans. Little's Law Applications**

Other ways in which Little's Law can be used are the following:

- **Estimate Waiting Times:** [W = Average Number of Customers/Average Throughput] (as the patient flow example above)
- **Planned Inventory Time:** Suppose a product is scheduled so that we expect it to wait for 2 days in finished goods inventory before shipping to the customer. These two days is called

planned inventory time and is sometimes used as protection against system variability to ensure high delivery service. Using Little's Law, the total amount of inventory in finished goods can be computed as [FGI = Throughput × Planned Inventory Time]

- **WIP Reduction:** Reducing WIP in a process without making any other changes will also reduce throughput. So, simply reducing inventory is not enough to achieve a Lean Manufacturing system. An integral part of any Lean Manufacturing implementation is an effort to reduce variability (often the domain of Six Sigma), to enable a line to achieve the same (or greater) throughput with less WIP.
- **Little's Law for Product Development**
- **Little's Law and Multi-tasking**

**Q.2. What is utilization factor of queuing model?**

**Ans.** The proportion of time, a server or a group of servers may be busy, is given by the server utilization

$$U = \frac{\lambda}{c\mu}$$

$$= \frac{\rho}{c}$$

whereas $c$ describes the number of servers in a queuing system. Please note, that a system with $U = 1$ is called a fully loaded system. Many common models are based on steady state concepts, which are comparable to the physical concept of equilibrium. As a consequence, they are not applicable to systems in overload, *i.e.*, $U > 1$. Due to statistical effects, they don't provide proper results in fully loaded systems as well. Thus $U < 1$ defines a necessary stability condition for commonly used models.

**Q.3. Consider a supermarket with eight cashiers open and a common queue. Customers arrive approximately every 1.25 sec and take on average 5 minutes to be served. Assuming exponential arrival and service processes following a first come, first serve discipline. Calculate traffic intensity and server utilization.**

**Ans.** The model input is given as follows:

$$\lambda = \frac{1}{1.25}$$

$$= 0.8$$

$$\mu = \frac{1}{5}$$

$$= 0.2$$

$$c = 8$$

From these parameters the traffic intensity $\rho$ and the server utilization $U$ may be calculated

$$\rho = \frac{\lambda}{\mu} = \frac{0.8}{0.2} = 4$$

$$U = \frac{\lambda}{c\mu} = \frac{\rho}{c} = \frac{4}{1} = 0.5$$

thus satisfying the stability condition $U < 1$.

**Q.4. Consider a database system with an average service time of 450 m/sec. As database requests are inititated by a large number of clients, a random arrival pattern may be assumed. Thus the arrival process is assumed to be Poisson. On the average a new database query arrives every 500 m/sec. Service times are assumed to be exponentially distributed, the queuing discipline is assumed to follow a first come, first serve pattern. Calculate various parameter of performance evaluation.**

**Ans.**

$$\lambda = \frac{1}{0.5} = 2 \text{ sec}$$

$$\mu = \frac{1}{0.4} = 5 \text{ sec}$$

$$U = \rho = \frac{\lambda}{\mu} = 0.9$$

$$N = \frac{\rho}{(1-\rho)}$$

$$= \frac{0.9}{0.1}$$

$$= 9$$

$$T = \frac{N}{\lambda} = \frac{9}{2} = 4.5$$

**Q.5. Write the name of model for which exact solution is possible.**

**Ans.** Exact solutions are possible for Exact solutions are possible for following model.

- *M/Er* /1 (Erlangian service)
- *M/D*/1 (special case of *M/G*/1)
- *M/M*/1 with bulk arrivals (restricted cases)

**Q.6. Write the name of queuing model for which analysis is difficult.**

**Ans.** Analysis is more difficulty for following queuing model:

- *G/M*/1
- *M/G*/1
- *G/G*/1

**Q.7. Differentiate between M/M/1 and M/G/1.**

**Ans.**

| | *M/M/1* | *M/G/1* |
|---|---|---|
| 1. | Service time is exponential distribution | Service time is general distribution |
| 2. | Structure is 1 dimensional | Structure is 2 dimensional |
| 3. | No concept of residual time | Concept of residual time |
| 4. | Service history does not matters | Service history matters |

**Q.8. Discuss some characteristics of all Markovian systems.**

**Ans.** The common characteristic of all *markovian systems* is that all interesting distributions, namely, the distribution of the interarrival times and the distribution of the service times are exponential distributions and thus exhibit the markov (memoryless) property. From this property we have two important conclusions:

1. The state of the system can be summarized in a single variable, namely, the number of customers in the system. (If the service time distribution is not memoryless, this is not longer true, since not only the number of customers in the system is needed, but also the remaining service time of the customer in service.)
2. Markovian systems can be directly mapped to a *continuous time markov chain* (CTMC) which can then be solved.

## EXERCISES

1. Consider an M/M/1 queue with an arrival rate of 60 customers per hour and a mean service time of 45 seconds. A period during which there are 5 or more customers in the system is called crowded, when there are less than 5 customers it is quiet. What is the mean number of crowded periods per day (8 hours) and how long do they last on average?
2. Consider a machine where jobs arrive according to a Poisson stream with a rate of 20 jobs per hour. The processing times are exponentially distributed with a mean of 1 μ hours. The processing cost is 16 μ dollar per hour, and the waiting cost is 20 dollar per order per hour. Determine the processing speed μ minimizing the average cost per hour.
3. A repair man fixes broken televisions. The repair time is exponentially distributed with a mean of 30 minutes. Broken televisions arrive at his repair shop according to a Poisson stream, on average 10 broken televisions per day (8 hours).
   (*i*) What is the fraction of time that the repair man has no work to do?
   (*ii*) How many televisions are, on average, at his repair shop?
   (*iii*) What is the mean throughput time (waiting time plus repair time) of a television?
4. One is planning to build new telephone boxes near the railway station. The question is how many boxes are needed. Measurements showed that approximately 80 persons per hour want to make a phone call. The duration of a call is approximately exponentially distributed with mean 1 minute. How many boxes are needed such that the mean waiting time is less than 2 minutes?

■■

# 11

# Arrival Patterns and Discrete System Simulation

**INSIDE THIS CHAPTER**

**11.1** Arrival Patterns; **11.2** Model of Arrival Process; **11.3** Service Times; **11.4** Simulation Programming Tasks; **11.5** Gathering Statistics; **11.6** Measuring Utilization and Occupancy; **11.7** Recording Distributions Transit Times; **11.8** Service Delay

## 11.1 ARRIVAL PATTERNS

It defines the way customers enter in system. Customers arrive in individually or batches. The arrival patterns can be defined in term of inter-arrival time. Arrival pattern may be constant and variable.

**Constant arrival pattern:** If there is no variability in inter-arrival time.

**Variable arrival pattern:** Inter-arrival time may vary stochastically. So it is necessary to define the probability function of the inter-arrival time.

Two or more arrivals may be simultaneous. Mean arrival time is given by following formula;

$$\lambda = 1/T_a$$

where $T_a$ is mean inter-arrival time.

The arrival time distribution can be approximated by following probability distribution.

(*a*) **Poisson Distribution:** It is discrete probability distribution that provides probabilities for the number of customers that may arrive in any specific interval of time.

(*b*) **Exponential Distribution:** It provides probabilities for time gap between two consecutive arrivals (inter-arrival time).

(*c*) **The Erlang Distribution:** The distribution functions named after A.K. Erlang, who found these distributions to be representative of certain types of telephone traffic. An arrival pattern governed by this type of function has the following density function and arrival distribution.

$$f(t) = (k\lambda)^k \left[\frac{e^{-k\lambda t}}{(k-1)!}\right] t^{k-1}$$

$$A_0(t) = e^{-k\lambda t} \sum_{n=0}^{k-1} \frac{(k\lambda t)^n}{\lfloor n} \qquad \ldots(11.1)$$

where $k$ is a positive integer greater than zero. Putting $k = 1$ in equation 11.1 and comparing with equation $f(t) = \lambda e^{-\lambda t}$ $(t \geq 0)$, it can be seen that the case of $k = 1$ is the exponential distribution.

$$A_0(t) = e^{-\lambda t}\,\frac{(\lambda t)^n}{\lfloor n}\qquad \leftarrow \text{(By } k = 1)$$

The standard deviation of the Erlang distribution is $T/k^{1/2}$, so that the coefficient of variation is $\dfrac{1}{k^{1/2}}$.

The coefficient has a maximum value of 1 for the case of the exponential distribution, and it decreases as $k$ increases. It follows that data represented by an Erlang distribution ($k > 1$) Cluster more closely to the mean value than exponentially distributed data, so that there are fever low values and high values. Ultimately, as $k$ tends to infinity, the Erlang distribution tends to the case of a contant arrival time, for which the arrival distribution is the step function.

**Poisson Arrival Pattern:** The Poisson process is most commonly used model for the arrival process of customers in a queueing theory. Poisson process is an arrival process for which interarrival time IATi's are IID exponential. It is discrete probability distribution that provides probabilities for the number of customers that may arrive in any specific interval of time.

Here are some important points of Poisson arrival pattern.

- Arrivals are said to be completely random.
- Arrival can occur at any point only to the restriction that mean arrival rate be some given value.
- The time of next arrival is independent of the previous arrival, and probability of an arrival in an interval $\Delta t$ is proportional to $\Delta t$.
- If $\lambda$ is the mean number of arrivals per unit time, then the probability of an arrival in $\Delta t$ is $\lambda \Delta t$.
- The probability density function of the inter arrival time is given by:

$$f(t) = \lambda e^{-\lambda t}\qquad (t \geq 0)\qquad \ldots(11.2)$$

It follows that the arrival distribution is:

$$A_0(t) = e^{-\lambda t}\qquad \ldots(11.3)$$

The actual number of arrivals in a period of time $t$ is a random variable. It can be shown that with an exponential distribution of inter-arrival times, the probability of n arrival occurring in a period of length $t$ is given by:

$$p(n) = (\lambda t)^n\, e^{-\lambda t}/n! \qquad (n = 0,1,2,\ldots)\qquad \ldots(11.4)$$

This distribution, which is called the Poisson distribution, is discrete. The exponential distribution is continuous because the inter-arrival time can take any non-negative value. Just because of this reason of two distribution, a random arrival pattern is often called a Poisson arrival pattern. Where this term is used, it will mean that the inter-arrival time is exponentially distributed.

## 11.2 MODEL OF ARRIVAL PROCESS

In simulation we required to generate a sequence of random points in time $0 = t_0 \leq t_1 \leq t_2 \leq \ldots$, such that the $i$th event of some kind occurs at time $t_i$ ($i = 1, 2, \ldots$) and distribution of the event time $\{t_i\}$ follows some specified form.

Let $N(t) = \max\ \{i : t_i) \leq t\}$ be the number of events to occur at or before time $t$ for $t \geq 0$. We say stochastic process $\{N(t), t \geq 0\}$ as arrival process because:

- The events of interest are usually arrivals of customers for any kind of service.
- Follows a interarrival time (IAT = $t_i - i - 1$) between ($i - 1$)th and $i$th customer.

There are following types of arrival process:

(*i*) Poisson process

(*ii*) Non-stationary poisson process

(*iii*) Batch process

(*i*) **Poisson Process:** The poisson process is most commonly used model for the arrival process of customers in a queueing theory. Poisson process is an arrival process for which interarrival time IATi's are IID exponential random variable. Let see why the interarrival times for many real-world systems closely resemble IID exponential random variables.

The stochastic process $\{N(t), t \geq 0\}$ is said to be a poisson process if:

(*a*) Customers arrive one at a time.

(*b*) $N(t + s) - N(s)$ (the number of arrivals in the time interval $(t, t + s)$ is independent of $\{N(u), 0 \leq u \leq t\}$.

(*c*) The distribution of $N(t + s) - N(s)$ is independent of $t$ for all $t, s \geq 0$.

Properties (*a*) and (*b*) are characteristics of many arrival processes. If customer arrived in batch form then property (*a*) would not be hold. Property (*b*) is define that the number of arrivals in interval $[t, t + s]$ is independent of the number of arrivals in the interval $[0, t]$ and also of the times at which these arrivals occur. This property will violate if large number of arrivals in $[0, t]$ caused some customers arriving in $(t, t + s)$ to balk (balk is defined as those customers who don't join the queue either by seeing the number of customers already in service system or by estimating excessive waiting time for desired service). Property (*c*) will be voilated by many real-life arrival process since it implies that arrival rate of customers does not depend of time.

(*ii*) **Non-stationary Poisson Processes:** The non-stationary poisson process is often used as a model of the arrival process to a system when the arrival rate varies with time. Let $\lambda(t)$ be the arrival rate of customers to some system at time $t$. If the customers arrive at system in accordance with a poisson process with constant rate $\lambda$, then $\lambda(t) = \lambda$ for all $t \geq 0$. If the arrival rate $\lambda(t)$, changes wih time, then interarrival times IAT1, IAT2, ... are not idenically distributed hence used non-stationary poisson process. The stochastic process $\{N(t), t \geq 0\}$ is said to be a non-stationary poisson process if:

(*a*) Customers arrive one at a time.

(*b*) $N(t + s) - N(t)$ is independent of $\{N(u), 0 \leq u \leq t\}$.

Thus, for a non-stationary poisson process, customers must still arrive one at a time, and the number of arrivals in disjoint intervals are independent, but now the arrival rate $l(t)$ is allowed to be a function of time.

**Batch Arrivals**

In real world, customers arrive in batches, or groups, so that property 1 of the poisson process and of the non-stationary poisson process is violated.

**Example:** People arriving at sporting event or cafeteria often come in batches.

Let $N(t)$ be the number of batches of individual customers that have arrived by time $t$. If the interarrival times of batches appear to be approximately IID exponential random variables, $\{N(t), t \geq 0\}$ can be modeled as a poisson process. Let assume batches of customers arrive in accordance with the arrival process $\{N(t), t \geq 0\}$ and that the number of customers in each batch is a random variable with the fitted discrete distribution.

If $X(t)$ is a total number of individual customers to arrive by time $t$, and if $B_i$ is the number of customers in the $i$th batch, then $X(t)$ is given by

$$X(t) = \sum_{i=1}^{N(t)} B_i \text{ for } t \geq 0$$

If the $B_i$'s are assumed to be IID random variables that are also independent of $\{N(t), t \geq 0\}$, and if $\{N(t), t \geq 0\}$ is a poisson process, then the stochastic process $\{X(t), t \geq 0\}$ is said to be a compound poisson process.

## 11.3 SERVICE TIMES

Service time is defined as time required by one server to serve one customer. The service time may be deterministic, or it may be random variable whose probability distribution is presumed known. Following notation will be used in service time.

- Mean service time: *Ts*
- Mean service rate: μ
- Probability that service time is >*t*: $S(t)$

If service time is completely random, it may be represented by an exponential distribution or, if the coefficient of variation is found to differ significantly from 1, an erlang, or hyper-exponential distribution may be used. Some time if service time is constant then there are random fluctuations due to uncontrolled factors. The normal, or Gaussian distribution is often used to represent the service time under these circumstances.

Service time also depends on whether customers are attended completely by one server or as given in figure, considers this in different cases:

**Case 1:** There is only one server and one queue.

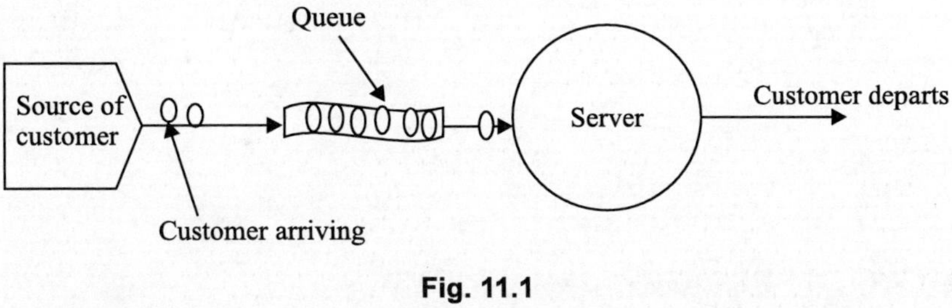


**Fig. 11.1**

**Case 2:** Single queue and multiple server in series

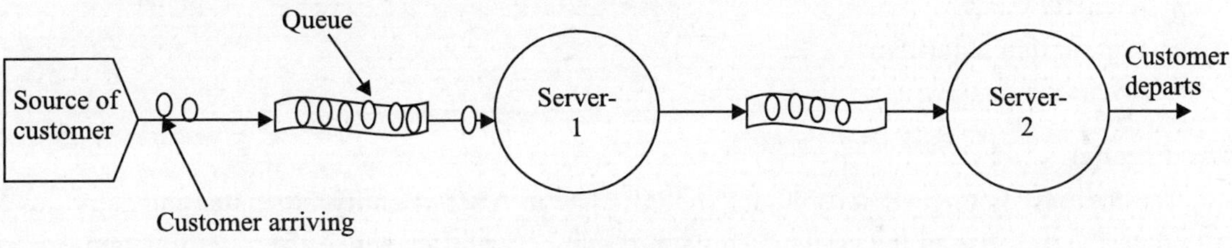


**Fig. 11.2**

**Case 3:** Single queue and multiple servers in parallel.

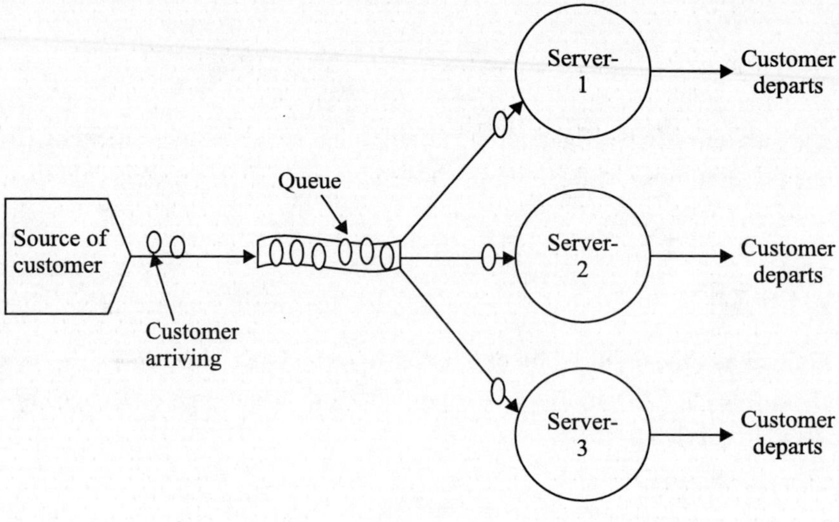


**Fig. 11.3**

**Case 4:** Multiple queues, multiple servers in parallel queuing system

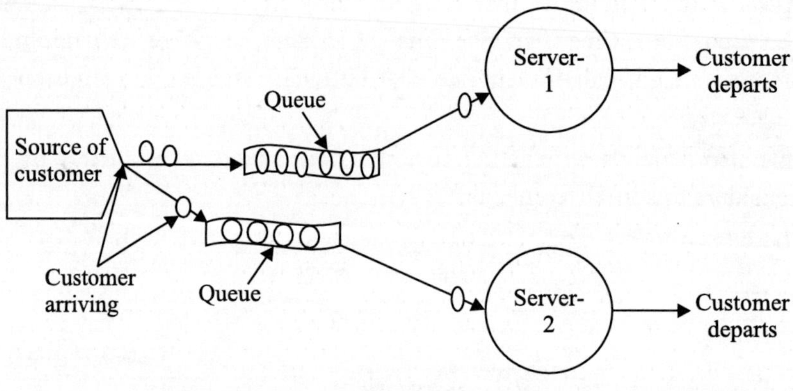


**Fig. 11.4**

## 11.4 SIMULATION PROGRAMMING TASKS

There are three main tasks to be performed.

- General model
- Simulation algorithm
- Report generator

**General Model**

The first task is to generate a model and initialize it. A set of entity, attributes and activities are used to describe the state of the system. This set of numbers will be called the system image since its purpose is to reflect the state of the system at all times. The activities of the system must be represented as routines that create the discrete events making the changes to the system image.

**Simulation Algorithm**

The second task is to program the procedure that executes the cycle of actions in carrying out the simulation. This procedure is called as the simulation algorithm. Simulation algorithm involves repeated execution of the five steps.These steps are:

(*a*) Find the next potential event

(*b*) Select an activity

(*c*) Test if the event can be executed

(*d*) Change the system image

(*e*) Gather the statistics.

**Report Generator**

The third task is the generation of report of simulation output, which is usually executed once at the end of simulation. The basic purpose of report generator is to print statistics gathered during the run.

## 11.5 GATHERING STATISTICS

So far we have discussed about task of report generator and found that report generator is use to print statistics gathered during simulation run. The exact statistics required from a model depend upon the study being performed. There are following commonly needed statistics.

**(*i*) *Counts*:** Represent the no. of entities or no. of events of particular type occur in system.

**(*ii*) *Summary measures*:** Represents as extreme values, mean values, and standard deviations.

**(*iii*) *Utilization*:** Defined as the fraction or percentage of a group of entities in use on the average.

**(*iv*) *Occupancy*:** Represent as the percentage of group of entities in use on the average.

**(*v*) *Distributions*:** Responsible for distribution of important variables, such as waiting time or queue length .

**(*vi*) *Transit time*:** Represent as the time taken for an entity to move from one part of the system to some other part.

## 11.6 MEASURING UTILIZATION AND OCCUPANCY

The simple measure is to determine what percentage of time the item (or entity) of equipment was engaged in simulation run. The term **utilization** will be used to describe this statistics. It is necessary to keep record of the time $T_{\text{busy}}$ at which the item last became busy. When the item becomes free at time $T_{\text{free}}$, the intervals $T_{\text{free}} - T_{\text{busy}}$ is derived and added to a counter. At the end of the simulation run, the utilization $U$ is derived by dividing the accumulated total time $T$, so that, if the entity is used $N$ times:

$$U = \frac{1}{T} \sum_{r=1}^{N} (T_{\text{free}} - T_{\text{busy}})r$$

to build up the count by counting the number of intervals.

- A discrete simulation program, update the time as events occur, will measure the intervals $(T_{\text{free}} - T_{\text{busy}})$ directly.

- A continuous simulation program, updating time in small intervals will require in which the item is busy.
- If group of entities are there then calculation is similar, requiring that information about the number of entities involved also be kept.
- If there is an upper limit on the number of entities, for example, if there was a limit on the number of links in the telephone system, the term occupancy is often used to describe the average number in use as a ratio to the maximum. If there are *M* links in a telephone exchange and the quantity *Nr* is the number busy in the interval $T_r$ to $T_{r+1}$, the average **occupancy**, assuming the number *Nr*, changes *N* times, is

$$B = \frac{1}{NM} \sum_{r=1}^{N} (T_{r+1} - T_r)\, Nr$$

- It is important to check the conditions at the beginning and end of simulation.

The basic difference between utilization and occupancy is that timing information must be kept for each individual entity for utilization while in case of occupancy only require keeping a count of a class of entities, and recording the last time the count changed, just two numbers.

## 11.7 RECORDING DISTRIBUTIONS AND TRANSIT TIMES

Determining the distribution of a variable requires counting how many times the value of the variable falls within specific intervals. A table is set in which to record the values defining the intervals and to accumulate each count. As each new observation is made, its value is compared with the limits established for the intervals, and counter is incremented by 1. The tabulation intervals are uniform in size and the specifications will be for:

(*i*) The lower limit of tabulation

(*ii*) The interval size

(*iii*) The number of intervals

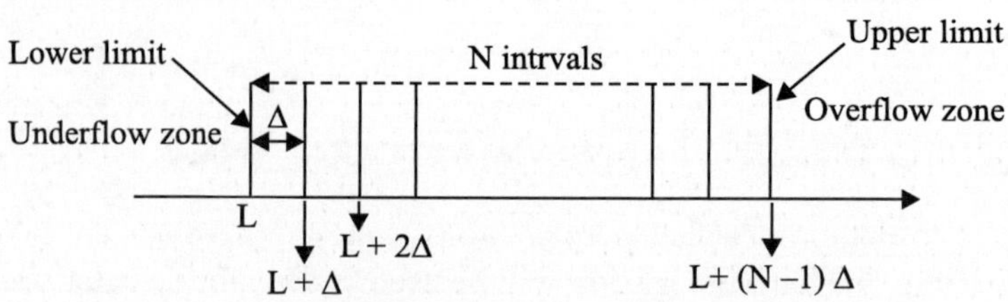


**Fig. 11.5** Definition of a distribution table.

It is necessary to count how many times an observation falls below the lower limit and beyond the upper limit. It is necessary to accumulate the number of observations and the sum of the squares, in order to calculate the mean value and the standard deviation, at the same time the distribution is being derived.

When the distribution derived is an approximation, since it matches the values of the observations to an interval, the mean or standard deviation will be accurate within the accuracy limits of the computer, even if some observations fall outside the table limits.

To measure transit times, the clock is used in the manner of a time stamp. When an entity reaches a point from which a measurement of transit time is to start, a note of the time of arrival is made. Later, when the entity reaches the point at which the measurement ends, a note of the clock time upon arrival is made and compared with the measurement ends, a note of the clock time upon arrival is made and compared with the first time to derive the elapsed interval.

## 11.8 SERVICE DELAY

A delay's duration is determined by system conditions (not specified by the modelar ahead of time). It is also called conditional wait.

**Example:** A customer's delay in a waiting line may be dependent on the number and duration of service of other customers ahead in line and whether a server has a failure during delay.

## SUMMARY

- The arrival patterns can be defined in term of inter-arrival time. Arrival pattern may be constant and variable.
- The arrival time distribution can be approximated by following probability distribution:
  (*a*) Poisson distribution, (*b*) Exponential distribution, (*c*) Erlang distribution
- The Poisson process is most commonly used model for the arrival process of customers in a queueing theory. Poisson process is an arrival process for which interarrival time IATi's are IID exponential.
- The Poisson process is most commonly used model for the arrival process of customers in a queueing theory. Poisson process is an arrival process for which interarrival time IATi's are IID exponential.
- There are following commonly needed statistics:
  (*i*) Counts, (*ii*) Summary measures, (*iii*) Utilization,
  (*iv*) Occupancy, (*v*) Distributions, (*vi*) Transit time,
- The utilization $U$ is derived by dividing the accumulated total time $T$, so that, if the entity is used $N$ times:

$$U = \frac{1}{T} \sum_{r=1}^{N} (T_{\text{free}} - T_{\text{Busy}})\, r$$

- To measure transit times, the clock is used in the manner of a time stamp.

## SOLVED QUESTIONS

**Q. 1. Where do we use batch process?**

**Ans.** When customers arrived in batches.

**Q. 2. Discuss hyper-expontential distribution.**

**Ans.** The hyper-exponential distribution defines a class of distributions that have a standard deviation greator than their mean value; consequently, they can be used to match data found to have a coefficient of variation greater than 1. They represent data where low and high values occur more frequently than with an exponential distribution, the data distribution may in fact be bimodal, showing a peak on either side of the mean value.

**Q. 3. What is transit time?**

**Ans.** Transit time is defined as the time taken for an entity to move from one part of the system to some other part.

**Q. 4. What do you mean by arrival pattern?**

**Ans.** The arrival pattern, which describes the statistical properties of the arrivals.

## EXERCISES

1. How do you gather statistics in simulation programming system?
2. Differentiate between poisson distribution and exponential distribution.
3. Differentiate between utilization and occupancy.
4. What do you mean by service time?

12

# Simulation Language

**INSIDE THIS CHAPTER**



## 12.1 CONCEPTS IN SIMULATION LANGUAGE

Computer simulation is essentially an experimental technique, used for studying a wide variety of systems. The purpose of the experiment is to observe the behaviour of a given system within a given environment which would eventually provide a basis for understanding the system and making decisions. The abstract model of the system being simulated takes the form of a computer program and the system behaviour is given by the output, as the program, runs. A complex simulation model can become difficult to program, to debug, and to modify.

A programmer who has to perform simulation frequently would be better off learning a higher-level special-purpose language, which facilitates simulation programming.

A simulation language often serves some very useful purpose, like:

- It provides a conceptual framework in which the system analyst can work while building a model of the system.
- It provides a terminology for describing a complex dynamic system.

There is a variety of simulation languages.

Simulation languages that have been designed implemented, and used, there is no single language which can be termed as the "best", "most useful", or "universally" available.

**Note:** A computer simulation language tells us how to perform simulation by using computer.

### 12.1.1 Classification of Simulation Languages

There are three main types of simulation languages:

1. Continuous simulation languages
2. Discrete simulation languages
3. Combined simulation languages

- Continuous simulation languages are designed for simulating continuous models. The model is viewed essentially as a set of differential equation. Examples:
  - **Advanced Continuous Simulation Language (ACSL):** It supports textual or graphical model specification.
  - Dynmo.
  - **Vissim:** A usually programmed block diagram language.
- Discrete simulation languages are designed for simulating discrete models. The model is viewed essentially as a sequence of random events, they are causing a change in state. Examples:
  - Auto mod
  - Simula
  - GPSS
  - SIMSCRIPT 11.5
  - GASP
  - SIMAN
- Some languages have been designed which are suitable for both discrete as well continuous models such a languages are known as combined simulation languages. Examples:

| | | |
|---|---|---|
| Sabes-Simulation | – | Provide continuous and discrete capability |
| Simulation K | – | Provide continuous and discrete capability |
| Flexsim 4.0 | – | Powerful interactive software for discrete event and continuous flow simulation |
| GASP IV | – | Best known example of combined simulation language |

CLASS (Composite language approach for system simulation)

### 12.1.2 Comparison of Simulation Packages with Programming Language

One of the most important decision in case of simulation study concerns the choice of software. If the selected software is not flexible enough or is too difficult to use, then the simulation project may produce erroneous results or may not even be completed.

**The following are some advantages of using a simulation package rather than a general purpose programming language.**

1. Simulation packages automatically provide most of the features needed to build a simulation model, resulting in a significant decrease in programming time and a reduction in overall project cost.
2. It provide a natural framework for simulation modeling.
3. Simulation models are generally easier to modify and maintain when written in a simulation package.
4. They provide better error detection because many potential types of errors are checked for automatically. The chance of making an error will probably be smaller in general-purpose program.

Simulation models are still written programming language. **Some advantages are as follows:**

- Analysis modelers already know a programming language this is often not the case with a simulation package.
- Simulation model efficiently written in C or C++ may require less execution time than a model developed simulation package.

- A simulation package is designed to address a wide of systems with one set of modeling constructs, a C program can be more closely tailored to a particular application.
- Programming languages may allow greater programming than certain simulation packages. Software cost is generally lower, but we can not say about cost of total project.

## 12.2 CONTINUOUS SIMULATION LANGUAGE

*[UPTU 2005-06, 2006-07]*

Before the digital computers come into wide spread use, **analog computers** were being used for simulating continuous dynamics systems.

As soon as the digital computer come into the picture some of its advantages over the analog computer became obvious like.

- Greater accuracy
- Freedom for scaling
- Easily stored for reuse

Therefore, special program packages were implemented on digital computers to make a digital computer appear like an analog computer. This is called as digital analog simulators. It checks the result of an analog simulation on a digital computer.

**Digital Analog Simulators:**

- It checks the result of an analog simulation on a digital computer.
- Such simulation language that makes a digital computer emulate an analog computer is called a BLOCK STRUCTURED CONTINUOUS SYSTEM SIMULATION LANGUAGE.
- It requires the user to prepare an analog-computer-type block diagram and then input this diagram.
- Later, more versatile expression-based simulation languages were developed.

### 12.2.1 Block-Structured Continuous Simulation Languages

They were designed to simulate only those systems that could be represented as analog block diagrams and thus their application area is restricted.

Following Era in Block-structured continuous simulation language is listed in Table 12.1.

**Table 12.1**

| Name of Simulator | Year | Computer (can perform their operation) | Integration method |
|---|---|---|---|
| 1. Digital analog simulator by R.G. Selfridge | 1955 | IBM 701 | Simpson's rule |
| 2. DEPI | 1958 | DATARON 204 Computer | Fourth-order Runge-Kutta method for integration |
| 3. DEPI-4 | 1959 | IBM 704 | Fourth-order Rungek-Kutta method for integration |
| 4. DYSAC | 1961 | CDC 1604 | Fourth-order Runge-Kutta method for integration |
| 5. DAS | 1962 | IBM 7090 | Euler's formula |
| 6. MIDAS | 1963 | For many computer initially it was an interpreter | Variable step 5th order predictor corrector method |

By the mid-1960's the use of digital computer for simulation of continuous systems was well-establishment and so were the block-oriented simulation language.

**Some Important Points of Block Structured Continuous Simulation Language**

1. It requires the user to first prepare a block diagram to represent the system to be simulated.
2. Then "patching" of instructions are required. Which specify how these blocks are to be connected together.
3. A typical block oriented language provides a number of standard functional blocks. These include the common simulation elements such as:
   - Integrators (*I*)
   - Multipliers (*X*)
   - Summers (+)

Some more specialized functional elements, such as:

- Limiter (*L*)
- Unit Delay (*V*)
- Hyteresis or Dead space (*D*)
- Zero order hold (*Z*)

For example, the IBM 1130 CSMP provides 25 different types of specified blocks and 5 blocks which the user can specify by supplying appropriate subroutines. As example Fig. 12.1 shows the diagrams for two blocks in the IBM 1130 CSMP.

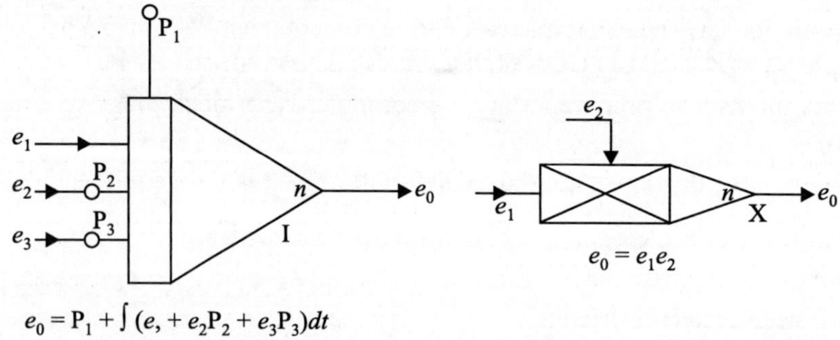


**Fig. 12.1** Integrator (*I*) and Multiplier (*X*) blocks n represents the block number.

- In order to simulate the given continuous system we first represent it by means of a block diagram.
- This block is then put into a computer program by means of three types of statements:

1. **Configuration Statements:** Configuration statements are the set of patching instruction which define the interconnections between the blocks.
2. **Parameter Statements:** Parameter statement associate the numerical values with block elements.
3. **Function Generation Statements:** Function generation statements define the input/output of the function generator elements.

**Advantages of Block-Oriented Language**

1. It is simple and natural with which a continuous dynamic system can be simulated.
2. A new user can learn to use such a language in an hour or two.
3. It provides greater accuracy and freedom from scaling.

**Disadvantages of a Block-Oriented Language**

1. It is relatively inflexible.
2. The system being simulated has first to be represented by means of a block diagram.

Consequently a block-oriented language is popular amongst those who naturally use "analog type" block diagrams to represent the system they are studying, often without explicity writing model equations.

**MOBSSL–UAF**

(Merrit and Millers Own Block Structured Simulation Language-Unpronounceable Acronym for). It was developed at the university of southern California in 1969. It is an interactive augmented block-structured simulation language for continuous systems. It is an extremely simple language to learn, and one need not have any prior programming knowledge or experience.

MOBSSL is a descendant of the IBM 1130 CSMP with some additional features such as:

- Disk Input/Output blocks
- A gradient processor
- A variable constant
- A storage block
- A large selection of trigonometric function
- Digital to analog and analog-to-digital blocks for hybrid computation.

Other block-structured simulation language are—IOAS, ACTOLUS 1130 CSMP. MOBSSL-UAF was implemented on PDP-11/40, IBM 360/44, CDC 3600 and IBM 5/370.

### 12.2.2 Expression-Based Language [*UPTU 2006-07*]

A purely block-oriented continuous system simulation language is severely limited because of its small specialized vocabulary. It does not allow any FORTRAN-type statement. Therefore no relationship or equation in the system can be expressed directly. The system must first be represented in an analog-block diagram so this languages are suited for analog system not for those who model their system as a set of equations this is a limitation of purely block-oriented language. After the success of MIDAS (1965, purely block-oriented language), this limitations was realized, and a successor of MIDAS called MIMIC was written for the IBM 7090 family of computers.

Continuous system simulation languages such as BED SOCS (Bradford Education Simulation Language for Continuous System) are called expression-based or statement-based continuous system stimulation languages.

Expression-based languages dont need to follow block structures of the analog patchboards but directly implement the differential equation of the model.

**Characteristics of Expression Based Language:** Following characteristics are there.

(*i*) **Methods of Integration:** In most expression-based languages several different routines are available for the user to choose from. They include both fixed step-size routines and variable integration step-size routines.

For example, in S/360 CSMP five fixed-size routines are available: fixed Runge-Kutta, Simpson's, trapezoidal, rectangular, and second order Adams. Two variable step-size routines are also available: fifth-order Milne predicter-corrector and fourth-order Runge-Kutta.

(*ii*) **Automatic Sequencing of Statements:** We know that an assignment statement in a program would be invalid if all the variables appearing on the right hand side had not already been defined by earlier statements. It is necessary to ensure in a simulation program that the statements appear in a correct sequence. In writing a simulation program for a large complex system, task of ensuring a correct sequence of statements may be too much of a burden on the programmers. Therefore, most of the expression-based simulation languages have a built in facility which automatically sorts the input statements.

(*iii*) **Intermediate Language:** Most of the expression-based languages are designed to be extensions of some well-known, existing high-level language. Example, S/360 CSMP are extensions of FORTRAN. Expressions in S/360 CSMP can be inter mixed with FORTRAN statements. A preprocessor first translates all the non-FORTRAN statements (of S/360 CSMP) into FORTRAN. These statements along with the FORTRAN statements constitute a single program, which then gets compiled by the FORTRAN compiler and run.

**Advantages of an Intermediate Language are:** The power, flexibility, ease of implementation. While disadvantages are: (*i*) First program has to be translated into the intermediate language, (*ii*) Programmer has also to become proficient at the intermediate language.

(*iv*) **Prepackaged Input Output:** In an expression-based continuous system simulation language the input is usually of free-form. The user does not have to format it. Similarly, the output is also displayed in a fixed format by means of a built-in output package. One has only to specify the variables to be displayed and the increments of the independent variable (time). This prepackaging of input or output relieves the programmer from the task of formatting them.

(*v*) **Macro Facilities:** An important feature available in many of the expression-based languages is the powerful function-defining capability called **macro**. It allows the user to build larger functional blocks from the basic functions available in the language, and thereby, to identify, a subsection of a simulation block diagram or the corresponding subset of structure statement. Macro functions may be used any number of times, as if they were operators of the language.

(*vi*) **Optional Block-Oriented Representation:** Many expression based languages have the facilities of block-oriented languages. This can have both type of users.

(*a*) Those who model their system as a set of equations.

(*b*) Those who model their system as a set of interconnected blocks.

Example: S/360 CSMP, CSSL, DSL have this facility.

Now we are highlighting some features of a few of the better known **expression based language** like DSL, S/360 CSMP, CSSL, DYNMO and MIMIC.

- **DSL (Digital Simulation Language)**

DSL is a classic and a very well known continuous system simulation language. This language was developed first in 1964 by IBM as DSL/90 for its 7090/7094 computers. DSL is an extension of FORTRAN.

Some DSL versions and its implementation on different computers are:

| (Version) | | | | Computer |
|---|---|---|---|---|
| DSL/40 | are | for | the | IBM 7040 |
| DSL/44 | are | for | the | IBM 360/44 |
| DSL/360 | are | for | the | IBM 360 |

- **S/360 CSMP (Continuous System Modeling Program)**

This language can handle either expressions or block oriented descriptions. It is a preprocessor with FORTRAN as an intermediate Language, means it accepts a simulation model description and run time instructions and produced a FORTRAN program which can be compiled and executed.

The capabilities of S/360 CSMP include:

➢ Seven integration methods

➢ 50 simulation functions

➢ Macros

➢ Excellent diagnostics and debugging aids

➢ Input/output formats are rich

➢ It include various tabular and plot arrangements

➢ It is expensive in computations.

**CSSL (Continuous System Simulation Language):** CSSL was defined in 1967, in accordance with the recommendations made by a simulation software committee formed by simulation councils, Inc. some salient features of a simulation languages are:

(*a*) It combined desirable properties like application areas, programming skills and digital computers.

(*b*) It has a powerful capability to be extended in order to satisfy user requirements.

(*c*) It has error control and debugging facilities.

(*d*) Integration by using several numerical methods.

Some CSSL versions and its implementation on different computers are:

| **Version** | | | | **Computer** |
|---|---|---|---|---|
| CSSL III | are | for | the | CDC 6400, XDS sigma 7 |
| CSSL IV | are | for | the | IBM 360/370, CDC 6600/7600 |
| RSSL | are | for | the | CDC 7600 |

HYTRAN and SL-1 languages are also versions of CSSL.

**DARE (Differential Analyzer Replacement):** DARE is a series of continuous system simulation languages for batch and online use. One version DARE P(Portable), was developed by the University of Arizona for the IBM S/360, CDC 6000 series. Some salient features of this language are following:

(*a*) The program is modulor.

(*b*) It uses system independent FORTRAN-based methods for writing and manipulating solution files.

(*c*) Users can add new integration algorithms or special library of subroutines.

(*d*) Problem equations are entered in a form resembling mathematical notations, and the associated procedural language is FORTRAN.

(*e*) Input data are provided by format free data cards.

(*f*) Outputs can be generated on a line-printer or plotter can be used.

(*g*) Following integration methods are incorporated like Runge-Kutta, 4th/3rd order Runge-Kutta, Adam's two point predictor, simple Euler one-point predictor.

**DYNAMO**

DYNAMO was developed in the late 1950's by J.W. Forrester and his industrial dynamics group at MIT. It uses first order difference equations to approximate the modeled continuous process. The level variables (commodities, money, population, etc.) characterize the state of the system at any particular time. These equations and variables define how quickly the levels are changing.

The DYNAMO compiler accepts the model written in the form of level, rate and auxiliary equations using special notation. The compiler functions are:

(*a*) Error control

(*b*) Sequencing equations according to the structural concepts of system dynamics.

(*c*) Compilation of the program defining the model.

(*d*) Execution of the program and generation of output in a tabular or graphical form.

The language has many computational facilities in the form of special statistical and mathematical functions.

DYNAMO has been extensively used to model and solve numerous information feed back systems. It is conceptually simple and easy to use.

But DYNAMO has one draw back that it is a very primitive and inaccurate integration scheme.

Available series of DYNAMO is on IBM 360, 370, 7090, B 5500, UNIVAC 1107, 1108. The latest version DYNAMO III extends the capabilities of DYNAMO II by the addition of arrays.

**MIMIC**

MIMIC was one of the first expression-based continuous system simulation languages, developed in 1965 for the IBM 7090 family of computers. This language has very rigid coding requirements, and has only one integration subroutine which uses fourth-order Rungekutta variable-step method. Some salient features of this language are:

(*a*) In MIMIC one could write FORTRAN-like statements.

(*b*) A MIMIC program is directly translated into a machine-language program.

(*c*) MIMIC is an inexpensive language to use in terms of the processing time.

(*d*) MIMIC eliminates the necessity of drawing a block diagram.

(*e*) In MIMIC, differential, algebric and logical equations can be written directly as statements or expressions in the language.

## 12.3 DISCRETE-SYSTEM SIMULATION LANGUAGES

**[*UPTU 2005-06, 2006-07*]**

Many large programs for simulating discrete systems have been written in general purpose languages, such as FORTRAN. These languages offer many convenient facilities such as:

- Automatic generation of streams of pseudorandom numbers for any desired statistical distribution.
- Automatic data collection
- Statistical analysis and report generation
- Good diagnostics
- Automatic handling of queues, etc.

A good simulation language also provides the model builder with a view of the world that makes model building easier. Thus a simulation language can be of help even at the model building stage. There are three major problems with such languages.

1. The programmer has to learn a new language.

2. Discrete-system simulation languages are highly problem oriented.
3. The language, a programmar is familiar with and is natural to his class of problems may not be available in his computing centre.

Every discrete-system simulation language must provide the concepts and statements for following :

(*a*) Representing the state of a system at a single point in time (Static modeling).

(*b*) Moving a system from state to state (Dynamic modeling).

(*c*) Performing relevant chores, such as, random number generation, data analyses, and report generation.

Points (*a*) and (*b*) which have been treated differently in different languages. From this point of view discrete-system simulation languages can be classified into three main categories:

1. Event-oriented languages.
2. Activity-oriented languages.
3. Process-oriented languages (including transaction-flow type).

**1. Event-Oriented Languages:** In an event-oriented language each event is represented by an instantaneous occurrence in simulated time and must be scheduled to occur (in advance) when a proper set of conditions exists. The state of the system changes at the occurrence of an event. The languages in this category are used to model process that are characterized by a large number of entities. The two well-known representatives of this group of languages are SIMSCRIPT and GASP.

**2. Activity-Oriented Languages:** In an activity-oriented language the discrete occurrences are not scheduled in advance. They are created by a program which contains descriptions of conditions under which any activity can take place. The activity-oriented languages have two major components:

(*i*) A test component

(*ii*) An action component

An activity-oriented language should be considered for use if the model has the following characteristics:

(*i*) The model uses a next event or variable time increment type of timing.

(*ii*) The process simulated is highly interactive but involves a fixed number of entities with events happening irregularly.

(*iii*) Event occurrence is controlled by cyclic scanning activity programs.

Examples: CSL (on IBM 7090/94, 1620).

**3. Process-Oriented Languages:** A key feature of a process orientation is that of a single process routine, composed of a number of segments describing a sequence of activities. Each segment behaves as an independently controlled program. On receiving control only the statements composing the segment are executed, and then control is returned.

Some of the languages belonging to this class are SIMULA, ASPOL (CDC 6000 series).

### Transaction-Flow Oriented Language

Transaction-flow oriented languages form a subcategory of process-oriented languages except that the flow of activities passes through specially defined blocks. The system model is represented by a flow chart consisting of the language blocks. The program creates transactions, executes them in the blocks and moves them along the flow chart.

We will now give a very brief description of three of the most popular discrete system simulation languages like: SIMSCRIPT, GPSS, and SIMULA.

### 12.3.1 Simula (Simulation Language)

SIMULA
- → Extension of ALGOL 60
- → General purpose programming language
- → Implemented on UNIVAC 1107, CDC 3000 and 6000, IBM-360 and 370 series
- → Design to provide simulation facilities.

In simula, their is addition of collection of a programs, called **processes** conceptually operating in parallel.

Process has two parts:

- Static part — A sequence of attribute declarations
- Dynamic part — A sequence of statement called an "Operation Rule".
  — Describe the dynamic behaviour of the process.

The processes perform their operations in groups called **active phase**.

In simulation we deal with several processes that have the same data structure and operation rule but differ in values of attributes such a group of alike processes is called the **process class.**

Path structure is defined as a part of the block head, and statement of the block define the operation rule.

**Salient Features of SIMULA**

1. Algorithmic capabilities is like to ALGOL.
2. Input output capabilities is like to SIMSCRIPT.
3. SIMULA is a complex language, but once mastered it is easy to use.
4. It has very good debugging features.
5. It is easy to read and structural.
6. Machine efficiency of simula 67 compiler appears to be on the average at least as good as that of a corresponding SIMSCRIPT compiler.
7. No automatic collection of statistics in SIMULA 67 (as is in GPSS).
8. It is possible to design experiment with SIMULA as in SIMSCRIPT.

### 12.3.2 GPSS (General Purpose Simulation System)

GPSS
- → Discrete simulation language
- → Developed by Geoffrey Gordon (1961)
- → Implemented on IBM 704, 709 and 7090 computers
- → Good debugging facilities

**Following Characteristics of GPSS**

1. GPSS was designed specially for analysts not for programmers.
2. GPSS is described as a block diagram in which the block represent the activities, and lines joining the blocks indicate the sequence in which the activities can be executed.

3. GPSS is a set of 48 specific block types. Each block has a name, and a specific task to perform.
4. Moving through the system of blocks are entities called **transaction**, *e.g.*, customers, messages, machine part, vehicles etc.
5. Transaction are created at GENERATE blocks and are removed for the simulation at TERMINATE blocks.
6. There can be many transactions simultaneously moving through the block diagram.
7. Blocks that can handle only one item at a time are called the **facilities**, *e.g.*, of a facility is a machine on a person.
8. A storage represents another block type that can be used by one or more transactions simultaneously, *e.g.*, storages are groups of machines or a warehouse.
9. GPSS handles the advancement of time by a block called ADVANCE when a transaction enters this block, an action time is computed and added to the current time to produce a block department time.
10. No language extension possible.
11. Restricted to simple queueing problems.
12. Machine efficiency is often poor since GPSS is an intepretates system.

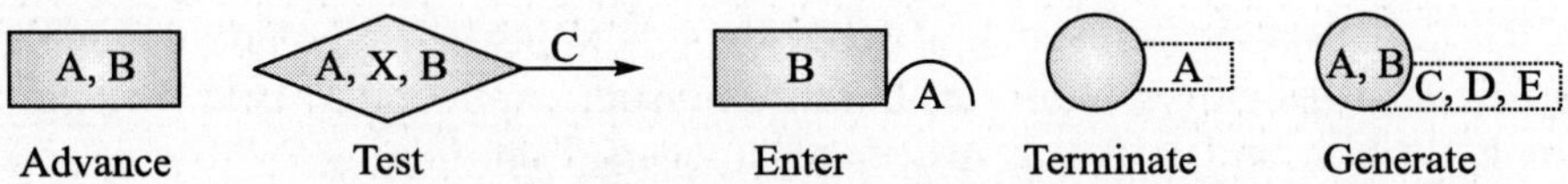


**Fig. 12.2** Some GPPS block diagram symbols.

### 12.3.3 SIMSCRIPT II.5

SIMSCRIPT II.5 was released by the RAND corporation in 1972 SIMSCRIPT II (or II. 5) can be represented as a general programming language with extra features for discrete-event simulation. It has been implemented on IBM 7090, 7094, 360, 370 series; CDC 3600, 3800, 6000, and 7000 series; UNIVAC 490, 1107, and 1108.

SIMSCRIPT II can be learned and used at different levels:

**Level 1**

- It is comparable to a very simple algorithmic language such as BASIC.
- It have format-free data and ease of programming, simplified output options and automatic mode conversion.

**Level 2**

- It is FORTRAN-plus language.
- Additional capabilities such as data structures is found.
- Extensive report generator is available.

**Level 3**

SIMSCRIPT-II provides ALGOL-like (or PL/I-like) statements

Level-1 + Level-2 + Level-3 $\Rightarrow$ Algebraic procedural language

The first three levels combinely represented SIMSCRIPT-II as an algebraic procedural language.

**Level 4**

It gives SIMSCRIPT to define and manipulates entities, attributes and sets. Individual entities have values, called attributes, which define a particular state of entity. Attributes are named not numbered

Entities can be of two types:

1. Permanent — (*a*) Specified for all time of program execution
   (*b*) Permanent entities have their attributes stored as arrays
2. Temporary — (*a*) Specified dynamically as the program proceeds
   (*b*) Temporary entities are physically created and destroyed through special statements.

Entities may be declared as temporary or permanent in the PREAMBLE of a SIMSCRIPT II program.

**Level 5**

- It keeps track of dynamics of the model.
- It describes system dynamics, for controlling system dynamics for modeling statistical phenomena.

SIMSCRIPT II requires a model to be specified only in terms of separate events, and scheduling an event means filing an "event notice" in the "events set".

SIMSCRIPT II provides eleven functions for generating pseuderandom samples from the following statistical distributions: uniform, normal, poison, exponential, beta, Erlang, lognormal, binomial, gamma, Weibull, and uniformly distributed integers. Contributing to the wide acceptance and success of SIMSCRIPT II.5 modeling, we have following outline.

**Design**

It consist entities and processes, provides a natural conceptual framework with which to relate real objects to the model.

**Programming**

SIMSCRIPT II.5 is a modern, free from language with structured programming constructs and all built in facilities needed for model development. The amount of program needed to model a system is typically 75% less than its FORTRAN or *C* counterpart.

**Debugger**

A well designed package of program debug facilities is provided. The required tools are available to detect errors in a complex computer program without reporting an error.

**Documentation**

You get a powerful, English like language supporting a modular implementation. Because each model component is readable and self-contained the model documentation is the model listing; it is never absolute or inaccurate.

## 12.4 SLAM

SLAM Simulation Language for Alternative Modeling, is a FORTRAN-based simulation language designed by Dennis Pegden in 1979 and subsequently supported and marketed by Pristker and Associates, Inc. SLAM was based on GASPIV and Q-GERT (an acronym for Queues—Graphical Evaluation and Review Techniques) and has three structures.

- Process
- Event
- Continuous

The process structure is similar to Q-GERT, and the event structure is similar to GASP-IV like a GPSS model, a SLAM model requires fewer statements than a FORTRAN model. SLAM-II, is the current version of SLAM, which is available in different forms. The basic SLAM-II does not include animation.

SLAM SYSTEM, the micro computer version of SLAM-II, has following capability:

- Animation
- Graphics
- User friendly environment

SLAM II/TESS is similar to SLAMSYSTEM but, in addition, has enhanced statistical features and an integrated database for model input/ output.

Using process structure of SLAM requires that the analyst first develops a graphical network representing the flow of an entity in the system.

The network model represents all possible paths that the entity can take as it flows through the system. The network is developed by combining a standard set of nodes and branches.

A node represent an event or system action such as an arrival event and queue, whereas a branch is used to represent an activity or the passage of time.

The analyst directly translates the network representation of the system into computer statements.

The analyst codes each discrete event as a FORTRAN subroutine. The subroutine SLAM is called by the user—written main program and serves as the executive control for discrete event simulation. It automatically handles event-handling algorithms, random variate generation and routines for gathering statistics.

## 12.5 OBJECT-ORIENTED SIMULATION

We know that there is strong interest in object-oriented programming in general so there has been a lot of interest in object-oriented simulation. Actually both object oriented simulation and programming originated from the object-oriented simulation language SIMULA; which was introduced in the 1960s.

In object-oriented simulation a simulated system is considered to consist of objects that interact with each other as the simulation evolves through time. These objects contain data and methods.

**Data:** data describe the state of an object at a particular point in time.

**Methods:** It describe the actions that the object is capable of performing.

Examples of true object-oriented simulation packages are MODSIM III and SIMPLE++. Three major features of such a simulation package are inheritance, polymorphism and encapsulation.

Many vendors claim that their simulation software is object-oriented, but in some case's the software may not include inheritance, polymorphism, or encapsulation. Furthermore, certain of the above three features are some times assigned different meanings.

The following are possible advantages of object-oriented simulation.

- It promotes code reusability because existing object can be reused.
- It manages complexity by breaking the system into different objects.
- It makes model changes easier.
- It facilitates large projects with several programmers.

Disadvantages of the object-oriented approach are:

- Some object-oriented simulation packages may have a steep learning curve.
- One must do many projects and reuse objects to achieve its full benefits.

## 12.6 GENERAL-PURPOSE VERSUS APPLICATION-ORIENTED SIMULATION PACKAGES

Simulation packages were divided in two major parts:

- Simulation languages
- Application-Oriented Simulators.

**Simulation Languages**

Simulation languages were general in nature, and model development was done by writing code. This language provides a great deal of modeling flexibility, but were often difficult to use.

Vendors of simulation languages have attempted to make their software easier to use by employing a graphical model building approach.

**Application Oriented Simulators**

Application oriented simulators were oriented toward a particular application, and a model was developed by using graphics, dialog boxes, and pull-down menus. Simulators were sometimes easier to learn and use, but might not have been flexible enough for some problems.

Vendors of simulators have attempted to make their software more flexible by allowing programming in certain model locations using an internal pseudo-language. In at least one simulator, it is now possible to modify existing modeling constructs and to create new ones.

We will now say that there are two types of simulation packages.

- A general-purpose and simulation packages.
- An Application-oriented simulation packages.

A General-purpose simulation can be used for any application, but might have special features for certain ones (*e.g.*, for manufacturing, communication or process engineering).

An Application-oriented simulation package is designed to be used for a certain class of applications such as manufacturing, health care or call centers.

## 12.7 WHAT IS MODSIM-III? [*UPTU 2006-07*]

It is an object oriented simulation tool, specifically designed for modeling of complex and large system.

It supports multiple inheritance, encapsulation, and polymorphism MODSIM III is used by no. of users for network simulation, supply chain management, business process modeling, manufacturing system, and other application. A MODSIM-III simulation model contains:

- A single module–"main".
- Definition module for each required object type.
- Implementation module for each required object type.

**Main Module**

Defines the model's data structures and starts the execution of simulation.

**Definition Module**

The definition block describes the object type by declaring its variable and methods. This object description is referred by other objects in the simulation and it provides the formal interface specification.

**Implementation Module**

The logic of what they will do and how they affect the state variables of the object are described in the implementation blocks.

There are two types of methods.

(*i*) ASK (*ii*) TELL

**ASK Method**

It is like a procedure in any other programming language, but is associated with a particular object type.

**TELL Method**

It describes the behaviours of an object that result in the passage of simulated time, and corresponds to a process routine or TELL methods are used to describe behaviour that elapse simulation time.

**What MODSIM-III Provides:**

(*a*) MODSIM-III provides a rich collection of simulation building block objects to fulfil many simulation modeling requirements.

(*b*) MODSIM-III provides an essentially unlimited no. of random-number streams.

(*c*) MODSIM-III provides you with a rich set of features to manage the complex scheduling, interaction and synchronizing of time-elapsing behaviours, it saves significant development time.

(*d*) MODSIM-III provides rich graphics libraries for graphical scenario, dynamics analysis, and animation (2D and 3D). Addition of graphics is easy. It also provides a graphical editor to configure the icons, menus, dialog boxes and presentation charts. If also provides histograms, bar charts and pie charts.

(*e*) MODSIM-IIFs object-oriented approach has the greatest benefits when one builds large and complex simulation models.

Unlike general purpose languages such as C++ and Java, MODSIM-III:

- ➢ Captures both concurrent and interacting behaviours of system components in simulation time.
- ➢ It provides built in statistical modeling and statistic functions.
- ➢ It comes with simulation specific graphics libraries, and animation.
- ➢ It provides support for the high level architecture (HLA) which is the U.S. Department of Defense's standard for interoperability among simulation model.

MODSIM is a product of CACI company. SIMPROCESS and COMNET-III are CACI's domain specific tools, that are developed by using MODSIM-III.

## 12.8 CSMP-III (CONTINUOUS SYSTEM MODELING PROGRAM VERSION III) [*UPTU 2006-07*]

CSMP-III programs have following three general type of statements:

- Structured statements

- Data statements
- Control statements

**1. Structured Statements**

- Structured statements define the structure of model.
- It consists of FORTRAN-like statements.
- It consists functional blocks in a model definition.
- It can make use of the arithmetic operations (+, –, ×) and exponentation using same notation and rules as used in FORTRAN.

**Some Rules are given here**

➢ For example, equation $Y = X^2 + 2X - 3X/Z$ will be represented by following equation

$$Y = X * X2 + 2 * X - 3.0 * X/Z$$

➢ 2.2 $E - 2$ represents 0.022, means exponent notations use $E$.

➢ Fixed value constants may also be declared.

➢ Variables can have up to six character.

➢ There are many functional blocks which, provide some operations specific to simulation.

Table 12.2 is a list of some of the functional blocks.

**Table 12.2**

| General form | Function |
|---|---|
| $Y$ = INTGRL ($IC$, $X$) $Y(0) = IC$ INTEGRATOR | $Y = \int_0^t X dt + IC$ |
| $Y$ = STEP ($P$) STEP FUNCTION | $Y = 0 \quad t < P$<br>$Y = 1 \quad t \geq P$ 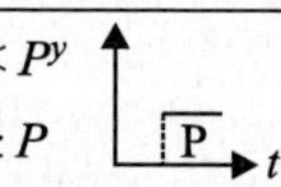 |
| $Y$ = EXP($X$) EXPONENTIAL | $Y = e^x$ |
| $Y$ = SIN($X$) TRIGONOMETRIC SINE | $Y = \sin(X)$ |
| $Y$ = SQRT($X$) SQUARE ROOT | $Y = X^{\frac{1}{2}}$ |

**2. Data Statements**

- It assign numerical values to parameters, constant and initial condition.
- One Data Statement INCON is used to set the initial values of integration function block.

**3. Control Statement**

- It specify options in assembly.
- It specify execution of the program.
- It specify, how we can get output of our choice.
- One control statement called PARAM can also be to assign values to individual parameters, but its main purpose is to specify a series of values for one parameter.
- One control statements is called TIMER, which must be present to specify certain time intervals.
- Output can be in the form of printed tables and/or print-plotted graphs for this purpose, control statements with words PRINT and PRTPLT are used.

- Two other control statements with the words TITLE and LABEL can be used to put headings on the printed and print plotted output, respectively.
- Set of structural, data and control statements for a problem can be assembled in any order but they must end with an END control statement.

## 12.9 HOW TO SELECT SIMULATION SOFTWARE

There are number of things that make an ideal simulation packages like:

- Properties related to package, such as support, reactivity to bug notification, etc.
- Properties related to user, such as what's their needs, their level of expertise.
- What purpose you need the software? Is it for education, teaching, student project or research.
- What are the important aspects to look for in a package?

However there are some criteria of selection :

- Input facilities, output facilities.
- Optimization capabilities.
- Environment including training and support services.
- Input-Output statistical data analysis capability.
- Cost factor.
- Recovery from the failure.

You must know which features are appropriate for your situation, your objective.

The following are some sample of software that is available in the market today:

- ACSL
- Med Model
- Automod
- Simscript
- Arena
- Simulate
- C + +SIM
- PowerSim
- CSIM
- QUEST
- JavaSim
- GPSS

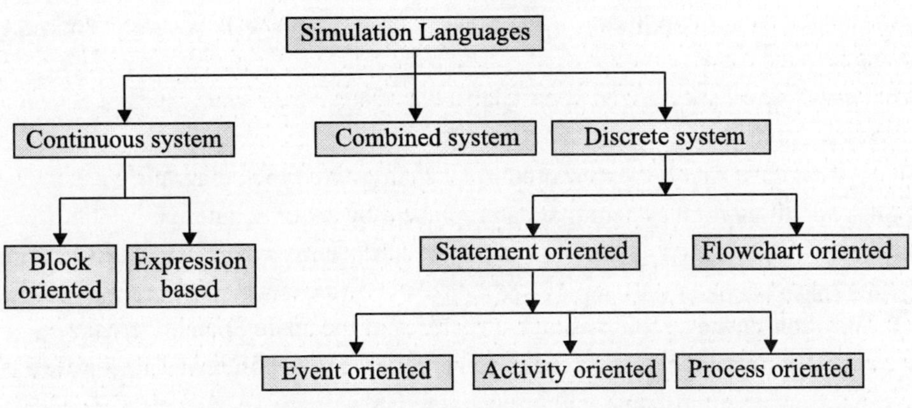


**Fig. 12.3** Classification of simulation languages.

## SUMMARY

- The efficiency of programming and execution of a simulation project depends upon the programming language used.
- Simulation packages provide a natural framework for simulation modeling while programming languages may allow greater programming flexibility.
- Continuous simulation languages are designed for continuous models. While discrete simulation languages are designed for discrete models.
- Some languages are suitable for both discrete as well as continuous models. These languages are called combined simulation languages.
- Simulation language that makes a digital computer emulate an analog computer is called a block-structured continuous system simulation language.
- If system can be expressed differential equation algebraic and logical equations directly as statements or expressions in the language that language is called expression-based languages.
- DSL, S/360 CSMP, CSSL, DARE, DYNAMO and MIMIC are examples of continuous system simulation languages.
- SIMSCRIPT, GPSS and SIMULA are most popular discrete system simulation languages.
- Object oriented simulation provide reusability because objects can be reused or easily modified.
- Application-oriented simulation package is designed to be used for a certain class of application such as manufacturing health care or call centers, while general purpose simulation package can be used for any application.

## SOLVED QUESTIONS

**Q.1. What is combined simulation language?**

**Ans.** Some languages have been designed which are suitable for both discrete as well as continuous models. This type of language is called as combined simulation language, it is written particularly for system models in which some of the variables change continuously and other variables change discretely. For example, GASP IV.

**Q.2. Write disadvantage of block-oriented continuous system simulation language.**

**Ans.** A pure block-oriented continuous system simulation language is severely limited because of following reason:

1. It has small set of specialized vocabulary.
2. It does not allow any FORTRAN-type statements.
3. The system must first be represented in an analog-like block diagram.
4. It does not allow for a system that can represent by set of equations.

**Q.3. What is expression based or statement-based continuous system simulation languages?**

**Ans.** Expression based simulation language can directly implement the differential equations of the model. They don't need to follow block-structures of the analog patchboards.

**Q.4. What are the important common characteristics of expression based simulation languages?**

**Ans.** There are following important common characteristics:

(*a*) Methods of integration

(*b*) Automatic sequencing of statements

(*c*) Intermediate

(*d*) Prepackaged input output

(*e*) Macro facilities

(*f*) Optional block-oriented representation.

**Q.5. Write some salient features of SIMULA.**

**Ans.** There are following salient features of SIMULA:

1. Algorithmic capabilities comparable to ALGOL.
2. It is a complex language, but once mastered it is easy to use.
3. It has good debugging features.
4. Good program readability and structuring.
5. Machine efficiency of simula 67 compiler appears to be on the average at least a good.

**Q.6. Discuss the methods of MODSIM-III.**

**Ans.** There are two types of methods in MODSIM-III:

1. ASK 2. TELL

**1. ASK:** An ask method is similar to a procedure in any other programming language, but is associated with a particular object type.

**2. TELL:** Tell method describes the behaviours of an object that result in the passage of simulated time, and corresponds to a process routine as defined.

**Q.7. Write some examples of application-oriented simulation packages.**

**Ans.** There are some example of the application-oriented simulation packages that are currently available.

- Manufacturing—Automod [Auto simulations (1999*a*)], Arena packaging edition [Systems 1999*b*)], Promodel [PROMODEL (1999*b*)].
- Communication Networks—COMNET III (CACI (1999)], IT Decision Guru (MIL 3(1999*a*)].
- Health care—Med Model [PROMODEL (1999*a*)]
- Call centers—Arena Call Center Edition [Systems (1999*a*)].

**Q.8. Write various advantages of object-oriented simulation.**

**Ans.** Following advantages are there:

1. It provides code reusability
2. It makes model changes easier
3. It manages complexity by breaking the system into different objects
4. It provides inheritance, polymorphis, encapsulation and various other object oriented features.

**Q.9. Write some limitation of block-structured continuous simulation language.**

**Ans.** Limitation of block-structured continuous simulation language—

- It is relatively inflexible
- The system being simulated has first to be represented by means of a block diagram.
- Their application area is restricted.

**Q.10. Differentiate continuous vs discrete simulation languages.**

**Ans.**

| Continuous simulation language | Discrete simulation language |
|---|---|
| 1. Simulation languages are designed for simulating continuous model. | Discrete simulation language are designed for simulating discrete model. |
| 2. It is written in special program packages. | It is written in general-purpose languages. |
| 3. Examples: DSL, S/360, CSMP, MIMIC, and DYNAMO. | Examples: SIMCRIPT, GPSS, and SIMULA. |

## EXERCISES

1. Differentiate between continuous simulation languages and discrete-event simulation.
2. Explain the classification of the discrete simulation language?
3. Differentiate general-purpose vs. application-oriented simulation packages.
4. Discuss object-oriented simulation.
5. Write short notes on:
   (*a*) DSL
   (*b*) MIMIC
   (*c*) CSSL
   (*d*) DYNAMO
   (*e*) Expression based language
   (*f*) GASP
   (*g*) SIMULA
6. Explain in brief, continuous and discrete simulation language and expression based language. [*UPTU 2006-07*]
7. Explain in brief, expression-based language.
8. Write some advantages of discrete-event simulation.
9. What are the criteria to select simulation software.
10. Write a short note on CSMP-III, and MODSIM-III.
11. Explain block structured continuous system simulation language with some advantages and disadvantages.
12. Discuss various characteristics of expression-based language.
13. Write short notes on :
    (*a*) MIMIC
    (*b*) CSSL
    (*c*) DARE
14. On what basis you can categories the discrete-system simulation languages and also explain these classification.
15. Explain and characterize GPSS.

13

# GPSS

**INSIDE THIS CHAPTER**



**GPPS (General Purpose Simulation System)**

- Discrete simulation language
- Developed by Geoffrey Gordon (1961)
- Implemented on IBM 704, 709 and 7090 computers
- Good debugging facilities

A short, easily understood GPSS model would require many pages of FORTRAN coding to accomplish a similar goal. The GPSS user is free to concentrate on the important issues in the model being developed since the language itself collects statistics, produces tabulated results and performs a host of mundane tasks one would prefer not to deal with. The purpose of this brief chapter is to present some basic terminology and concepts which will enable the person meeting GPSS for the first time to understand the broad range of applications, power and ease of use that this language brings to the simulationist.

## 13.1 GPSS ENTITIES

GPSS is designed around number of elementary abstractions called entities. GPSS entities are abstract objects that exist in a simulation so it is most important to understanding of these entities and their rules of manipulation. The collection of all these entities is called the simulation. There are two most prominent types of entity.

(*i*) Transactions (*ii*) Blocks

**Fig. 13.1**

Simulation, consists of many transactions moving from one block into the another. Transactions are the only entity types that can be deleted from the simulation. There are a number of entity types, and a simulation may contain many instances of any entity type. To be effective in creating simulations first one has to understand the properties of each of the GPSS entities and how to use GPSS blocks in order to cause interactions among the entities.

There are following properties of GPSS entities:

(*a*) GPSS entities are numbered. The integer value associated with the name is used to find the entity. EQU statement automatically assign these integer values to names.

(*b*) GPSS World normally assigns a unique value greater than or equal to 10,000.

(*c*) GPSS entities are created automatically when needed. This convenience can sometimes cause your simulation to use a lot of virtual memory due to a bug in your GPSS model.

(*d*) The use of virtual memory provides for simulations taking up to half a gigabyte (512 Megabytes) and hence reduces the simulation size.

(*e*) Entities must be specifically declared before they can be used.

**The following entities must be declared before they can be used:**

- Storage entities must be declared in STORAGE Statements.
- Arithmetic Variables must be declared in VARIABLE statements.
- "Floating point" Variables must be declared in FVARIABLE statements.
- Boolean Variables must be declared in BVARIABLE statements.
- Matrices must be declared in MATRIX statements, or Temporary Matrix PLUS Declarations.
- Tables must be declared in TABLE statements.
- Qtables must be declared in QTABLE statements.
- Functions must be declared in FUNCTION and Function Follower statements.
- Transaction Parameters must be declared in ASSIGN, MARK, READ, SELECT, SPLIT, COUNT or TRANSFER SUB Blocks before they are referenced.

**GPSS Entity Categories**

There are following four categories of GPSS entity.

1. **Basic Category:** This category includes Transaction and Block Entities.
2. **Structural Category:** This category represents structural aspect of simulation and includes Facility, Storage, User Chain, LogicSwitch, Numeric Group and Transaction Group entities.
3. **Data and Procedures Category:** This category used for generating and storing data. These include the Matrix, Save Value, Variable, and Function entities.
4. **Output Category:** This category is used for collecting and presenting statistical data and includes the Queue and Table Entities.

## 13.2 TRANSACTIONS

A basic simulation needs to have transactions generated pass through some blocks and then terminated. Transactions move from Block to Block in a simulation in such a manner which represents the real-world system which you are modeling. The transactions are tracked and placed on different holding areas for processing dependent on whether they are OK to proceed now (current event chain CEC), or need to wait for some future event or time (future event chain FEC). When all the current events are processed, then the time is incremented and the simulation proceeds to work through the new set of CEC.

Once a Transaction starts to move in the simulation, it continues to enter blocks as long as it can. During a simulation, the Transaction which is attempting from block to block is called the *Active Transaction*. If transactions cannot move further then another Transaction is selected to be the Active Transaction and at a time there can be only one Active Transaction.

Transactions are numbered sequentially throughout a session starting with 1. A CLEAR statement begins the numbering of transactions at 1 again. The behaviour of a transaction is determined by transaction attributes. There are following important attributes of the transactions.

**Parameters:** Transaction parameters are a set of values associated with a transaction. Each Transaction may have any number of parameters. Each parameter has a parameter number, by which it is referenced, and a value. The parameter number is a positive integer. The value of any parameter of the active transaction may be returned by the SNA P*Parameter* where *Parameter* is the name or number of the Parameter. If a block operand specifies parameter name or number as the desired value, P$ *Parameter* or P*Parameter* should not be used, use only the Parameter Name or Number by itself.

Transaction Parameters must be created and assigned values before they can be referenced. ASSIGN, MARK, and TRANSFER SUB, SELECT, SPLIT and COUNT blocks create a transaction parameter if one does not exist.

**Related SNAs of Transaction**

There are following SNAs associated with Transactions.

1. **A1:** Assembly Set. A1 returns the Assembly Set of the Active Transaction.
2. **MB*Entnum*:** Match at Block. MB*Entnum* returns a 1 if there is a Transaction at Block *Entnum* which is in the same Assembly Set as the Active Transaction. MB*Entnum* returns a 0 otherwise. This SNA class should not be used in a Refuse Mode GATE or TEST Block condition test, you should use MATCH Blocks instead.
3. **MP*Parameter*:** Transit Time, Parameter. Current absolute system clock value minus value in Transaction Parameter *Parameter*.
4. **M1:** Transit Time. M1 returns the absolute system clock minus the "Mark Time" of the Transaction.
5. **P*Parameter* or ****Parameter*:** Parameter value. P*Parameter* or **Parameter* returns the value of Parameter *Parameter* of the Active Transaction.
6. **PR:** Transaction priority. The value of the priority of the Active Transaction.
7. **XN1:** Active Transaction. The transaction number of the Active transaction.

## 13.3 BLOCKS IN GPSS

In **GPSS,** the structure of the system being simulated is described in the form of a block diagram drawn with a fixed set, of predefined *block types.* The GPSS Block entity is the basic structural element of the simulation. Each block type represents a specific action that is characteristic of some basic operation that can occur in a system. Blocks that can handle only one item at a time are called the **facilities** *e.g.*, of a facility is a machine on a person.

Connections between the blocks of the diagram indicate the sequence of actions that occur in the system. Where there is a choice of actions, more than one connection is made from a block to indicate the choice. Moving through the system being simulated are certain basic units that depend upon the nature of the system. For example, a communication system is concerned with the movement of messages, a traffic system with vehicles or a data processing system with data records, and so on. These units are identified with entities called transactions. Transactions have a priority level and they can carry items of data called parameters.

**Consider following basic block with figure:**

**Advance**

ADVANCE block is concerned with representing the expenditure of time. The program computes an interval of time called an action time for each transaction as it enters an ADVANCE block.

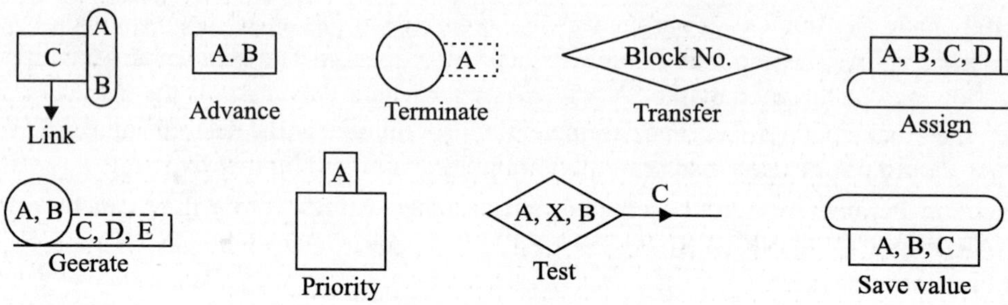


**Fig. 13.2. GPSS block diagram.**

**Generate**

This block controls the interval between successive arrivals of transactions and normally creating transaction from zero time.

**Terminate**

Transactions are removed from the simulation at TERMINATE block.

**Transfer**

The transfer block allows some location other than the next sequential location to be selected.

**Adopt:** Change Assembly Set.

**Advance:** Place Transaction on Future Events Chain.

**Alter:** Test and modify Transactions in a Group.

**Assemble:** Wait for and destroy related Transactions.

**Assign:** Modify Transaction Parameter.

**Buffer:** Place Transaction on the Current Events Chain behind its priority peers.

**Close:** End the Data Stream.

**Count:** Place count of entities into a Transaction Parameter.

**Depart:** Decrement content of a Queue Entity.

**Displace:** Change the Next Sequential Block of a Transaction.

**Enter:** Occupy or wait for storage units in a Storage Entity.

**Examine:** Test group membership.

**Execute:** Perform action specified by a different Block.

**Favail:** Change status of a Facility Entity to "available".

**Funavail:** Change status of a Facility Entity to "not available".

**Gate:** Test entity and modify Transaction flow.

**Gather:** Wait for related Transactions.

**Generate:** Create Transaction and place on Future Events Chain.

**Index:** Modify Transaction Parameter.

**Integration:** Turn the integration of a User Variable On or Off.

**Join:** Place a member into a Numeric or Transaction Group.

**Leave:** Release storage units of a Storage Entity.

**Link:** Move Transaction to Userchain Entity.

**Logic:** Modify Logicswitch Entity.

**Loop:** Decrement Parameter, jump to different Block if result is nonzero.

**Mark:** Place value of system clock into Transaction Parameter.

**Match:** Wait for related Transaction to reach conjugate MATCH Block.

**Msavevalue:** Assign value to Matrix Entity element.

**Open:** Initialize a Data Stream.

**Plus:** Evaluate PLUS Expression and save result in Parameter.

**Preempt:** Displace Facility owner.

**Priority:** Modify Transaction priority.

**Queue:** Increment content of a Queue Entity.

**Read:** Bring the next line of data from a Data Stream.

**Release:** Free Facility Entity.

**Remove:** Take a member out of Numeric or Transaction Group.

**Return:** Free Facility Entity.

**Savail:** Change status of Storage Entity to "available".

**Save Value:** Assign a value to Save Value Entity.

**Scan:** Test Transaction group, place value in Parameter.

**Seek:** Change the line pointer in a Data Stream.

**Seize:** Assume ownership of or wait for a Facility Entity.

**Select:** Place selected entity number into Transaction Parameter.

**Split:** Create related Transaction.

**Sunavail:** Change status of Storage Entity to "not available".

**Tabulate:** Update Table Entity.

**Terminate:** Destroy Transaction, decrement Termination Count.

**Test:** Test arithmetic condition and modify Transaction flow.

**Trace:** Set Trace Indicator of the Active Transaction.

**Unlink:** Remove Transaction from Userchain Entity.

**Untrace:** Turn off Trace Indicator in the Active Transaction.

**Write:** Send a value to a Data Stream.

## 13.4 PROCESS ORIENTED PROGRAMMING

GPSS is a process-oriented simulation. A process begins when we start generating transactions. As transactions percolate through system blocks, the process advances. Process orientation is communicated in a FORTRAN type simulation platform. Process oriented languages are based on the fact, that events are not independent. An events typically a consequence of the previous events. It is possible to define sequence of events that may be viewed as entities of simulation model at the hierarchy. A sequence of events is called **process**. Process based abstract system are very close to reality. Process way of viewing system dynamics is very natural.

Mostly a process models an activity of a real object. It is believed, that process oriented discrete simulation is the best way how to create discrete simulation models.

A key feature of process-orientation is that a single programs is made to act as thought it is several programs, independently controlled either by activity type scan or event scheduling. Each process can has several points at which it interacts with other processes. Each process can have several active phases; each active phase of a process is an event. This is different from pure event or acitivity approaches that allow an interaction only when all the actions associated with an event or activity have been completed.

## 13.5 ACTION TIMES

The sequence of actions occurring in the system in real time is reflected in the movement of transactions from block to block in simulated clock time. Clock time is represented by an integral number, with the interval of real time corresponding to a unit change of clock time chosen by the program user. The program computes an *action time* for each transaction entering a block to represent the time taken by the system action simulated by the block. The transaction remains at the block for this interval of simulated time before attempting to proceed. The action time may be a fixed interval (including zero), a random variable, or it can be made to depend upon conditions in the system in various ways. An action time is defined by giving a *mean* and *modifier* for the block. If the modifier is zero, the action time is a constant equal to the mean. If the modifier is a positive number (≤ mean), the action time is an integer random variable chosen from the range, mean of modifier, with equal probabilities given to each number in the range. It is possible to introduce into the simulation, a number of functions. The action time is derived by multiplying the mean by the value of the function.

## 13.6 SUCCESSION OF EVENTS

Once the program has begun moving a transaction, therefore, it continues to move the transaction through the block diagram until one of several circumstances arises.

(*i*) The transaction may enter an ADVANCE block with a non-zero action time, in which case, the program will turn its attention to other transactions in the system and return to that transaction when action time has been expended.

(*ii*) The conditions in the system may be such that the action the transaction is attempting to execute by entering a block cannot be performed at the current time.

(*iii*) The transaction enters a TERMINATE block, in which case it is removed from the simulation.

(*iv*) The transaction may be put on a chain.

The program maintains a record of each transaction in the system is due to move. If there is more than one transaction due to move, the program processes transactions in the order of their priority class such as FCFS.

## 13.7 SYSTEM NUMERICAL ATTRIBUTES (SNA)

System Numerical Attributes, or SNAs, are simulation "state variables" that are available for use through a simulation. Parameters, attributes, and transactions are collectively called as system Numerical attributes. They return numeric or string values, and may be used in GPSS Statement operands and in expressions. Some important points of SNA:

- Each type of SNA is identified by one or two letter code and a number. For example, the contents of storage number 5 is denoted by S5 and the length of queue number 15 is Q15.
- Some SNAs are maintained automatically, and others require a calculation when they are called.
- Many SNAs, such as PR, are evaluated on behalf of the active Transaction. This means that some attribute of a Transaction is necessary in order to evaluate the SNA.
- Other SNAs, such as the system clock C1, can be evaluated without referring to a Transaction.

SNAs are available for use in the operand fields and Expressions of commands and statements. In all cases *Entnum* must be replaced by any entity specifier. The entity specifier could be a name (preceded by a $ separator) or number, or for indirect addressing, it could be an asterisk,*, followed by a name or number.

**SNA Format:**

SNA is divided into following parts:

- SNA Class
- Atomic SNA
- An operand which is typically an Entity Identifier.

Most SNAs may be specified in one of many forms, beginning with the SNA class. For example, the SNA W22 returns the number of Transactions waiting in Block number 22. The Block identifier in this case is the number 22, but there are several other possibilities. Examples with SNA Class **W** provides number of Transactions in Block.

- **W22:** Represents the number of Transactions in Block 22.
- **W$CompA:** Represents the number of Transactions in Block labelled CompA (GENERATE block in the example of the previous section).
- **W*5:** Represents the number of Transactions in the Block identified by the value of the current Transaction's parameter 5.
- **W*BLKNUM:** Represents the number of Transactions in the Block identified by the value of the current Transaction's parameter named BLKNUM.
- **W*$BLKNUM:** Same as W*BLKNUM.
- **W*j*:** Where *j* is a positive integer, the number of the Block in the simulation.
- **W$*Name*:** Where *Name* is the location of the desired Block.
- **W*j:** Where *j* is a positive integer, the number of the Parameter of the active Transaction which contains the number of the desired Block. This is indirect addressing.
- **W*Name:** Where *Name* is the name of the Parameter of the Active Transaction which contains the number of the desired Block. This is indirect addressing.
- **W*$Name:** Where *Name* is the name of the Parameter of the Active Transaction which contains the number of the desired Block. The $ is not needed and is used only as a separator. Essentially, this is identical to W**Name*. This is indirect addressing.
- **W*Parameter:** Denotes that either the **W*j**, the **W***Name, or the **W*$**Name may be used.

The Matrix Entity SNA class MX is a special case. It may contain up to 3 indirect addresses. For example the SNA MX*Sales (*Partnumber,*January) locates the Matrix Entity whose number is in the Transaction parameter named Sales, and then retrieves the element whose row and column numbers are in the Parameters named Partnumber and January, respectively. Normally, the Active Transaction would have already passed through three ASSIGN blocks initializing the Parameters Sales, Partnumber, and January with the numbers or name values desired.

**Atomic SNA:**

Some SNA classes, such as A1, AC1, C1, M1, MP, PR, and TG1, are "atomic" SNAs. Atomic SNAs are complete and do not need an entity name or number to complete the evaluation of the SNA. Atomic SNA do not have an operand. Some atomic SNA are:

- Parameters of the active Transaction
  - **A1:** Assembly Set
  - **M1:** Transit Time which equals to the AC1–mark time (time that the transaction entered the simulation).
  - **PR:** Priority
  - **XN1:** Identifer.
- **AC1:** Absolute value of the system clock (since the start of the simulation).
- **C1:** Relative value of the system clock, since the last RESET Command.
- **TG1:** Remaining termination count. See the TERMINATE Block.

When you are in any operand field, the valid SNA forms are available in an online help message.

**Uses of SNA's:**

1. SNA provides the inputs to the functions by allowing a great variety of functional relationships to be introduced into the model.
2. Transit time is one of the SNA's and M1 is its symbol.
3. The current value of any SNA may also be used as the value of almost any field of a block.
4. The values of the SNA's change as the simulation proceeds.
5. The program does not continuously maintain current value, it computes the value of SNA's at the time they are needed.

## 13.8 LOGICSWITCH

A Logicswitch Entity is the simplest entity that is made available to represent two-state conditions in the system. Each switch is either "Set" or "Reset" and a block type called LOGIC is used to change the status of the switch. There are BLOCKS which alter a Logicswitch and SNAs which return the state of a Logicswitch.

Logicswitches are given the value of 0 when created, or when a CLEAR Command (without the OFF option) is sent to the Simulation Object. The program keeps no statistics about logic switches. However, it prints the numbers of switches that are set at the time the run ends.

**Related Block**

LOGIC Blocks set, reset, or invert the state of a Logicswitch Entity. In coding the LOGIC block, the letters S,R, or I indicating set, reset, or invert.

**Related SNA**

The SNA associated with Logicswitches is:

LS*Entnum* - Logicswitch set. LS*Entnum* returns 1 if Logicswitch is in the "set" state, 0 otherwise.

**Related Command**

INITIAL – Used to initialise a LogicSwitch.

*Example*

(*a*) In model of factory system, a logic switch can be used to represent that whether the machine is working order or not.

(*b*) In model of vehicular traffic system, a logic switch can be used to represent that whether the traffic light is ON or OFF.

## 13.9 SAVE LOCATIONS

The program does not continuously maintain current value, it computes the value of SNA's at the time they are needed. It is good to save values computed at one time and use it after some another time. This can be done by the use of block type called SAVEVALUE. The block indicates in field A the number of one of many save Value locations and, in field B, gives the SNA to be saved. A SaveValue Block changes the value of a Save Value Entity.

*Example:* SAVE VALUE A, B, C

**Operands**

**A:** Save value Entity number. must be required. May be followed by + or - to indicate addition or subtraction to existing value. (Required). The operand must be *Name*, *PosInteger*, *ParenthesizedExpression, SNA*, or *SNA*Parameter*, followed by +, –, or *Null*.

**B:** The value to be stored, added, or subtracted. (Required). The operand must be *Name, Number*, *String*, *ParenthesizedExpression, SNA*, or *SNA*Parameter*.

**C:** The C field of the savevalue block must carry one of the symbols XF, XH, XB, or XL to indicate a fullword, halfword, byte-sized, or floating-point type of savevalue. Example:

```
SAVEVALUE        10,S6,XF
```

The above block will save the current contents of storage number 6 in fullword savevalue number 10.

The current content of a savevalue location *n* is available as an SNA indicated by Xxn, where x denotes the type of savevalue. A control statement, called INITIAL can set the value of a savevalue at the beginning of the simulation so that savevalues can be used to introduce data and set initial condition. The value of initial conditions is put in field B and the notation Xxn is put in field A. The values of the non-zero savevalues are printed at the end of the simulation run.

## 13.10 USER CHAINS

A User chain Entity contains a special Transaction Chain, called a User Chain. Userchain can be manipulated by LINK and UNLINK Blocks. Userchains are useful for modeling Queuing algorithms and complex scheduling algorithms. They provide closer control of Transaction queuing than is available with entity Delay Chains. A "Link Indicator" is part of each Userchain Entity and, is useful for using a User Chain to control the queuing of Transactions on a resource.

Following SNAs are associated with Userchain Entities.

- CA Entnum: Average Userchain content. The time weighted average number of chained Transactions for Userchain Entity.
- CC Entnum: It counts all transactions that have been chained to the User Chain of Userchain Entity.

- CH *Entnum*: It defines the current number of Transactions chained at Userchain Entity.
- CM *Entnum*: It defines the maximum number of Transactions chained at Userchain Entity.
- CT *Entnum*: It defines the average duration of Transactions at Userchain Entity.

**Related Blocks of userchain:**

There are two GPSS Blocks used in Userchain Entities:

**1. LINK Blocks:** Link block is used to conditionally place a Transaction on a User Chain.

A LINK Block controls the placement of the Active Transaction on the User Chain of a Userchain Entity.

*Example:* LINK A,B,C

**A: Userchain number.** The Userchain Entity which may receive the entering Transaction. (Required). The operand must be *Name*, *PosInteger*, *Parenthesized Expression, SNA*, or *SNA*Parameter*.

**B: Chain ordering.** The placement of new Transactions on the Userchain. (Required). The operand must be LIFO, FIFO, *Parenthesized Expression, SNA*, or *SNA*Parameter*.

**C: Next Block location.** The destination Block for Transactions which find the Link Indicator of the Userchain in the off state (reset). (Optional). The operand must be *Null*, *Name*, *PosInteger*, *Parenthesized Expression, SNA* or *SNA*Parameter*.

**2. UNLINK Blocks:** It is used to remove Transactions from a User Chain.

*Example:* **UNLINK O A, B, C, D, E, F**

**Where:**

**O: Relational operator.** Relationship of D to E for removal to occur. These choices are explained below. (Optional). The operator must be *Null*, E, G, GE, L, LE or NE.

**A: User Chain number.** User Chain from which one or more Transactions will be removed. (Required). The operand must be *Name*, *PosInteger*, *ParenthesizedExpression, SNA*, or *SNA*Parameter*.

**B: Block number.** The destination Block for removed Transactions. (Required). The operand must be *Name*, *PosInteger*, *ParenthesizedExpression, SNA*, or *SNA*Parameter*.

**C: Removal limit.** The maximum number of Transactions to be removed. If not specified, ALL is used. (Optional). The operand must be ALL, *Null*, *Name*, *PosInteger*, *Parenthesized Expression, SNA*, or *SNA*Parameter*.

**D: Test value.** The member Transaction Parameter name or number to be tested, a Boolean variable to be tested, or BACK to remove from the tail of the chain. (Optional). The operand must be *Null*, *Name*, *PosInteger*, *ParenthesizedExpression, SNA*, *SNA*Parameter* or BACK.

**E: Reference value.** The value against which the D Operand is compared. (Optional). The operand must be *Null*, *Name*, *Number*, *String*, *ParenthesizedExpression, SNA*, or *SNA*Parameter*. Operand E is not used if Operand D is a Boolean Variable or BACK.

**F: Block number.** The alternate destination for the entering Transaction. (Optional). The operand must be *Null*, *Name*, *PosInteger*, *ParenthesizedExpression, SNA*, or *SNA*Parameter*.

*Example:* UNLINK OnHold,Reentry,1

This is the simplest way to use the UNLINK Block. The first Transaction at the head of the Userchain Entity named on Hold, if any, is taken off the chain and is directed to the Block labeled Reentry. It is put on the CEC behind Transactions of the same priority. The Transaction entering the UNLINK Block proceeds to the Next Sequential Block.

## 13.11 TABLE ENTITY

The Table entity is used to accumulate data for creating a histogram.

**Related SNAs:**

- TB: Average of the entries in the entity.
- TC: Number of entries in the entity.
- TD: Standard deviation of the entries in the entity.

**Related Block:**

- TABULATE: Use by a Transaction to add entries to a Table entity.

**Related Commands:**

- TABLE: Used to declare a Table entity.
- QTABLE: Declares a Table Entity that can be used to create a histogram of the times that Transactions spend in a Queue entity.

**Tabulate**

As we know that tabulate is a related block in table entity. A TABULATE Block triggers the collection of a data item in a Table Entity.

**TABULATE A,B**

**Where**

| A | B |
|---|---|
| • Table Entity name or number.<br>• Required.<br>• The operand must be *Name*, *PosInteger*, *ParenthesizedExpression, SNA*, *SNA*Parameter*. | • Weighting factor.<br>• Optional.<br>• The operand must be *Null, Name*, *Number, Parenthesized Expression, SNA*, or *SNA*Parameter*. |

*Example:* TABULATE Accounts

When a Transaction enters this TABULATE Block, the Table Entity named Accounts is found. Accounts must have been defined in a TABLE Command. Then the statistics associated with the table are updated with no weighting.

When a Transaction enters a TABULATE Block, Operand A is evaluated and used to find a Table Entity. If there is no such entity an Error Stop occurs.The Table Entity is then updated according to the operands in the TABLE Statement. If the B Operand is used, it is evaluated and used as a weighting factor. Otherwise the factor is taken to be 1.

**Some Special Restrictions**

1. A must be positive.
2. B, if specified, must be positive.
3. A must be the name or number of a predefined TABLE entity.

A Table Entity must be defined in a TABLE Command before it can be updated by a TABULATE Block. The TABLE Command must exist in the model, or must be sent to the Simulation Object interactively, before a Transaction can enter the TABULATE Block. Any attempt to do so before the Table Entity is defined, cases an Error Stop. A Table Entity can be redefined by an interactive TABLE Command.

## 13.12 FUNCTIONS

To introduce functional relationships in a model, a GPSS model can include a number of functions. Each function is defined by giving two or more pairs of numbers that relate a input $x$ to an output $y$. the function can be in a continuous or discrete mode.

In a continuous mode, the program will interpolate linearly between the defined points, allowing the user to approximate a continuous function by a series of straight line segments. In a discontinuous mode, the function is treated as a "stair case function". Fig. 13.3 illustrates these two modes of using functions.

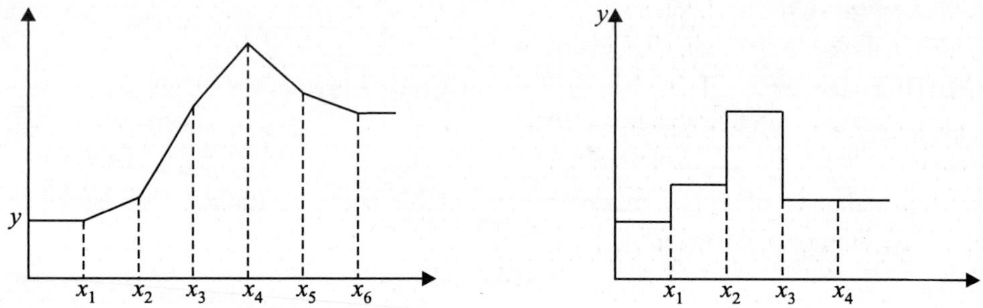


**Fig. 13.3.** Continuous and discrete functions.

If $x_i$ and $x_{i+1}$ are two successive points at which the function has been defined, an input value in the range $x_i < x \leq x_{i+1}$ will result in the value $y_i$ being produced.

Any number of points (> 1) can be used to define a function; and the size of the intervals between successive points can vary freely, except that a function defined to have a random number as input must have values of $x$ in the range 0 to 1. The values of $x$ and $y$ can be fractional and negative. A reference to a value of $x$ below the lowest defined value $x_1$ results in the value $y_1$. Similarly, reference beyond the highest defined value of $x$ produces the value of $y$ at that highest value.

## 13.13 SIMSCRIPT

In 1962, Harry Markowitz, Bernard Hausner, and Herb Karr at the RAND dorporation developed the SIMSCRIPT-I simulation programming language for the U.S. Air Force. Philip Kiviat, Richard Villanueva, and George Benedict later collaborated with Markowitz and Hausner to create SIMSCRIPT II, also at RAND. A commercial version known as SIMSCRIPT II.5 has been marketed and improved by CACI since the 1970s SIMCRIPT I language had "considerable impact" on the desion of SIMULA "through its list processing and time scheduling mechanisms" and its "random drawing and other utility routines". Hausner spent the summer of 1963 in Norway, which increased the communication between the SIMSCRIPT and SIMULA teams. The influence was reciprocal: Nygaard is given credit in the SIMSCRIPT II manual for his design suggestions, and in the mid-1970s, SIMSCRIPT II.5 incorporated SIMULA's "quasi-parallel" timeelapsing process. SIMSCRIPT II.5 has been widely used in the programming of large and small simulation models, at least in part because of language features that greatly facilitate simulation programming. The language includes support for discrete-event simulation, combined discrete-event/continuous simulation, interactive graphics and animation, database access, and an integrated development environment.

**Concept of SIMSCRIPT Program**

A system that is to be simulated considered as collection of entity, attributes that interact with activities. In SIMCRIPT, following main concepts are used:

- Entities (Permanent/ Temporary)
- Sets
- Event routines.

In describing the system, SIMSCRIPT uses the terms entities and attributes and it distinguishes between temporary and permanent entities and attributes. The **temporary entities** are created and destroyed during the execution of a simulation while **permanent entities** are those that remain during the run. Temporary entities form **sets.** The user can define sets, and facilities are provided for entering and removing entities into and from sets.

Activities are considered as extending over time with their beginning and their end being marked as events occurring instantaneously. Each type of is described by an **event routine, each** of which is given a name and programmed as a separate closed subroutines. **Event can be endogenous (internal event) and exogenous (external event).**

## 13.13.1 Organization of SIMSCRIPT Program

Event routines are called as closed routines. Some events are created for transferring control between them. The event notices when it is determined that an event is scheduled. Each event notice records the time of event when it is due to occur and event routine that is execute the event.

There are following points to be consider in which the simulation proceeds.

- The event notices are filed in chronological order.
- All events that can be executed at a particular time have been processed.
- The clock is updated to the time of the next event notice and control is passed to the event routine identified by the notice.
- Event notices do not usually go to more than one activity in the manner of a GPSS transaction.
- It is also possible to some and rescheduled an event notice.
- If the event executed by a routine results in another event, either at the current clock time or in the future, the routine must create a new event notice and file it with the other notice.
- In case of exogenous events, a series of exogenous event statements are created.
- All exogenous event statements are sorted into chronlogical order and they are read by the program when the time for the event is due.

## 13.13.2 Names and Labels

**Name**

Each entities and attributes are described and identified by its **name.** There are following rules to assign the name of entity and attribute:

- Names can be combination of letters and digits and there is at least one letter.
- Some names are reserved for system use.
- Use of periods (.) in name is permissible. By this compound names can be constructed.

**Labels**

Labels for identifying programming statements similarly consist of any combination of letters and numbers, without the restriction that at least one be a letter. They also may be made compound by using periods. Labels are identified by being enclosed between single quotation marks.

**SIMSCRIPT Syntax Rules**

- There are no statement continuation marks in Simscript.
- Statements begin with a key word.
- Key words are ***not*** reserved words.
- No variable name, key word or literal may be broken at the end of a line.
- Statements may extend over as many lines as necessary.
- There are no statement delimiters in Simscript.
- Comments begin with two apostrophes (")-- not a quote mark, and end either:
  1. at the end of the record
  2. or at the next pair of apostrophes (on the same line)
- System Variables

All predefined Variables have names ending in ".letter",, for example,

**pi.c** (constant) **hours. v** (variable)
**sqrt. f** (function) **time. r** (routine)

- All System-Generated Variables have names beginning with "letter."

*Examples*

**N.MACHINE L.BAKERY.QUEUE**

- User-Defined Variables

Any combination of letters, digits, periods, and underscored, which is not a number (Terminal periods are ignored)

*Example*

**AIRPLANE NO.OF.TERMINALS**

### 13.13.3 SIMSCRIPT Statements

SIMSCRIPT statements are written in a form closely resembling the English language. To enhance the similarity, there are many alternative terms or modes of expression which will be interpreted as equivalent statements.

**The Standard "IF" Statement**

Simple "either or" logic is supported by the **IF** statement

The general form is:

**If** *logical expression*
    *group of statements 1*
**Else**
    *group of statements 2*
**Always**

Sometimes it is convenient to simplify this to:

**If** *logical expression*
    *group of statements 1*
**Always**

**The "Select Case" Statement**

The ***SELECT CASE*** statement provides for choosing one among many.

The general form is:

**Select Case** *expression*
**Case** *constant list 1*
*group of statements 1*
**Case** *constant list n*
*group of statements n*
**Default**
*group of statements*
**Endselect**

**Simple Decision Logic**

Logical Expressions may be simple or compound. A simple logical expression is a single comparison, such as:

**A > B**

or **B**2 - 4* A* C is positive**

Compound Logical Expressions are just simple logical expressions connected by AND, OR and parentheses

**If PRODUCTION.RATE (MACHINE) > 0 and DUE.DATE(JOB) <= time. v**

OR

**If A < X <= B and (C = D or E = F)**

**Other Logical Control phrases**

**While/Until**

Standing alone, these phrases serve as alternative iteration phrases

*Example*

| | | |
|---|---|---|
| **While M < 10000** | | **Until M >= 10000** |
| **Do** | **OR** | **Do** |
| **Let M = M * X** | **equivalently,** | **Let M = M * X** |
| **Let X = X + 1** | | **Let X = X + 1** |
| **Loop** | | **Loop** |

**Termination Phrases**

***While*** and ***Until*** have an entirely different meaning when used to modify a ***For*** phrases.

*Example*

**For I = 1 to N, while SUM <= MAX**

**Do**

**........**

### 13.13.4 Defining Telephone System

Telephone system is the example of discrete on system, illustrated in Fig. 13.4.

The system has a number of telephone (only the first eight are shown), connected to a switchboard by lines. The switchboard has a number of links which can be used to connect any two lines, subject to the condition that only one connection at a time can be made to each line. It will be assumed that the system is a lost call system, that is, any call that cannot be connected at the time it arrives is immediately abandoned.

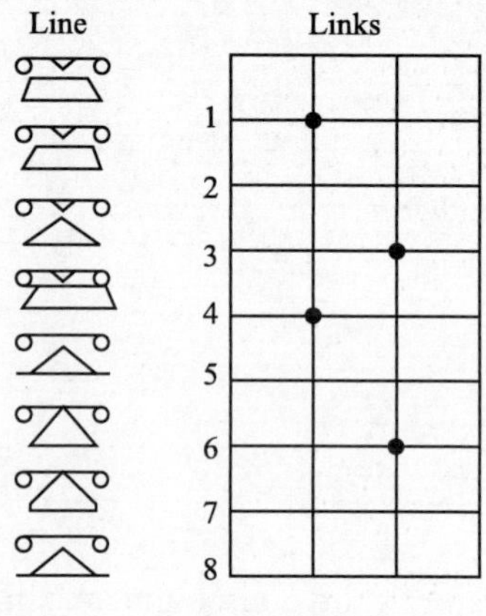


**Fig. 13.4** Simple telephone system.

A call may be lost because the called party is engaged, in which case the call is said to be a **busy call;** or it may be lost because no link is available, in that case it is called to be a **blocked call.** The object of the simulation will be to process a given number of calls and determine what proportion are successfully completed, blocked, or found to be busy calls.

According Fig. 13.4, line 1 is connected to line 3 and line 4 is connected to line 6.

To illustrate the use of SIMSCRIPT, the telephone system will be programmed as a SIMSCRIPT model. The telephone lines are obviously permanent entities, and represented by TLINE. The attribute will be called STATE. It is an integer variable and can takes only 0 or 1. Values 0 means free and 1 means busy state. Calls are to be represented by temporary entities, each with two attributes for carrying the origin and destination.

Three important section of the SIMSCRIPT model that define telephone system.

**1. Preamble**

A SIMSCRIPT program begins with preamble section that defines the system structure and establish the condition under which the simulation will be run.

The preamble for model 1 of the telephone system is shown in Fig. 13.5.

1. Telephone System-Model I
2. Preamble
3. Normally, Mode is Integer
4. Event Notices Include Arrival and Closing
5. Every Disconnect Has a Dis.call
6. Temporary Entities
7. Every Call Has an Origin and a Destination
8. Define Links.in.use, Max. Links, Blocked, Busy, Finished
9. And Stop. Time as Variables
10. Define Inter. Arrival. Time and Mean. Length as Real Variables
11. Permanent Entities
12. Every Tline has a State
13. Define Secs To Mean/60 Minutes
14. End

**Fig. 13.5.** Preamble for model I.

2. **A main program** where execution begins. Resources must be created and initialized before they can use by processes. This is usually accomplished in MAIN. see Fig. 13.6.
3. **A process/event** routine for each process/event notice declared in the preamble.

   Here we need to write routine for:

   Arrival Event

   Disconnect Event

   Closing Event

```
Main
Read N.Tline, max.Links, Inter arrival.time, mean.length and stop.time
Create every Tline
If n Tline < 2* man* links +2
Print 1 line thus
    too few lines specified, simulation abandoned
Stop
Else
Schedule an arrival now
Schedule a closing in stop.time minutes
Start simulation

"Report to be written at end of simulation"

Print 1 line thus
    simulation of telephone system = model
Skip 3 output lines
Print 5 lines with N.Tline, max.Links, linter.arrival.time,
mean.length, stop. time thus
    NUMBER OF LINES                 *
    NUMBER OF LINES                 *
    MEAN INTER = arrival time       *. * /seconds
    MEAN CALL LENGTH                * /seconds
    Simulated time                  * /minutes
SKIP 2 output lines
Print 3 lines with finished, busy, blocked thus
    Calls processed                 *
    Busy calls                      *
    Blocked calls                   *
stop
end
```

**Fig.13.6** Telephone system-model 1-Main routine.

## SUMMARY

- GPSS is a discrete event simulation language/package which operates not primarily by stepping through time, but by generating transactions (people to serve, jobs to process, etc.) and then progresses them through various stages (or BLOCKS to use the correct term).
- GPSS is designed around number of elementary abstractions called entities.
- GPSS is described as a block diagram in which the block represents the activities, and lines joining the blocks indicate the sequence in which the activities can be executed.
- System Numerical Attributes (SNAs) are simulation "state variables" that are available for use through a simulation. Parameters, attributes, and transactions are collectively called as system Numerical attributes.
- Moving through the system of blocks are entities called **transaction,** ***e.g.***, customers, messages, machine part, vehicles, etc.
- Transaction are created at GENERATE blocks and are removed for the simulation at TERMINATE blocks.
- A **Logicswitch** Entity is the simplest entity that is made available to represent two-state conditions in the system.
- A storage represents another block type that can be used by one or more transactions simultaneously, *e.g.* storages are groups of machines or a warehouse.
- SAVEVALUE block is use to save values computed at one time and use it after some another time.
- Userchains are useful for modeling Queuing algorithms and complex scheduling algorithms. User chain can be manipulated by LINK and UNLINK Blocks.
- A TABULATE Block triggers the collection of a data item in a Table Entity.
- SIMSCRIPT, considered as collection of entity attributes that interact with SIMSCRIPT. main concepts are; Entities, sets and Event routines.

## SOLVED QUESTIONS

**Q. 1.** Discuss the storage block of GPSS with example.

**Ans.** A similar result to the restricted resource of facility can be achieved using STORAGE. This is also useful for defining stocks or resources with limited availability and has a structure of STORAGE/ENTER/LEAVE. These set up the amount available in the start of the programme and then are used up and then possibly replaced as the simulation progresses.

```
-----------------------------------------------------------------------------------------------
tottime TABLE M1,30,30,20 ; setup table to capture total time in system
crew STORAGE 3 ; make number of crew
;*****************************************************************************
GENERATE 60; arrival of customer 1 per minute
ENTER crew; use a crew
ADVANCE 19,2; do something
```

LEAVE crew; let a crew go back into stock
TABULATE tottime; record time in the system
TERMINATE 1; customer leaves and
; termination count increases

---

If the storage is not released, then it is a way of setting a finite limit for the simulation, *e.g.* when the crew have all been used and not replaced, the simulation will stop.

**Q. 2.** Discuss the various characteristics of GPSS.

**Ans. Following Characteristics of GPSS**

1. GPSS was designed specially for analysts not for programmers.
2. GPSS is described as a block diagram in which the block represents the activities, and lines joining the blocks indicate the sequence in which the activities can be executed.
3. GPSS is a set of 48 specific block types. Each block has a name, and a specific task to perform.
4. Moving through the system of blocks are entities called **transaction, e.g**., customers, messages, machine part, vehicles, etc.
5. Transaction are created at GENERATE blocks and are removed for the simulation at TERMINATE blocks.
6. There can be many transactions simultaneously moving through the block diagram.
7. Blocks that can handle only one item at a time are called the **facilities**, *e.g.*, of a facility is a machine on a person.
8. A storage represents another block type that can be used by one or more transactions simultaneously, *e.g.*, storages are groups of machines or a warehouse.
9. GPSS handles the advancement of time by a block called ADVANCE when a transaction enters this block, an action time is computed and added to the current time to produce a block department time.
10. No language extension possible.
11. Restricted to simple queuing problems.
12. Machine efficiency is often poor.

**Q. 3.** Discuss various GPSS commands.

**Ans. GPSS Commands:**

GPSS commands can serve two purposes

- Declaring and Initialising Entities
- Controlling a simulation
  - GPSS World offers a GUI for controlling a simulation

**Commands: Declaring and Initialising Entities:**

- BVARIABLE: Declare Boolean Variable Entities
- VARIABLE or FVARIABLE: Used to declare Variable Entities
- FUNCTION: Used to declare a Function entity.
- MATRIX: Used to declare a Matrix Entity.

- QTABLE: Used to declare a Table entity
- TABLE: Used to declare a Table entity
- STORAGE: Used to declare and configure a Storage entity.
- EQU: Used to assign a value to a Name. Useful for assigning names to constants.
- INITIAL: Used to initialise a Matrix Entity, a SaveValue Entity, or a LogicSwitch Entity.
- RMULT: Used to set the seeds for the random number generators.

**Commands: Controlling a Simulation:**

- CLEAR: Completely re-initialises the simulation
- RESET: Resets the statistical accumulators, but leaves Transactions in the simulation.
- START: Command that starts a simulation.
- STOP: Used to set or remove a stop condition (break point) at a Block.
- HALT: Used to halt the execution of the simulation. The hot key Cntrl-Alt-H can be used to invoke HALT.
- STEP: Used to execute a specified number of Blocks.
- CONTINUE: Used to continue the simulation after hitting a stop condition.
- SHOW: Used to display the state of the simulation. Can be used to examine SNAs or calculate expressions.

**Q. 4.** What is storage entity in GPSS? Discuss its various blocks and related SNA.

**Ans. Storage Entity**

A Storage Entity is associated with a number of storage units which are allocated and returned by Transactions. Storage Entities can be used as "token pools" for controlling the flow of Transactions in the model.

When a Transaction ENTERs a Storage Entity, it utilizes or occupies one or more storage units of the Storage Entity. A Transaction is denied entry into an ENTER Block if its storage demand cannot be met. Such a Transaction comes to rest on the Delay Chain of the Storage Entity. Then it must wait until other Transactions free enough storage by entering LEAVE Blocks.

Storage capacity may be released by any Transaction, even if it had not previously ENTERed the Storage Entity. However, if more capacity is released than was declared in the STORAGE Command an Error Stop occurs.

**Storage Entity SNAs**

1. SNAs related to the Storage Entity state:
    - R: Provides the amount of storage units available within the Storage entity.
    - S: Provides the amount of storage units in use within the Storage entity.
    - SE: Returns 1 if the Storage entity is empty and 0 otherwise.
    - SF: Returns 1 if the Storage entity is full and 0 otherwise.
    - SV: Returns 1 if the Storage entity is available and 0 otherwise.
2. SNA related to statistics:
    - SA: Returns the time weighted average of storage capacity in use.
    - SC: Returns the total number of storage acquired.

- SR: Returns the fraction of total utilization as expressed in parts per thousand using a value between 0 and 1000.
- SM: Returns the maximum storage value used within the entity, i.e. the high water mark.

ST – Returns the average holding time per unit storage.

**Blocks and Commands Related to Storage Entity:**

- Blocks
  - ENTER: Enter and occupy space in the entity.
  - LEAVE: Release space and exit the entity.
  - SAVAIL: Make entity available.
  - SUNAVAIL: Make entity unavailable
- Commands
  - STORAGE: Must be used to declare the capacity of the Storage entity.

**Q. 5.** Discuss various arithmetic operators in GPSS.

**Ans.** Arithmetic operators in GPSS:

^ Exponentiation. A^B returns A raised to the power of B.

# (or *) Multiplication. A # B returns the numeric product of A and B.

/ Division. A / B returns the quotient of A divided by B.

\ Integer Division. A \ B returns the result of integer division of A by B.

@ Integer Remainder. A @ B returns the modulo division of A by B.

– Subtraction. A – B returns the difference of A and B.

+ Addition. A + B return the sum of A and B.

*Examples.*

```
VARX        VARIABLE    C1*5
QTY         VARIABLE    (RN1@96+5)
FINANCE     VARIABLE    (RN1@3+1)#50+200
CHKOUT      VARIABLE    (P$ITEMS)#20+P$PMNT
```

**Q. 6. Discuss the queue in GPSS.**

**Ans.** Queues are an integral part of the GPSS world and are a simple way of capturing information on the transaction processing times, particularly when awaiting facilities. Data on queue entities are captured automatically and can be examined at the end of a run in the report which is automatically generated, or interactively in the queues window (do the menu command: window > simulation window > queues window). Queues have two parts a transaction being placed on the queue and then removed:

*Example:*

```
QUEUE order,1; make a queue called order
SEIZE crew1; use a facility
ADVANCE 10; do something for 10 sec
DEPART order; record transaction leaving order queue
RELEASE crew1; give up the facility.
```

The queues window and report shows the size of the queue and the average time that transactions were in the queue.

Information on the progress of transactions can also be captured easily using the TABLE/TABULATE pairs seen earlier above, and also with SAVEVALUE blocks.

Similar results to QUEUE can be obtained using LINK/UNLINK, which allow different types of rules to be applied to the transactions to sort the order in which they are processed, such as FIFO-first in first out, LIFO-last in first out, by priority of a certain attribute.

## EXERCISES

1. Discuss the important GPSS block.
2. Why GPSS is so important in simulation.
3. How do you gather statistics in GPSS.
4. Discuss SIMSCRIPT syntax Rule.
5. Discuss the purpose of TABULATE block.

■■

14

# Estimation and Output Analysis of Simulation

**INSIDE THIS CHAPTER**

**14.1** Introduction; **14.2** Confidence Intervals; **14.3** Standard Error; **14.4** Interpretation of Confidence Interval of a Parameter: Example; **14.5** Output Analysis of Simulation

## 14.1 INTRODUCTION

In statistics, estimation refers to the process by which one makes inferences about a population, based on information obtained from a sample.

There are two types of estimates we will find:

(*i*) Point estimates (*ii*) Interval estimates

(*i*) **Point estimate:** A point estimate of a population parameter is a single value of a statistic. For example, the sample mean $x$ is a point estimate of the population mean $\mu$. Similarly, the sample proportion $p$ is a point estimate of the population proportion $P$.

(*ii*) **Interval estimate:** An interval estimate is defined by two numbers, between which a population parameter is said to lie. For example, $a < x < b$ is an interval estimate of the population mean $\mu$. It indicates that the population mean is greater than $a$ but less than $b$. It provides a confidence level for the estimate. Such interval estimates are called **confidence intervals.** It involves a range and a probability.

**Confidence Coefficient:** The probability that the interval contains the parameter that the interval contains the parameter, *e.g.*, 0.90, 0.95, 0.99.

**Error Probability ($\alpha$):** The probability that an interval does not contain the parameter

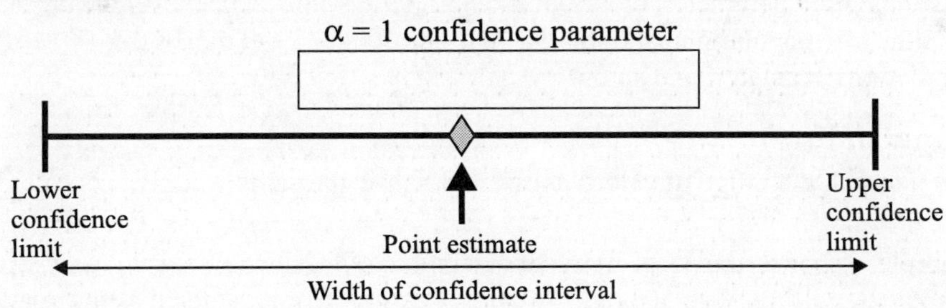

A good estimator must satisfy three conditions:

- *Unbiased:* The expected value of the estimator must be equal to the mean of the parameter.
- *Consistent:* The value of the estimator approaches the value of the parameter as the sample size increases.
- *Relatively Efficient:* The estimator has the smallest variance of all estimators which could be used.

### Sampling Method

A sampling method is a procedure for selecting sample members from a population. Three common sampling methods are: *simple random sampling, stratified sampling*, and *cluster sampling.*

## 14.2 CONFIDENCE INTERVALS

Statisticians use a confidence interval to express the precision and uncertainty associated with a particular *sampling method.* A confidence interval consists of three parts.

- A confidence level
- A statistic
- A margin of error

The confidence level describes the uncertainty of a sampling method. The statistic and the margin of error define an interval estimate that describes the precision of the method. The interval estimate of a confidence interval is defined by the *sample statistic ± margin of error*.

For example, suppose we compute an interval estimate of a population parameter. We might describe this interval estimate as a 95% confidence interval. This means that if we used the same sampling method to select different samples and compute different interval estimates, the true population parameter would fall within a range defined by the *sample statistic ± margin of error* 95% of the time.

Confidence intervals are preferred to point estimates, because confidence intervals indicate: (*a*) the precision of the estimate, and (*b*) the uncertainty of the estimate.

### 14.2.1 Confidence Level

The probability part of a confidence interval is called a **confidence level.** The confidence level describes the likelihood that a particular sampling method will produce a confidence interval that includes the true population parameter.

Here is how to interpret a confidence level. Suppose we collected all possible samples from a given population, and computed confidence intervals for each sample. Some confidence intervals would include the true population parameter; others would not. A 95% confidence level means that 95% of the intervals contain the true population parameter; a 90% confidence level means that 90% of the intervals contain the population parameter; and so on.

### 14.2.2 Margin of Error

In a confidence interval, the range of values above and below the sample statistic is called the **margin of error**.

For example, suppose the local newspaper conducts an election survey and reports that the independent candidate will receive 30% of the vote. The newspaper states that the survey had a 5% margin of error and a confidence level of 95%. These findings result in the following confidence interval: We are 95% confident that the independent candidate will receive between 25% and 35% of the vote.

**Note:** Many public opinion surveys report interval estimates, but not confidence intervals. They provide the margin of error, but not the confidence level. To clearly interpret survey results you need to know both! We are much more likely to accept survey findings if the confidence level is high (say, 95%) than if it is low (say, 50%).

### How to Compute the Margin of Error

The margin of error can be defined by either of the following equations:

Margin of error = Critical value × Standard deviation of the statistic ...(14.1)

Margin of error = Critical value × Standard error of the statistic ...(14.2)

Equation (14.1) is used when you know the standard deviation of the statistic to compute the margin of error. And the second equation is used when standard error of the statistics. Hence in next section we described how to compute the critical value, standard deviation and standard error.

### How to Find the Critical Value

The critical value is a factor used to compute the margin of error. This section describes how to find the critical value, when the *sampling distribution* of the statistic is *normal* or nearly normal.

The *central limit theorem* states that the sampling distribution of a statistic will be normal or nearly normal, if any of the following conditions apply.

- The population distribution is normal.
- The sampling distribution is *symmetric, unimodal*, without *outliers*, and the sample size is 15 or less.
- The sampling distribution is moderately *skewed*, unimodal, without outliers, and the sample size is between 16 and 40.
- The sample size is greater than 40, without outliers.

When one of these conditions is satisfied, the critical value can be expressed as a *t score* or as a *z score*. To find the critical value, follow these steps:

(*i*) Compute alpha ($\alpha$): $\alpha = 1 - (\text{confidence level}/100)$

(*ii*) Find the critical probability ($p^*$): $p^* = 1 - \alpha/2$

(*iii*) To express the critical value as a *z* score, find the *z* score having a *cumulative probability* equal to the critical probability ($p^*$).

(*iv*) To express the critical value as a *t* score, follow these steps.

➢ Find the *degrees of freedom* (DF). When estimating a mean score or a proportion from a single sample, DF is equal to the sample size minus one. For other applications, the degrees of freedom may be calculated differently. We will describe those computations as they come up.

➢ The critical *t* score ($t^*$) is the *t* score having degrees of freedom equal to DF and a *cumulative probability* equal to the critical probability ($p^*$).

Should you express the critical value as a *t* score or as a *z* score? There are several ways to answer this question. As a practical matter, when the sample size is large (greater than 40), it doesn't make much difference. Both approaches yield similar results. Strictly speaking, when the population standard deviation is unknown or when the sample size is small, the *t* score is preferred. Nevertheless, many introductory statistics texts use the *z* score exclusively.

You can use the *normal distribution calculator* to find the critical *z* score, and the *t distribution Calculator* to find the critical *t* score. You can also use a *graphing calculator* or standard statistical table (found in the appendix of most introductory statistics texts.)

## 14.3 STANDARD ERROR

The standard error is an estimate of the standard deviation of a *statistic*. This lesson shows how to compute the standard error, based on sample data.

The standard error is important because it is used to compute other measures, like *confidence intervals* and *margins of error.*

### Notation

The following notation is helpful, when we talk about the standard deviation and the standard error.

| Population parameter | Sample statistic |
|---|---|
| $N$: Number of observations in the population | $n$: Number of observations in the sample |
| $Ni$: Number of observations in population $i$ | $n$: Number of observations in sample $i$ |
| $P$: Proportion of successes in population | $p$: Proportion of successes is sample |
| $P_i$: Proportion of successes in population $i$ | $p_i$: Proportion of success in sample $i$ |
| $\mu$: Population mean | $\bar{x}$ : Sample estimate of population mean |
| $\mu_i$: Mean of population $i$ | $\bar{x}_i$ : Sample estimate of $\mu_i$ |
| $\sigma$: Population standard deviation | $s$: Sample estimate of $\sigma$ |
| $\sigma_p$: Standard deviation of $p$ | $SE_p$: Standard error of $p$ |
| $\sigma_{\bar{x}}$: Standard deviation of $x$ | $SE_{\bar{x}}$: Standard error of $\bar{x}$ |

### 14.3.1 Standard Deviation of Sample Estimates

Statisticians use sample statistics to estimate population parameters. Naturally, the value of a statistic may vary from one sample to the next.

The variability of a statistic is measured by its standard deviation. The table below shows formulas for computing the standard deviation of statistics from *simple random samples*. These formulas are valid when the population size is much larger (at least 10 times larger) than the sample size.

| *Statistic* | *Standard Deviation* |
|---|---|
| Sample mean, $\bar{x}$ | $\sigma_{\bar{x}} = \sigma/\text{sqrt}(n)$ |
| Sample proportion, $p$ | $\sigma_p = \text{sqrt}\ [P(1 - P)/n\ ]$ |
| Difference between means, $\bar{x}_1 - \bar{x}_2$ | $\sigma_{\bar{x}_1 - \bar{x}_2} = \text{sqrt}\ [\sigma_1^2\ /n_1 + \sigma_2^2/n_2]$ |
| Difference between proportions, $\bar{p}_1 - \bar{p}_2$ | $\sigma_{p_1 - p_2} = \text{sqrt}[\ P_1(1 - P_1)/n_1 + P_2(1 - P_2)/n_2]$ |

**Note:** In order to compute the standard deviation of a sample statistic, you must know the value of one or more population parameters.

### 14.3.2 Standard Error of Sample Estimates

Sadly, the values of population parameters are often unknown, making it impossible to compute the standard deviation of a statistic. When this occurs, use the standard error.

The standard error is computed from known sample statistics, and it provides an unbiased estimate of the standard deviation of the statistic. The table below shows how to compute the standard error for simple random samples, assuming the population size is at least 10 times larger than the sample size.

| *Statistic* | *Standard Error* |
|---|---|
| Sample mean, $\bar{x}$ | $SE_{\bar{x}} = s/\text{sqrt}(n)$ |
| Sample proportion, $p$ | $SE_p = \text{sqrt}\,[p(1-p)/n]$ |
| Difference between means, $x_1 - x_2$ | $SE_{\bar{x}_1-\bar{x}_2} = \text{sqrt}\,[s_1^2/n_1 + s_2^2/n_2]$ |
| Difference between proportions, $p_1 - p_2$ | $SE_{p_1-p_2} = \text{sqrt}\,[p_1(1-p_1)/n_1 + p_2(1-p_2)/n_2]$ |

The equations for the standard error are identical to the equations for the standard deviation, except for one thing—the standard error equations use statistics where the standard deviation equations use parameters. Specifically, the standard error equations use $p$ in place of $P$, and $s$ in place of $\sigma$.

**Example 14.1:** *Nine hundred (900) high school freshmen were randomly selected for a national survey. Among survey participants, the mean grade-point average (GPA) was 2.7, and the standard deviation was 0.4. What is the margin of error, assuming a 95% confidence level?*

**Solution:** To compute the margin of error, we need to find the critical value and the standard error of the mean. To find the critical value, we take the following steps:

- Compute alpha ($\alpha$): $\alpha = 1 - (\text{confidence level}/100) = 1 - 0.95 = 0.05$
- Find the critical probability ($p^*$): $p^* - 1 - \alpha/2 - 1 - 0.05/2 - 0.975$
- Find the critical $z$ score. Since the sample size is large, the sampling distribution will be roughly normal in shape. Therefore, we can express the critical value as a $z$ score. For this problem, it will be the $z$ score having a cumulative probability equal to 0.975. Using the *Normal Distribution Calculator*, we find that the critical value is 1.96.

Next, we find the standard error of the mean, using the following equation:

$$SE_{\bar{x}} = s/\ \text{sqrt}(n) = 0.4/\text{sqrt}(900) = 0.4/30 = 0.013$$

And finally, we compute the margin of error (ME).

$$ME = \text{Critical value} \times \text{Standard error} = 1.96 * 0.013 = 0.025$$

This means we can be 95% confident that the mean grade point average in the population is 2.7 plus or minus 0.025, since the margin of error is 0.025.

## 14.4 INTERPRETATION OF CONFIDENCE INTERVAL OF A PARAMETER: EXAMPLE

### Parameter and Statistics

The mean $\mu$ and standard deviation $\sigma$ of a population random variable $X$ are called **parameter**; and the mean $\bar{x}$ and standard deviation $s$ of a random sample are called **statistics**. Statistics are used to estimate parameter.

### 14.4.1 Confidence Interval [For a Single Population]

In addition to point estimates of a parameter there are interval estimates which stipulate, with a certain degree of confidence that the parameter lies between two values of the estimating statistics. Suppose a population random variable $X$ has mean $\mu$ whose value is unknown. From a random sample of size $n$, the value $\bar{x}$ of the sample mean $\bar{X}$ can be used to estimate $\mu$ at 95% confidence level as follows:

First, determine the value $ME$ for which

$$P(\mu - ME \leq \bar{X} \leq \mu + ME) = 0.95$$

which is equivalent to the equation

$$P(\bar{X} - ME \leq \mu \leq \bar{X} + ME) = 0.95$$

where $[\bar{X} + ME, \bar{X} + ME]$ is called a random 95% confidence interval for μ and *ME* is called the Margin of Error.

This means that the probability is 0.95 that a random sample will result in a value $\bar{x}$ of $\bar{X}$ for which the numerical interval $[\bar{x} - ME, \bar{x} + ME]$ contains μ.

### 14.4.2. Confidence Interval Estimation of Population Mean μ, when σ is Known (*z*-interval)

There are following assumption to estimate population mean μ, when σ is known:

1. Population standard deviation σ is known
2. Population is normally distributed
3. If population is not normal, use large sample

Confidence interval estimate:

$$\mu = \bar{X} \pm Z\frac{\sigma}{\sqrt{n}}$$

where *z* is the normal distribution's critical value for a probability of α/2 is each tail.

Consider a 95% confidence interval:

$$1-\alpha = 0.95 \quad \alpha = 0.05 \quad \alpha/2 = 0.025$$

The condition that need to be net for this procedure are:

1. The sample is a simple random variable
2. The population is normal or $n \geq 30$

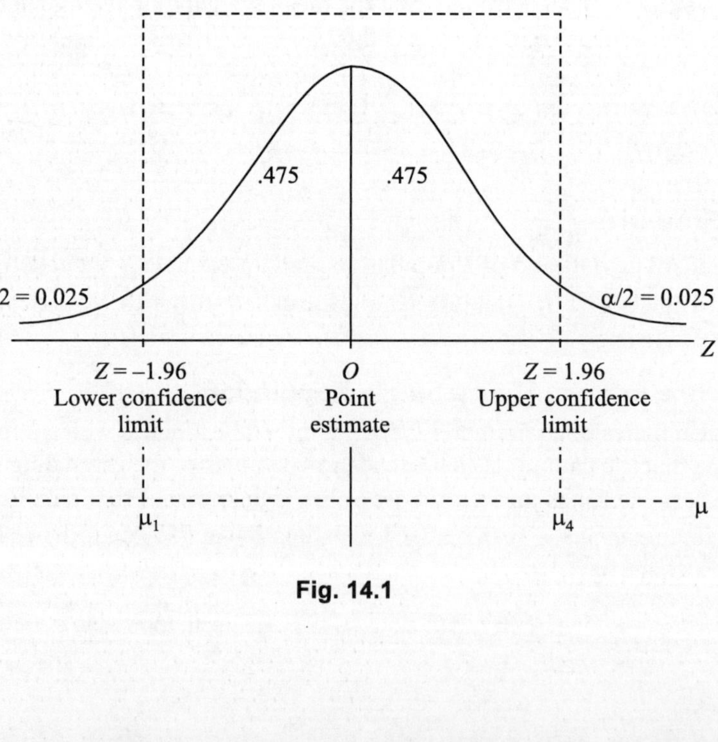


Fig. 14.1

### 14.4.3. Confidence Interval Estimation of Population Means μ, with σ is Unknown (*t*-interval)

The general formula will use the sample mean $\bar{X}$, the sample standard deviation ($s$), the sample size ($n$), and degree of freedom ($n - 1$).

$$\mu = \bar{X} \pm (t^*)\left(\frac{S}{\sqrt{n}}\right)$$

The condition that need to be net for this procedure are:

1. The sample is a simple random variable
2. The population is approximately normal (graphical support required) or $n \geq u_0$.
3. σ is unknown.

**Example 14.2:** *Suppose X is a normal random variable with mean μ, which is unknown, and standard deviation σ, which is known to be 2. A random sample (with replacement) of 25 values of X results in a sample mean $\bar{x} = 10$. Determine the margin of error ME for a 95% confidence interval for μ and find the confidence interval. Give an interpretation of the result.*

**Solution:** To determine *ME*, we first convert to standard units. The random variable

$$Z = \frac{\bar{X} - \mu}{\sigma/\sqrt{n}}$$

has mean 0 and standard deviation 1. Since *X* is normal. So is *Z*. Therefore, *Z* is a standard normal random variable.

Now $$\frac{\sigma}{\sqrt{n}} = \frac{2}{5} = 0.4 \quad \text{so} \quad Z = \frac{\bar{X} - \mu}{0.4}$$

The margin of error ME satisfies the equation

$$P(\mu - ME \leq \bar{X} \leq \mu + ME) = 0.95$$

equivalent to

$$P\left(-\frac{ME}{0.4} \leq \frac{\bar{X} - \mu}{0.4} \leq \frac{ME}{0.4}\right) = 0.95,$$

or $$P\left(-\frac{ME}{0.4} \leq Z \leq \frac{ME}{0.4}\right) = 0.95$$

From standard table (Table of standard normal distribution)

We get $P(-1.96 \leq Z \leq 1.96) = 0.95$

or $$P(0 \leq Z \leq 1.96) = \frac{0.95}{2} = 0.475$$

That, is 1.96 is the critical value of *Z* corresponding to probability 0.95. Therefore,

$$\frac{ME}{0.4} = 1.96 \quad \text{or} \quad ME = 0.4 \times 1.96 = 0.784$$

The corresponding 95% confidence interval is

$$[\bar{x} - ME, \bar{x} + ME] = [10 - 0.784, 10 + 0.784] = [9.216, 10.784]$$

We are 95% confidence that the mean μ of *X* is some value in this interval, which means that, as $\bar{x}$ ranges through all possible values of $\bar{X}$, 95% of all the intervals $[\bar{x} - 0.784, \bar{x} + 0.784]$ will contain μ.

Although different random samples of size 25 can give different values of $\overline{x}$, the value of ME is the same for each sample.

The probability theory for deriving confidence intervals for μ requires that the sample mean $\overline{X}$ be normally distributed. For large samples, the central limit theorem is use to conclude that $\overline{X}$ is approximately distributed, which then implies that $\overline{X}$ is also approximately normal. In applications, $X$ will always be a random variable defined on a finite population, so that $X$ and therefore $\overline{X}$ can never be exactly normal. Hence, the confidence intervals obtained in applications are only approximation. The theory provides a model for dealing with real populations, but our conclusions will be valid only to the extent that the real $\overline{X}$ approximates a normal distribution.

## 14.5 OUTPUT ANALYSIS OF SIMULATION

A model can derive by establishing the structure and supplying the data. The data provide the values to the attributes. These attributes can be called as variable that is used to define relationships involved in the activities. Activities can be stochastic or deterministic. In case of stochastic activities, activities can be Represented in form of stochastic variables. Hence we need many observations of the variable's value must be made, order to make a statistical estimate of its true value. Following objective of developing new statistical methodology:

(*a*) Statistical estimates should be consistent, meaning that, as the sample size increases, the estimates tends to the true value

(*b*) Should be control bias in measures of both mean values and variances.

(*c*) Develop sequential testing methods that will allow automatic controls to determine how long a simulation should be run in order to obtain a given level of confidence.

Here we describe some of methods being used to analyze simulation results.

### 14.5.1 Estimation Methods

Some statistical methods are commonly used to estimate parameters from observations on random variables. A random variable is drawn from an infinite population that has a stationary probability distribution with a finite mean, μ, and finite variance, $\sigma^2$. This means that the population distribution is not affected by the number of samples already made, nor does it change with time. If, further, the value of one sample is not affected in any way by the value of any other sample, the random variables are mutually independent. Random variables that meet all these conditions are said to be independently and identically distributed, usually abbreviated to i.i.d. Under broad conditions that can be expected to hold for simulation data, the central limit theorem can be applied to i.i.d. data. The theorem states that the sum of $n$ i.i.d. variables, drawn from a population that has a mean of μ and a variance of $\sigma^2$, is approximately distributed as a normal variable with a mean of $n\mu$ and a variance of $n\sigma^2$.

As we know that, any normal distribution can be transformed into a standard normal distribution, that has a mean of 0 and a variance of 1.

$a_i$ ($i = 1, 2, ..., n$) be the $n$ i.i.d. random variables. By using the central limit theorem, and applying the transformation, gives the following (approximate) normal variate.

$$Z = \frac{\sum_{i=1}^{n} a_i - n\mu}{\sqrt{n}\,\sigma} \qquad ...(i)$$

Dividing top and bottom of this equation (*i*) by *n*.

Let $$\bar{a} = \frac{1}{n}\sum_{i=1}^{n} a_i \qquad ...(ii)$$

$$z = \frac{\sum_{i=1}^{n} \frac{a_i}{n} - n\mu/n}{\sqrt{n}\,\sigma/n}$$

$$z = \frac{\sum_{i=1}^{n} \frac{a_i}{n} - \mu}{\sigma/\sqrt{n}}$$

By equation (*ii*)

$$z = \frac{\bar{a} - \mu}{\sigma/\sqrt{n}}$$

The variable $\bar{a}$ is the sample mean and the sample mean is the sum of random variable and it is itself a random variable. So confidence interval about its computed value, needs to be established.

The probability density function (*Pdf*) of the standard normal variate is given in Fig. 14.2. The integral from $-\infty$ to a value $u$ is the probability that $z$ is less or equal to $u$. The integral is generally represented by $\phi\,(u)$ and tables of its value are available. Let the value of $u$ is chosen so that $\phi(u) = 1 - \alpha/2$, where $\alpha$ is some constant less than 1, and denote this value of $u$ by $u_{m/2}$. The probability that $z$ is greater than $u_{\alpha/2}$ is then $\alpha/2$. The probability that $z$ is less than $-u_{\alpha/2}$ is also $\alpha/2$.

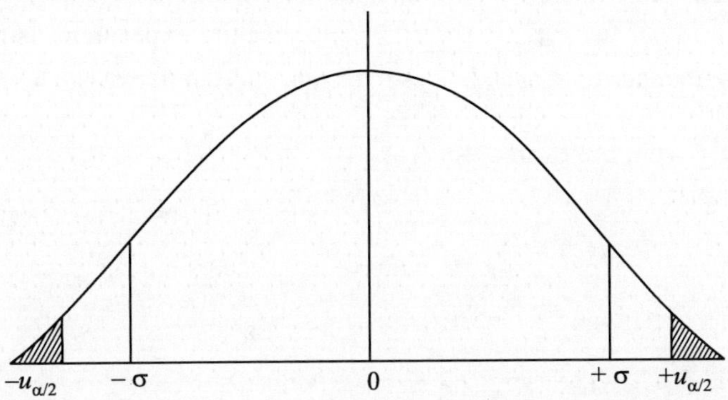


**Fig. 14.2.** Probability density function of standard normal variate.

The probability that $z$ lies between $-u_{\alpha/2}\ u_{\alpha/2}$ is $1 - \alpha$. We can say,

$$\text{probability } [-u_{\alpha/2} \le z \le u_{\alpha/2}] = 1 - \alpha$$

In terms of the sample mean, this probability statement can be written

$$\text{probability}\left[\bar{a} + \frac{\sigma}{\sqrt{n}}u_{a/2} \ge \mu \ge \bar{a} - \frac{\sigma}{\sqrt{n}}u_{a/2}\right] = 1 - \alpha$$

The constant $(1 - \alpha)$ is the confidence level and the interval

$$\bar{a} \pm \frac{\sigma}{\sqrt{n}} u_{\alpha/2}$$

is called as confidence interval, generally confidence level is represented in percentage. Typically, the confidence level might be 90%, in which case $u_{\alpha/2}$ is 1.65.

The $\mu$ will be covered by the confidence interval $\bar{a} \pm 1.65\, \sigma / \sqrt{n}$ with probability 0.9 or 90%.

In practice, the population variance $\sigma^2$ is not usually known; in which case it is replaced by an estimate calculated from the formula

$$s^2 = \frac{1}{n-1} \sum_{i=1}^{n} (a_i - \bar{a})^2$$

The normalized random variable based on $\sigma^2$ is replaced by a normalized random variable based on $s^2$.

Expressed in terms of the estimated variance, $s^2$, the confidence interval for $\bar{a}$ is defined by

$$\bar{a} \pm \frac{s}{\sqrt{n}} t_{n-1}, \alpha/2$$

This has a student-$t$ distribution, with $n - 1$ degrees of freedom. The quantity $u_{\alpha/2}$ used in the definition of a confidence interval given above, is replaced by a similar quantity, $t_{n-1}$, $\alpha/2$, based on the student-$t$ distribution, for which tables are also available.

### 14.5.2 Replication of Runs

Repeating the experiment with different random number for the same sample size $n$ gives a set of independent determinations of the sample mean $\bar{x}$. Suppose the experiment is repeated $p$ times with independent random number series. Let $x_{ij}$ be the $i^{th}$ observation in the $j^{th}$ run be denoted by $\bar{x}_j(n)$ and $S_j^2$ $(n)$, respectivel. For $j^{th}$ run, the estimates are:

$$\bar{x}_j(n) = \frac{1}{n} \sum_{i=1}^{n} x_{ij} \qquad \text{...(1)}$$

$$S_j^2\ (n) = \frac{1}{n-1} \sum_{i=1}^{n} \left[ x_{ij} - \bar{x}_j(n) \right]^2$$

Combine the results of $p$ independent measurements gives the following estimates for the mean waiting time, $\bar{x}$, and variance $S^2$ of the population.

$$\bar{x} = \frac{1}{p} \sum_{j=1}^{p} \bar{x}_j(n) \qquad \text{...(2)}$$

$$S^2 = \frac{1}{p} \sum_{j=1}^{p} S_j^2(n) \qquad \text{...(3)}$$

The value of $\overline{x}$ is an estimate for the mean waiting time, and, $S^2$ can be used to establish a confidence interval, according to the expression of equation (2) in chapter–5, sec. 5.2 with ($p$–1) degree of freedom.

- Increasing the number of repetitions decreases the size of the confidence interval, since the variance estimate is approximately inversely proportional to $p$. But this is a desirable effect, since it reduces the range of uncertainty about estimate of the mean.
- So assumes the estimate of the mean is unbiased.
- It is preferable to keep the number of repetition as low as possible, bearing in mind the need to approximate a normal distribution with the sample mean.
- In this method there is tradeoff between number of repetition and the run length.

### 14.5.3 Batch Means

- In the method of batch means the model is run only once but for a considerable period. It uses a single long run, preferably with the initial bias removed.
- The run is divided into a number segments to separate the measurements into batches of equal size.
- A complete run consists of $N$ observations that are broken into p batches of size n and N is exactly divisible by $p$.
- The experiment is equivalent to repeating an experiment of length n of $p$ times, with the final state of one run becoming the initial condition for the next.
- If representing ith observation in the jth batch by $X_{ij}$. Equation (1 to 3) can be used to estimate the mean and standard deviation of the variable being measured. And equation (2) (In chapter–5, see 5.2) can then be used to establish a confidence interval, with ($p$-1) degree of freedom.
- The connection between the batches introduces correlation. The effect of autocorrelation is that the value of one piece of data affects the value of following data.
- By making the batch sizes multiples of each other, it is possible to perform the operation in single run.
- This method is unreliable since the sub-periods are clearly not independent.
- However it has the advantage that only one set of observations needs to be discarded to overcome the warm-up effects in steady state analysis.
- In batch method there is tradeoff between batch size and number of batches.

### 14.5.4 Regenerative Techniques

It is sometimes possible within the run of a simulation model to identify points in the trajectory where the model returns to exactly equivalent states. These are called regeneration points. Between successive regeneration points, the system is said to execute a tour. The time the system spends in a tour is called an epoch. Periods between regeneration points are genuinely independent sub-runs. Examples of regenerative processes:

1. **G/G/1 queue:**

   In the G=G=1 queue the moments that a customer arrives at an empty queue are regeneration points.

2. **Two-machine production line:**

   In the two-machine production line, there are two types of regeneration points. The first type of regeneration points are the moments that machine 1 is repaired and the buffer is empty. The second type of regeneration points are the moments that machine 1 breaks down and the buffer is completely filled.

3. **Periodic review inventory system:**

   Consider a periodic review inventory system with independent, identicallydistributed demands per period and instantaneous delivery of replenishmentorders. Demand that cannot be satisfied directly from stock is lost. The stock levelis controlled by an (*s*; *S*) policy. For this system, the regeneration points are themoments just after an arrival of a replenishment order.

   - Notice that, unlike the previous two methods, the method of regeneration does not require any transient observations to be removed.
   - Unfortunately not all models have easily defined regeneration states, and even when they exist they can be computationally expensive to identify.
   - Another disadvantage is that it is not possible to define the length of a simulation run beforehand.

   However, research suggests that this method gives the most accurate results and so it is the method of choice when it is feasible.

### 14.5.5 Time Series Analysis

Time series analysis accounts for the fact that data points taken over time may have an internal structure (such as autocorrelation, trend or seasonal variation) that should be accounted for. Previous discussed topics such as: replication of runs, batch means, and the regenerative methods are based on making estimates of the variance of the sample mean. While time series analysis is based on autocorrelation.

An autocorrelation is the correlation between the target variable and lag values for the same variable. Correlation values range from –1 to +1. A value of +1 indicates that the two variables move together perfectly; a value of –1 indicates that they move in opposite directions. When building a time series model, it is important to include lag values that have large, positive autocorrelation values. Sometimes it is also useful to include those that have large negative autocorrelations.

Autocorrelation is measured by a series of autocovariances that show the extent to which values separated by s time units affect each other. The autocovariance between values of a variable made at two different times, $X_t$ and $X_u$, is defined as

$$R_{tu} = E\,[X_t - \mu\,)\,(X_u - \mu)]$$

Assume μ is independent of the independent of the indices *t* and *u*, and that $R_{tu}$ depends only with the separation ($s = t–u$) is said to be covariance stationary. The autocovariance has the property of symmetry that $R_{tu} = R_{ut}$.

- The autocovaiance usually represented by $R_s$, and property of symmetry says that $R_s = R_{-s}$ and s take all the value 0, ± 1 , ± 2, …, but if $s = 0$ is actually the definition of the variance of $X_t$.
- The autocovariance are estimated from the equations at T uniform intervals of time.

$$R_s = \frac{1}{T}\sum_{t=1}^{T-s}(X_t - \bar{X})(X_{t+s} - \bar{X}) \quad (s = 0, 1, \ldots, T–1) \qquad \ldots(6)$$

$$\bar{X} = \sum_{t=1}^{T} X_t$$

- The estimation of an unbiased sample means, $V(\bar{X})$.

$$V(\bar{X}) = \frac{T}{T-k}\{R_0 + 2\sum_{s=1}^{k-1}(1 - s/T)R_s\} \quad ...(7)$$

The estimate is taken to have 3 $k/2n$ degrees of freedom. Only the first $k - 1$ autocovariance estimates are involved.

- The value used for k should be large enough to include the autocovariances of significant size, but, as s increases, the accuracy of the autocovariance estimates decreases.

**Limitations:**

- This method involves a considerable amount of calculation.
- It required that all the data be simultaneously available.

**Time Series Analysis is used for many applications such as:**

- Economic Forecasting
- Sales Forecasting
- Budgetary Analysis
- Stock Market Analysis
- Yield Projections
- Process and Quality Control
- Inventory Studies
- Workload Projections
- Utility Studies
- Census Analysis

### 14.5.6 Spectral Analysis

The definition the spectral methods used to establish confidence intervals to analyze data generated by simulation experiments. Spectral analysis considers data arranged in a series according to historical time. When one so plots the movement of many economic variables through time, he often observes a remarkable degree of smoothness in the curve; that is, although a variable assumes values that are at different times relatively high and low, its path, nevertheless, progresses evenly enough to suggest that the current value of that variable is related to its past values.

In short, the data of such a time series appear to be autocorrelated. It is essentially the quantification and evaluation of this autocorrelation at which spectral analysis is aimed, after the data have been transformed into the frequency domain. The application of spectral analysis to a time series (whether it be actual or simulated) yields two types of information:

(*i*) The magnitudes of deviations from the average level of a given activity.

(*ii*) The period or length of these deviations, both of which require the transformation of the time series into the frequency domain. To obtain this information from a time series we make use of the following notation. Denote a particular stochastic generating process or ensemble by $\{X_t, t \in T\}$, from which a "sample" time series $\{x_t, t = 1, 2, ..., n\}$ is taken. It is important to note that $\{X_t\}$ will indicate the manner in which $\{x_t\}$ is formed for all t; but, due to its stochastic nature, $\{X_t\}$ cannot determine exactly the value of the series at any particular $t$. By studying the series one attempts to approximate the structure of the generating process.

A spectral analysis can provide more information than is contained in the estimate of a mean value. Comparing two systems on the basis of mean values of such factors as waiting time or queue length. Two systems may show no significant difference in their mean values, but their transient behavior may be significantly different. So spectral analysis can distinguish the difference by showing whether the spectrum emphasizes low or high frequencies.

Spectral analysis has been use to test a simulation model by comparing the spectral pattern of the simulation output against the corresponding pattern derived from the system itself.

### 14.5.7 Autoregressive Process

Autoregressive process. Most time series consist of elements that are serially dependent in the sense that you can estimate a coefficient or a set of coefficients that describe consecutive elements of the series from specific, time-lagged (previous) elements. This can be summarized in the equation:

$$X_t = \xi + \phi_1 * X_{(t-1)} + \phi_2 * X_{(t-2)} + \phi_3 * X_{(t-3)} + \ldots \phi_p * X_p + \varepsilon \qquad \ldots(8)$$

Where:

$\xi$ : is a constant (intercept), and $\phi_1$, $\phi_2$, $\phi_3$ are the autoregressive model parameters.

$\varepsilon$ : is a random variable

The process is determined by p coefficients $\phi_i$ $(i = 1, \ldots, p)$, together with the first $(p-1)$ values of Xt, which must be given as initial conditions. Equations states that the tth value of $X_t$ $(t \geq p)$ is a linear sum of the previous $(p-1)$ values, plus a random component, . The conditions state that the random variable, has a zero mean, finite variance, and all its autocovariance are equal to zero and the random component is not correlated with $X_t$. The definition can be written in following form.

$$\sum_{s=0}^{p} \phi_s (X_{t-s} - \mu) = \varepsilon \quad \phi_0 = 1$$

It can be shown that the autocovariances, which are still defined by equation (6), are related by the following set of equations, known as the Yule-Walker equations:

$$\sum_{s=0}^{p} \phi_s R_{s-r} = \begin{cases} \sigma^2 & r = 0 \\ 0 & r \neq 0 \end{cases} \qquad \ldots(9)$$

This is a set of $p + 1$ linear equations involving the $p$ coefficients $\phi_s (s = 1, \ldots, p; \phi_0 = 1)$, $2p + 1$ autocovariances, $R_s$ $(s = -p, \ldots, -1, 0, 1, \ldots, p)$, and $\sigma^2$.

Here very important question is that what value of p should be assumed. The order of the autoregressive process is left open for researchers. There are following way to select the value of $p$:

- In equation (9) small value of $p$ should be suggested.
- In time series approach, where no assumption was made about the nature of the stochastic process, the values needed for a good approximation appear to be much less than the large number of auto covariance.
- It is also possible to establish statistical tests for deciding whether to accept a particular value of $p$.

**Stationarity requirement** Note that an autoregressive process will only be stable if the parameters are within a certain range; for example, if there is only one autoregressive parameter then is must fall within the interval of $-1 < \phi < 1$. Otherwise, past effects would accumulate and the values of successive xt' s would move towards infinity, that is, the series would not be stationary. If there is more than one autoregressive parameter, similar (general) restrictions on the parameter values can be defined (*e.g.*, Ref. Box & Jenkins, 1976; Montgomery, 1990).

**Moving average process.** This process is independent from the autoregressive process, each element in the series can also be affected by the past error (or random shock) that cannot be accounted for by the autoregressive component, that is:

$$X_t = \mu + \varepsilon_t - \theta_1 * \varepsilon_{(t-1)} - \theta_2 * \varepsilon_{(t-2)} - \theta_3 * \varepsilon_{(t-3)} - \ldots$$

Where:

$\mu$ : is a constant, and $\phi_1$, $\phi_2$, $\phi_3$ are the moving average model parameters.

Put into words, each observation is made up of a random error component (random shock, $\varepsilon$) and a linear combination of prior random shocks.

## SUMMARY

- An estimator is a formula or process for using sample data to estimate a population parameter.
- An estimate is a specific value (or a range of values) used to approximate a population parameter.
- A confidence interval is a range of values used to estimate the true value of a population parameter.
- Probability part of confidence interval is called confidence level.
- The range of values above and below the sample statistics is called the margin of error.
- Estimation methods, Replication Runs, Regenerative Technique and Time series analysis are the methods used to analyze simulation results.

## SOLVED QUESTIONS

**Q.1. What is margin of error discuss with example.**

**Ans. Margin of Error**

In a confidence interval, the range of values above and below the sample statistic is **called** the **margin of error.**

For example, suppose we wanted to know the percentage of adults that exercise daily. We could devise a sample design to ensure that our sample estimate will not differ from the true population value by more than, say, 5 per cent (the margin of error) 90 per cent of the time (the confidence level).

**Q.2. How do you construct a confidence interval?**

**Ans.** There are four steps to constructing a confidence interval.

- Identify a sample statistic. Choose the statistic (*e.g.*, sample mean, sample proportion) that you will use to estimate a population parameter.
- Select a confidence level. As we noted in the previous section, the confidence leavel describes the uncertainty of a sampling method. Often, researchers choose 90%, 95%, or 99% confidence levels; but any percentage can be used.
- Find the margin of error. If you are working on a homework problem or a test question, the margin of error may be given. Often, however, you will need to compute the margin of error, based on one of the following equations.

Margin of error = Critical value * Standard deviation of statistic

Margin of error = Critical value * Standard error of statistic

- Specify the confidence interval. The uncertainty is denoted by the confidence level. And the range of the confidence interval is defined by the following equation.

  Confidence interval = Sample statistic ± Margin of error

**Q.3. Suppose we want to estimate the average weight of an adult male in Mumbai County, India. We draw a random sample of 1,000 men from a population of 1,000,000 men and weigh them. We find that the average man in our sample weighs 180 pounds, and the standard deviation of the sample is 30 pounds. What is the 95% confidence interval?**

**Ans.** To specify the confidence interval, we work through the four steps below.

- Identify a sample statistic. Since we are trying to estimate the mean weight in the population, we choose the mean weight in our sample (180) as the sample statistic.
- Select a confidence level. In this case, the confidence level is defined for us in the problem. We are working with a 95% confidence level.
- Find the margin of error. Previously, we described how to compute the margin of error. The key steps are shown below.
- Find standard error. The standard error (SE) of the mean is:

  $$SE = s/\text{sqrt}(n) = 30/\text{sqrt}(1000) = 30/31.62 = 0.95$$

- Find critical value. The critical value is a factor used to compute the margin of error. To express the critical value as a $t$ score ($t^*$), follow these steps.
  - Compute alpha ($\alpha$): $\alpha = 1 - (\text{confidence level}/100) = 0.05$
  - Find the critical probability ($p^*$): $p^* = 1 - \alpha/2 = 1 - 0.05/2 = 0.975$
  - Find the degrees of freedom ($df$): $df = n - 1 = 1000 - 1 = 999$
  - The critical value is the $t$ score having 999 degrees of freedom and a cumulative probability equal to 0.975. From the $t$ distribution calculator, we find that the critical value is 1.96.

**Note:** We might also have expressed the critical value as a z score. Because the sample size is large, a $z$ score analysis produces, the same result – a critical value equal to 1.96.

- Compute margin of error (ME): ME = critical value * standard error = 1.96 * 0.95 = 1.86
- Specify the confidence interval. The range of the confidence interval is defined by the *sample statistic* ± *margin of error*. And the uncertainty is denoted by the confidence level. Therefore, this 95% confidence interval says that the population mean falls within the interval 180 ± 1.86.

**Q.4. Suppose that a 90% confidence interval states that the population mean is greater than 100 and less than 200. How would you interpret this statement?**

**Ans.** Some people think this means there is a 90% chance that the population mean falls between 100 and 200. This is incorrect. Like any population parameter, the population mean is a constant, not a random variable. It does no change. The probability that a constant falls within any given range is always 0.00 or 1.00.

The confidence level describes the uncertainty associated with a sampling method. Suppose we used the same sampling method to select different samples and to compute a different interval estimate for each sample. Some interval estimates would include the true population parameter and some would not. A 90% confidence level means that we would expect 90% of the interval estimates to include the population parameter; A 95% confidence level means that 95% of the intervals would include the parameter; and so on.

**Q.5. What are the requirements of Confidence Interval?**

**Ans.** There are following three basic requirements of confidence interval:

1. Confidence level
2. Statistic
3. Margin of error

**Explain each by Yourself**

**Q.6. The values 2, 7, 3, 8 were obtained as a random sample of a random variable *X*. Give unbiased point estimates of the mean μ and variance $\sigma^2$ of *X*.**

**Ans.**

$$\bar{X} = \frac{2+7+3+8}{4} = \frac{20}{4} = 5 \text{ is an unbiased point estimate of } \mu.$$

$$S^2 = \frac{(2-5)^2+(7-5)^2+(3-5)^2+(8-5)^2}{3}$$

$$= \frac{9+4+4+9}{3} = \frac{26}{3} = 8.67$$

is an unbiased point estimate of $\sigma^2$.

**Q.7. Show that the sample mean $\bar{X}$, for random sample of size *n*, is an unbiased estimator of the population mean μ.**

**Ans.** Definition $\bar{X} = \dfrac{X_1 + X_2 \ldots X_n}{n}$

where $X_1, X_2, \ldots X_n$ are independent random variable , each with mean μ and standard deviation σ. The expected value of a constant times a random variable is that variable is equal to the sum of the expected values of the random variables. Therefore,

$$E\left(\frac{X_1+X_2+\ldots+X_n}{n}\right) = \frac{1}{n}E(X_1+X_2+\ldots+X_n)$$

$$= \frac{1}{n}(E(X_1)+E(X_2)+\ldots+E(X_n))$$

$$= \frac{1}{n}(\mu+\mu+\ldots\mu) = \frac{1}{n}(n\mu) = \mu$$

**Q.8. Suppose a sample of 100 students with mean SAT score of 1020, standard deviation of 200. How do we find the 95% confidence interval for the population mena?**

**Ans.** We know that:

1. The sample distribution is normally distributed.
2. Therefore 95% of samples will yield a mean estimate within two standard deviations of the population (μ).

Thus, 95% of estimates we make are within two "standard errors" of $\bar{Y}$.

95% CI: $\bar{Y} \pm 2\,(\hat{\sigma}_{\bar{Y}})$ $\hat{\sigma}_Y = \dfrac{S_Y}{\sqrt{n}}$

$$\bar{Y} \pm 2\left(\frac{S}{\sqrt{n}}\right) = 1020 \pm (2)\left(\frac{200}{\sqrt{100}}\right)$$
$$= 1020 \pm 2 \times 20$$
$$= 1020 \pm 40$$

95% CI : (980, 1060)

We are 95% confident that the mean SAT for all students fall within the ranger of (980, 1060).

**Q.9. How do you read *Z*–score table for 95% confidence level?**

**Ans.**
- Compute $\alpha = 1 - 95/100 \Rightarrow 0.05$
- Find $Z_{\alpha/2} \Rightarrow \alpha/2 = 0.025$

Search 0.025 in *Z*–table, which is at *Z* = 1.9 and from 1.9 now we will get 0.06, column as 0.025. So *Z*–score = 1.9 + 0.06 ⇒ 1.96.

## EXERCISES

1. A biology student at a major university is writing a report about bird watchers. She has developed a test that will score the abilities of a bird watcher to identify common birds. She collects data from a random sample of people that classify themselves as bird watchers (data shown below). Find a 90% confidence interval for the mean score of the population of bird watchers. At a 90% CI, the critical value is $t^* = 1.729$.

| 4.5 | 9.1 | 8 | 5.9 | 7.0 | 5.2 | 7.3 | 7.0 | 6.6 | 5.1 |
|---|---|---|---|---|---|---|---|---|---|
| 7.5 | 8.2 | 6.4 | 4.8 | 5.8 | 6.2 | 8.5 | 7.3 | 7.8 | 7.4 |

2. If $\sigma = 20$, what sample size is needed to estimate the mean within ± 4 margin of error with 95% confidence?
3. An asbestos removal company places great importance on the safety of their employees. The protective suits that the employees wear are designed to keep asbestos particles off the employee's body. The owner is interested in knowing the average amount of asbestos particles left on the employee's skin after a days work. A random sample of 100 employees had skin tests after removing their protective suit. The average number of particles found was 481 particles per square centimeter. Assuming that the population standard deviation is 0.35 particles per square centimeter, calculate a 95% confidence interval for the number of particles left on the employee's skin.
4. The owner of a popular chain of restaurants wishes to know if completed dishes are being delivered to the customer's table within one minute of being completed by the chef. A random sample of 75 completed dishes found that 60 were delivered within one minute of completion. Find the 95% confidence interval for the true population proportion.
5. How large of a random sample is required to insure that the margin of error is 0.08 when estimating the proportion of college professors that read science fiction novels with 95% confidence.
6. Explain types of simulation with respect to output analysis.
7. Explain point estimator and interval estimator.
8. Explain output analysis for steady state simulation.

15

# Simulation of PERT Network

**INSIDE THIS CHAPTER**



The evolution of CPM (critical path method) and PERT (Program Evaluation and Review Technique) in 1957-59 provided the great help in planning and scheduling of large projects. CPM and PERT show the interrelationships between various 'activities' of a project and its identify critical activities which need to paid most attention because any delay on the critical path will delay the project.

**Table 15.1.** Difference between CPM and PERT

| CPM | PERT |
|---|---|
| 1. Duration of each activity is deterministic. | 1. There is uncertainty regarding the time taken by various activities in project. |
| 2. The CPM analysis is activity oriented. | 2. PERT is event oriented. |
| 3. CPM is based on a single time estimate. | 3. PERT is based on three time estimate for each activity:<br>• most optimistic time<br>• most pessimistic time (longest time)<br>• most likely time for completion of each activity. |
| 4. Best for simple project. | 4. Best for research and development type project. |

## 15.1 NETWORK MODEL OF A PROJECT

There are two basic elements in network model of a project: Activity and event.

**Activity:** The activity represents for time-consuming part of a project. It represents a job.

**Event:** The event, also called a node, it is either the beginning or the end of the job. The events are represented by circles or rectangles.

The activities are denoted by arrows or edge $(i, j)$ cannot be started before all activities leading to node $i$ have been completed some activities cannot be started before the completion of some other activities that means there are precedence relationships between activities. Each activity will require certain time to complete, called its **duration.**

The activities and their precedence relationships, and duration can be conveniently represented by means of an arrow diagram called an **activity network model** or activity graph.

Fig. 15.1 shows how the events are connected by activities.

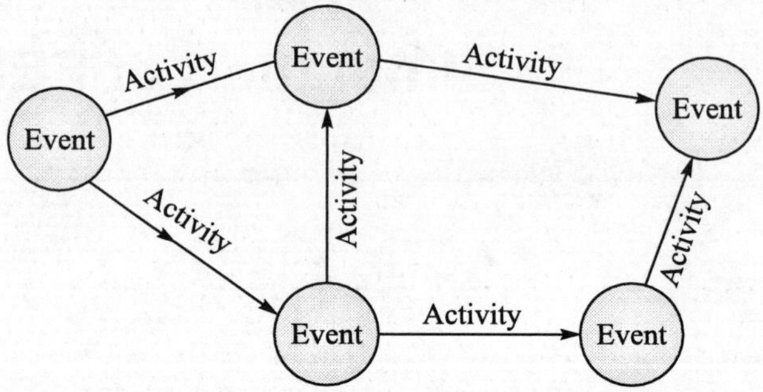


**Fig. 15.1**

Let consider a simple example, we have a project consisting of six well defined, non-overlapping individual jobs, $A, B, C, D, E, F$ with the restriction that $A$ must precede $C$ and $D$, $B$ and $D$ must precede $E$ and $C$ must precede $F$. The durations of activities are 5.6, 7.7, 6.5, 5.0, 16.3 and 3.0 days. The activity network of this project is shown in Fig. 15.2.

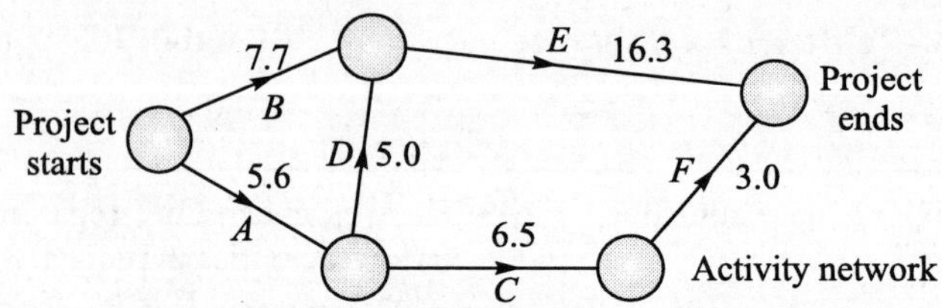


**Fig. 15.2**

**Note:** Activity network should not have a 'cycle', otherwise, we would have an impossible situation in which no activity lying in the cycle could be initiated.

### 15.1.1 Loop Network

We have to avoid a loop in a network this is what we have discuss above about cycle generally it occur in complex network. For example, event $A$ cannot started until activity $d$ is over which in turn cannot take place until event $D$ has taken place.

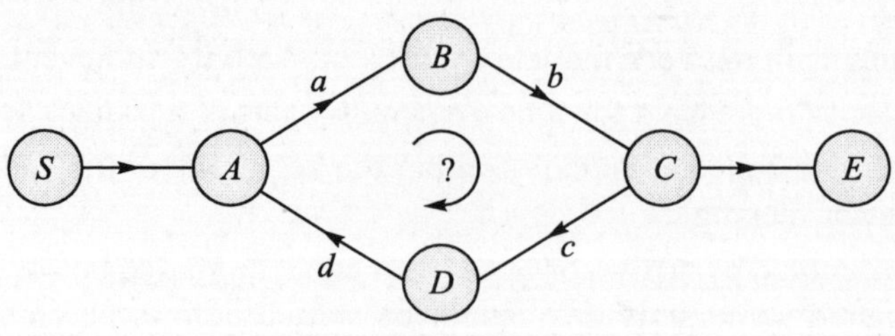


**Fig. 15.3**

### 15.1.2 Dummy Activity

Sometime there is a situation when a certain event *j* cannot occur until another event *i* has taken place, while we are connecting events by activities showing their interdependencies; activity connecting *i* and *j* will not involve any real time resources. In such case an edge, which represents only a precedence relationship and not any other job in the project is called as **Dummy Activity.**

- All dummy activities are of zero duration.
- Generally it is shown in broken lines.

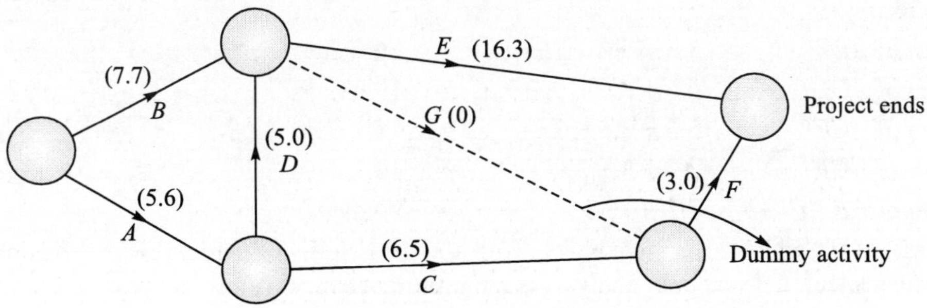


**Fig. 15.4.** Dummy activity in a network.

## 15.2 ANALYSIS OF THE ACTIVITY NETWORK

As we know that CPM and PERT are providing great help in planning and scheduling of large project. This is possible when we analysis our activity network and identified critical path. So analysis of the activity network is accomplished by managers or executive person for studying optimizing the execution of projects. They make their analysis by two most popular names–CPM and PERT.

### 15.2.1 Path

A sequence of adjacent edges from the source node to the sink node is called a path in a network, *e.g.*, *ADE*, *BE* and *ADGF*.

**Length** of a path is the sum of the duration of all the activities along that path.

Activity network is analyzed by following points:

(*i*) Identify critical activities (*ii*) Project completion time

Any delay in execution of critical activities will delay overall project so management must pay special attention to critical activities. To identify critical activities we have to identify critical path.

**Some Important Points for Analysis of Activity Network:**

- The number of edges entering a node is called the **in-degree** of that node.
- The number of edges going out of a node is called its **out degree.**
- There will be at least one node with zero in-degree and at least one node with zero out-degree in an activity network.
- If the network has more than one node with zero in-degree, we can arbitrarily select one of the nodes and draw dummy edges from this node to all others node.
- This will result exactly one node with zero in-degree. This node is called the **source node.**

- Similarly we can made to have exactly one node with zero out-degree, called the **sink node.**
- The source node denote start of the project and sink node denote end of the project.
- Activity network cannot have a cycle.
- There is a real number, associated with every edge called **duration.**

**How to Identify Critical Path**

1. Calculate length of all possible path.
2. The longest duration path through the network is called critical path.

**Example:**

Length of *ADE* = 5.6 + 5.0 + 16.3 = 26.9 days
Length of *ADGF* = 5.6 + 5.0 + 0 + 3.0 = 13.6 days
Length of *ACF* = 5.6 + 6.5 + 3.0 = 15.1 days
Length of *BGF* = 7.7 + 0 + 3.0 = 10.7 days
Length of *BE* = 7.7 + 16.3 = 24 days

Activities *A*, *D*, and *E* are critical in Fig. 15.5 and the length of a critical path is the time required to complete the project. Critical path analysis is an important aspect of project planning.

Theoretically we can trace all their lengths and identify critical path as longest these. But in reality this method is impractical because a typical activity network will have too many paths from source to sink. So we have given one practical or formal method. The critical path can be formally identified by determining the following four parameters for each activity:

- **EST–Earliest Start Time:** The earliest time at which the activity can start given that its precedent activities must be completed first.
- **EFT–Earliest Finish Time** : Earliest finish time to complete the activity.

$$EFT(k) = EST(k) + T(k)$$

**T(k):** is duration of $k^{th}$ activity

- **LFT–Latest Finish Time:** The latest time at which the activity can be completing without delaying the project.
- **LST–Latest Start Time:** It is equal to the latest finish time minus the time required to complete the activity (duration).

$$LST(k) = LFT(k) - T(k)$$

## Slack Time

Slack time for an activity is the time between its earliest and latest start time, or between its earliest and latest finish time. For a node that falls on the critical path this slack will be zero.

**Points to be Remember (While Analysing the Activity Network)**

Arrange all the activities in topological order in a activity network.

## Explanation

Project is said to be in topological order if no activity appears the list before all of its immediate predecessors have appeared, *e.g.*, Fig. 15.5. list *A*, *B*, *C*, *D*, *E*, *G*, *F* is in topological order not *A*, *B*, *C*, *D*, *E*, *F*, *G* because activity *F* appears before *G* · *G* is a predecessor of *F*.

*A*, *B*, *C*, *D*, *E*, *G*, *F* is not only one topological order of given activity network, we can have more than one topological order, *e.g.*, *B*, *A*, *D*, *C*, *E*, *G*, *F* is also in topological order.

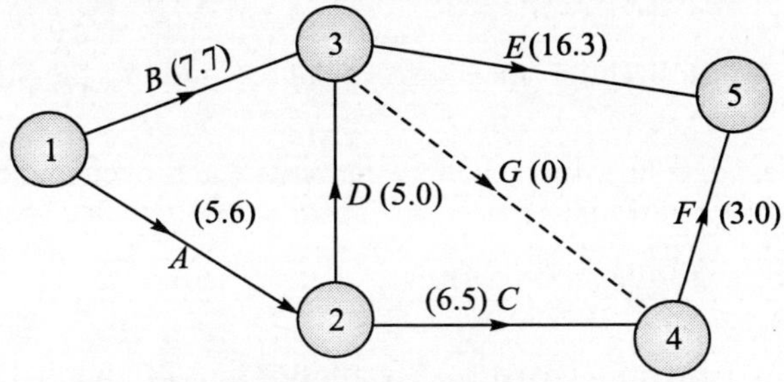


**Fig. 15.5.** Topologically ordered.

Topological order is possible only when there is no cycle in the network.

**How we can arrange list of activity in topological order?**

(*i*) We start by taking an activity which is not preceded by any other in the network. This is put first in the list.

(*ii*) Now we remove this activity or node from the network and select another activity which is not preceded by any other in the remaining network.

(*iii*) We place this activity as the second no. in the list and also remove it from the network.

(*iv*) We repeat this process till then all the activities are removed from the network and placed on the list.

(*v*) The final list is in topological order.

**Table 15.2.** Representation of topological ordered activity of Fig. 15.5

| *Activity No. (k)* | *SN(k)* | *FN(k)* | *T(k)* |
|---|---|---|---|
| 1. | 1 | 2 | 5.6 days |
| 2. | 1 | 3 | 7.7 days |
| 3. | 2 | 4 | 6.5 days |
| 4. | 2 | 3 | 5.0 days |
| 5. | 3 | 5 | 16.3 days |
| 6. | 3 | 4 | 0 days |
| 7. | 4 | 5 | 3.0 days |

## 15.3 CRITICAL PATH COMPUTATION

As we have already discussed in: How to identify critical path–that we have to calculate EST: earliest start time, EFT: earliest finish time, LFT: latest finish time and LST: latest start time. We can calculate these four term by traversing the network in two directions:

- Forward pass method : For *EST* and *EFT*, *ENT*
- Backward pass method : For *LST* and *LFT*, *LNT*

### 15.3.1 Forward Pass

Network is traversed in forward direction means from source node to the sink node passing through every edge once.

It calculate the earliest date on which each event may be achieved and the earliest date on which each activity may be started and completed.

- *EST*(*k*) – Earliest start time of activity *k*
- *EFT*(*k*) – *EST*(*k*) + *T*(*k*) ...(1)

Let the given activity network consist of *N* activities and *M* nodes. Let activities are listed in topological order and be labelled as 1, 2, ..., *N* (including dummy activity).

- The nodes are labelled as 1, 2, ..., *M*.
- Node 1 is source node.
- Node 5 is sink node.
- Let each activity be specified by its starting node *SN*, finishing node *FN* and its duration *T*.

For a *k*th activity

| | | |
|---|---|---|
| *SN*(*k*) | : | Starting node of *k*th activity |
| *FN*(*k*) | : | Finishing node of *k*th activity |
| *T*(*k*) | : | Duration of *k*th activity |

For example, the network of Fig. 15.6 (with its activities in topological order 1, 2, ..., 7 and nodes relabelled 1, 2, ..., 5) is represented as shown in Fig. 15.6. The activity durations are shown in parentheses.

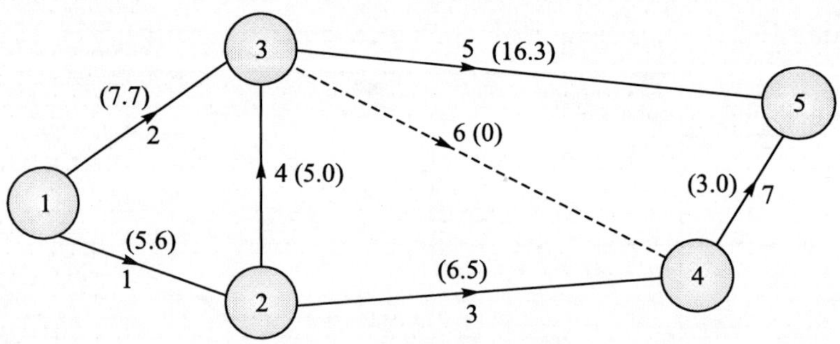


**Fig. 15.6**

**ENT (*i*): Earliest Node Time**

These nodes are also called milestones. A milestone *i* is said to be achieved when all activities terminating in node *i* are completed. Let *ENT* (*i*) denote the earliest time when node *i* is achieved. Therefore

$$ENT\ (i) = \text{Max}\{EFT\ (\text{all activities terminating in } i)\} \quad ...(2)$$

Earliest an activity *k* can begin is the time when node *SN*(*k*) achieved

*i.e.*,
$$EST(k) = ENT(SN(k)) \quad ...(3)$$

**At Source Node**

$ENT(1) = 0$ Since activity 1 begins at node 1 then $EST(1) = 0$

and $EFT(1) = EST(1) + T(1)$ or $0 + T(1)$

Similarly, we will compute EFT of all activities originating at node 1.

By using equation (2) we can compute *ENT* of the next node. Then we can compute the *EST*'s and *EFT* of some more activities and repeat this process till then all the activities in the project have considered. We cannot take any activity until all its immediate predecessors had been process.

Following steps are given for **FORWARD PASS method:**

(*i*) Read following values from the topological order activities (Table 15.2).

- $N$ = number of activities
- $M$ = number of node
- $SN(k)$: Start node, $FN(k)$: Finish node and $T(k)$ duration of all activities from $k = 1$ to $N$.

(*ii*) Initialize, for all $M$ values of node

- $ENT(M) = 0$
- Initialize $k = 1$
- Repeat this process while $(k \leq N)$

(*iii*) Initialize, snode and fnode as

- snode = $SN(k)$
- fnode = $FN(k)$

(*iv*) Initialize $EST(k)$, $EFT(k)$ as

- $EST(k) = ENT(\text{snode})$
- $EFT(k) = EST(k) + T(k)$

(*v*) If $(EFT(k) > ENT(\text{fnode}))$

- Yes then set:

  $ENT(\text{fnode}) = EFT(k)$

- No then:

  go to line no. (*vi*)

(*vi*) Increment the value of $k$ as

$$k = k + 1$$

(*vii*) Check if $k \leq n$

- Yes then:

  go to line no. (*viii*)

- No, go to line no. (*iii*)

(*viii*) $T_{min} = ENT(M)$

(*ix*) (*a*) Print value of $T_{min}$

(*b*) Print values of *EST*, *EFT* from $k = 1$ to $N$.

(*x*) Print values of *ENT* for all the values of $i = 1$ to $M$

On applying this sequence of forward pass on Fig. 15.6.

Set $N = 7, \quad M = 5$

Read values of $S(k)$, $FN(k)$ and $T(k)$ from Table 15.2

Initialize $ENT(1) = ENT(2) = ENT(3) = ENT(4) = ENT(5) = 0$

**For $K = 1$** $1 \leq 7$ Yes then

snode = $SN(1) = 1$

fnode = $FN(1) = 2$

$T(1) = 5.6$

$EST(k) = ENT(\text{snode})$

$EST(1) = ENT(1) = 0$

$EFT(1) = EST(1) + T(1)$

$= 0 + 5.6 \Rightarrow 5.6$

$EFT(1) > ENT(2)$ means $5.6 > 2$ yes then

$ENT(2) = EFT(1) = 5.6$

$EST(1) = 0 \quad EFT(1) = 5.6$

$ENT(1) = 0 \quad ENT(2) = 5.6$

**For K = 2** $2 \leq 7$ Yes then

snode = $SN(2) = 1$

fnode = $FN(2) = 3$

$T(2) = 7.7$

$EST(2) = ENT(1) = 0$

$EFT(2) = EST(2) + T(2) = 0 + 7.7$

$EFT(2) = 7.7$

$EFT(2) > ENT(3)$

So $ENT(3) = EFT(2) = 7.7$ (temporary)

**Note:** Because two activities 2nd and 4th are terminating on $ENT(3)$ so we have to take only one, *i.e.*,

$ENT(3) = \max(EFT$ all activities terminating in $i))$

$EST(2) = 0 \quad EFT(2) = 7.7$

$ENT(3) = 7.7$ (temporary)

**For K = 3** $3 \leq 7$ Yes

snode = $SN(3) = 2$

fnode = $FN(3) = 4$

$T(3) = 6.5$

$EST(3) = ENT(2) = 5.6$

$EFT(3) = EST(3) + T(3) = 5.6 + 6.5 = 12.1$

$EFT(3) > ENT(4)$

$ENT(4) = EFT(3) = 12.1$

So $EST(3) = 5.6 \quad EFT(3) = 12.1$

$ENT(4) = 12.1$

**For K = 4** $4 \leq 7$ then

snode = $SN(4) = 2$

fnode = $FN(4) = 3 \quad T(4) = 5.0$

$EST(4) = ENT(2) = 5.6$

$EFT(4) = EST(4) + T(4) = 5.6 + 5.0 \Rightarrow 10.6$

$EFT(4) > ENT(3)$ then

$ENT(3) = ENT(4) = 10.6$

**Note:** We are having one temporary value $ENT(3) = 7.7$

$$ENT(3) = \max \Rightarrow (10.6, 7.7) \Rightarrow 10.6 \text{ (conceptually)}$$

By computer program it is automatically assign maximum value.

So $EST(4) = 5.6 \quad EFT(4) = 10.6$

$ENT(3) = 10.6$

**For K = 5** $5 \leq 7$ Yes

snode $= SN(5) = 3$

fnode $= FN(5) = 5$

$T(5) = 16.3$

$EST(5) = ENT(3) = 10.6$

$EFT(5) = EST(5) + T(5) = 10.6 + 16.3 \Rightarrow 26.9$

$EFT(5) = 26.9$

$EFT(5) > ENT(5)$ then

$ENT(5) = EFT(5) = 26.9$

So $EST(5) = 10.6 \quad EFT(5) = 26.9$

$ENT(5) = 26.9$

**For K = 6** $6 \leq 7$ yes then

snode $= SN(6) = 3$

fnode $= FN(6) = 4$

$T(6) = 0$

$EST(6) = ENT(3) = 10.6$

$EFT(6) = EST(6) + T(6) \Rightarrow 10.6 + 0 \Rightarrow 10.6$

$EFT(6) \neq ENT(4)$ then {*i.e.*, $10.6 \not> 12.1$, Not)

Then simply $K = 6 + 1 \Rightarrow 7$

$EST = 10.6 \quad EFT(6) = 10.6$

**For K = 7**

snode $= SN(7) = 4$

fonde $= FN(7) = 5$

$T(7) = 3.0$

$EST(7) = ENT(4) = 12.1$

$EFT(7) = EST(7) + T(7) \Rightarrow 12.1 + 3.0 \Rightarrow 15.1$

$EFT(7) \neq ENT(5)$ (not)

So as above $k = 7 + 1 \Rightarrow 8$

**For K** = 8

$8 \leq 7$ Not true then

Termination of loop

$EST(7) = 12.1 \quad$ EFT(7) = 15.1

**EST and EFT in Tabular Form:**

**Table 15.3**

| *Activity* | *Duration* | *EST* | *EFT* |
|---|---|---|---|
| 1. | 5.6 | 0 | 5.6 |
| 2. | 7.7 | 0 | 7.7 |
| 3. | 6.5 | 5.6 | 12.1 |
| 4. | 5.0 | 5.6 | 10.6 |
| 5. | 16.3 | 10.6 | 26.9 |
| 6. | 0 | 10.6 | 10.6 |
| 7. | 3.0 | 12.1 | 15.1 |

$$ENT \text{ values} = (0, 5.6, 10.6, 12.1, 26.9)$$

$$T_{min} = ENT(5) = 26.9$$

So the project completion time is 26.9.

## 15.3.2 Program of Forward Pass

```
# include <stdio.h>
# include <conio.h> void main()
{
        int i, n m, fnode, snode, j, l, k;
        float sn[10], fn[10], t[10], ent[10], est[10], eft[10];
        clrscr();
        printf("Enter total no. of activities");
        scanf("%, d" & n);
        printf("Enter total no.of node");
        scanf("%d", & m);
        printf ("Enter source node of all activity");
                for(k = 1; k < = n; k + +)
                        {
                        scanf("%f", & sn[k]);
                        }
        printf("\n Enter Finish node of all activity");
        for(k = 1; k < = n:k ++)
                {
                scanf("% f", &fn[k])
                }
        printf("Enter time duration for each activity");
         for(k = 1; k < = n; k++)
                {
                scanf("% f", & t[k])
                }
        for (i = 1; i < = m, i + +)
                {
```

```
        ent[i] = 0;
        }
        k = 1;
   while(k < = n)
   {
        snode = sn[k];
        fnode = fn[k];
        est[k] = ent[snode];
        eft[k] = est[k] +t[k];
        if(eft[k] > ent [fnode])
        {
        ent[fnode] = eft[k];
        }
        k = k + 1;
        }
        tmin = ent[m];
   printf("Minimum Time need to complete this project = %2.2f", tmin);
   printf("in All the Ests are as follows {");
        for(j = 1; j < = n; j + +)
        {
        printf("%2.2f", est [j]);
        }
   printf ("}\n");
   printf ("All the ESTs are as follows {");
        for(j = 1; j < n; i = n; j + +)
        {
        printf("% 2.2 f", eft[j]);
             }
        printf("}/n")
        printf("All the ENTs are as follows {");
             for (l = 1; l < m; l + +)
             {
             printf("% 2.2 F"; ent[l]);
        }
        printf("}");
        getch();
        }
```

**Output**

Enter total No. of activities: 7

Enter total no. of node: 5

Enter source node of all activity

1

1

2

2
3
3
4
Enter Finish node of all activity
2
3
4
5
5
4
5
Enter time duration for each activity
5.6
7.7
6.5
5.0
16.3
0
3.0
Minimum time required to complete this project = 26.90
All the ESTs are as follows {0.00 0.00 5.60 5.60 10.60 10.60 12.10}
All the EFTs are as follows {5.60 7.70 12.10 10.60 26.90 10.60 15.10}
All the ENTs are as follows {0.00 5.60 10.60 12.10 26.90}

### 15.3.3 Backward Pass

This technique is used for identifying the critical path by traversing the network in backward direction.

**Rule of Backward Pass**

- The latest time for an event is the latest start time (LST) for all the activities that may starting from that event.
- LST($k$) is the latest start time of $k$th activity so that there is no delay in project.
- LST($k$) = LFT (k) – T($k$)
- Latest node time (LNT($i$)) ...(4)

where more than one activity commences at a common event we take minimum of the latest start time for those activities.

$$LNT(i) = MIN\{LST \text{ (all activities orignating in node } i)\} \quad ...(5)$$

- The completion of any activity terminating in node $i$ can be delayed up to (not beyond) $LNT(i)$.

$$LFT \text{ (every activity terminating in } i) = LNT(i) \quad ...(6)$$

In backword pass we traverse the network from sink node to source node. So we set its latest node time eqaul to the project completion time $T_{min}$ (compute in forward pass) *i.e.*, $LNT(M) = T(MIN) = ENT(M)$

**LST**

- By using equation (6) we set this time to be the *LFT* of each activity terminating in *M*. From these *LFT* we will calculate *LST*s using equation (4).

**LNT**

- We will have at least one node (other than *M*), say *A*, such that the *LST* of all activities originating from *A* have been calculated. This is possible by calculating *LNT*(*A*) by using equation (5).

**LFT**

We get LFT's is of all activities terminating in a using equation (6).

**Consider following step for the calculation of backward pass.**

(*i*) Read following values.

(*a*) $N$ = No. of activities

(*b*) $M$ = No. of node

(*c*) $T_{min}$ = Minimum time to complete the project (by forward pass)

(*d*) Start node $SN(k)$, $FN(k)$, $T(k)$ from $k = 1$ to $N$

(*e*) $EST(k)$ from $k = 1$ to $N$ (Result of forward pass)

(*ii*) Initialize *LNT* for all the values of *M*

$$LNT(i) = T_{min} \quad \text{from } \boldsymbol{i} = 1 \text{ to } M$$

(*iii*) Set $$k = N$$

While $k > 0$ repeat following step:

(*iv*) Initialize snode and fnode as:

$$\text{snode} = SN(k)$$
$$\text{fnode} = FN(k)$$

(*v*) Initialize $LFT(k)$ and $LST(k)$ as:

$$LFT(k) = LNT(\text{fnode})$$
$$LST(k) = LFT(k) - T(k)$$

(*vi*) If $LST(k) < LNT$ (snode) then

(a) If Yes, then compute

$$LNT\,(\text{snode}) = LST(k)$$

(*b*) If No, then go to step no. (*vii*)

(*vii*) If $LST(k) = EST(k)$ then

(*a*) If Yes, then

Print *k* as a critical activity

(*b*) If No, then goto step no. (*viii*)

(*viii*) $k = k - 1$

(*ix*) If $k > 0$?

(*a*) Yes then

goto step (*iv*)

(*b*) No then

goto step (*x*)

(*x*) Print all critical activities

(*xi*) Print all the values of

(*a*) *LST* and *LFT* from $k = 1$ to $N$

(*b*) *LNT* from $k = 1$ to $M$

On applying this step of backward pass on Fig. 15.6 we get

$$LNT(5) = 26.9 = LNT(4) = LNT(3) = LNT(2) = LNT(1)$$

**For $K = 7$**

$$\text{snode} = SN(7) = 4$$
$$\text{fnode} = FN(7) = 5$$
$$T(7) = 3.0$$
$$LFT(7) = LNT(5) = 26.9$$
$$LST(7) = LFT(7) - T(7) = 26.9 - 3.0 \Rightarrow 23.9$$
$$LST(7) = 23.9$$

(*a*) If $LST(7) < LNT(4)$ then (*i.e.*, $23.9 < 26.9$ Yes)

$$LNT(4) = LST(7) = 23.9$$

(*b*) Is $LST(7) = EST(7)$

$$23.9 \neq 12.1$$

**Note:** Values of *EST*(7) is taken from Table 15.2

$$LFT(7) = 26.9 \qquad LST(7) = 23.9$$
$$LNT(4) = 23.9 \qquad LNT(5) = 26.9$$

**For K = 6**

$$\text{snode} = SN(6) = 3$$
$$\text{fnode} = FN(6) = 4$$
$$T(6) = 0$$
$$LFT(6) = LNT(4) = 23.9$$
$$LST(6) = LFT(6) - T(6) \Rightarrow 23.9$$
$$LST(6) = 23.9$$

(*a*) If $LST(6) < LNT(3)$ then **Yes**

$$LNT(3) = LST(6) = 23.9$$

(*b*) Is $LST(6) = EST(6)$

$$23.9 \neq 10.6$$

$$LST(6) = 23.9 \quad LFT(6) = 23.9$$
$$LNT(3) = 23.9$$

**For K = 5**

$$\text{snode} = SN(5) = 3$$
$$\text{fnode} = FN(5) = 5$$
$$T(5) = 16.3$$
$$LFT(5) = LNT(5) = 26.9$$
$$LST(5) = LFT(5) - T(5) = 26.9 - 16.3 \Rightarrow 10.6$$
$$LST(5) = 10.6$$

(*a*) If $LST(5) < LST(3)$ then

$10.6 < 26.9$

$LNT(3) = LST(5) = 10.6$

(*b*) Is $LST(5) = EST(5)$

$10.6 = 10.6$ yes

**Mark Activity 5 is critical activity**

$LST(5) = 10.6 \quad LFT(5) = 26.9$

$LNT(3) = 10.6$

**For K = 4**

snode $= SN(4) = 2$

fnode $= FN(4) = 3$

$T(4) = 5.0$

$LFT(4) = LNT(3) = 10.6$

$LST(4) = LFT(4) - T(4) = 10.6 - 5.0 \Rightarrow 5.6$

$LST(4) = 5.6$

(a) If $LST(4) < LNT(2)$, then

$LNT(2) = LST(4) = 5.6$

(*b*) Is $LST(4) = EST(4)$

$5.6 = 5.6$ Yes

**Mark 4 as a critical activity**

$LST(4) = 5.6 \quad LFT(4) = 10.6$

$LNT(2) = 5.6$

**For K = 3**

snode $= SN(3) = 2$

fnode $= FN(3) = 4$

$T(3) = 6.5$

$LFT(3) = LNT(4) = 23.9$

$LST(3) = LFT(3) - T(3) \Rightarrow 23.9 - 6.5 \Rightarrow 17.4$

$LST(3) = 17.4$

(*a*) If $LST(3) < LNT(2)$ $17.4 \nless 5.6$

No.

(*b*) If $LST(3) = EST(3)$ *i.e.,* $17.4 \neq 5.6$

$LFT(3) = 23.9 \quad LST(3) = 17.4$

**For K = 2**

snode $= SN(2) = 1$

fnode $= FN(2) = 3$

$T(2) = 7.7$

$LFT(2) = LNT(3) = 10.6$

$LST(2) = LFT(2) - T(2) = 10.6 - 7.7 \Rightarrow 2.9$

$$LST(2) = 2.9$$

(*a*) $LST(2) < LNT(1)$

*i.e.,* $2.9 < 26.9$ Yes

Then $LNT(1) = LST(2) = 2.9$

(*b*) Is $LST(2) = EST(2)$

From Table 15.2 2.9 ≠ 0

$$LFT(2) = 10.6 \quad LST(2) = 2.9$$

**For K = 1**

$$\text{snode} = SN(1) = 1$$
$$\text{fnode} = FN(1) = 2$$
$$T(1) = 5.6$$
$$LFT(1) = LNT(2) = 5.6$$
$$LST(1) = LFT(1) - T(1) = 5.6 - 5.6$$
$$LST(1) = 0$$

(*a*) If $LST(1) < LNT(1)$ then

$0 < 2.9$ Yes

$$LNT(1) = LST(1) = 0$$

(*b*) Is $LST(1) = EST(1)$

*i.e.,* $0 = 0$ Yes

**Mark Activity 1 is critical activity**

$$LFT(1) = 5.6 \quad LST(1) = 0$$
$$LNT(1) = 0$$

Activity 1, 4, 5 is **critical activity**, the arcs are (*A*, *D*, *E*).

The values of *LNT* are as follows: (0, 5.6, 10.6, 23.9, 26.9)

**Table 15.4:** Table for *LST* and *LFT*

| *Activity K* | *Duration* | *LST* | *LFT* |
|---|---|---|---|
| 1. | 5.6 | 0 | 5.6 |
| 2. | 7.7 | 2.9 | 10.6 |
| 3. | 6.5 | 17.4 | 23.9 |
| 4. | 5.0 | 5.6 | 10.6 |
| 5. | 16.3 | 10.6 | 26.9 |
| 6. | 0 | 23.9 | 23.9 |
| 7. | 3.0 | 23.9 | 26.9 |

**Slack:** The slack is given by the difference between earliest node time (*ENT*) and latest node time (*LNT*), this represent the latitude available at each node

$$\text{Slack} = LNT(i) - ENT(i) \quad \ldots(7)$$

Slack is the amount of time by which achieving of node $i$ can slip and still the project can be completed in time. So, a node that falls on the critical path, this slack value is equal to zero. Similarly activity that falls on the critical path will be equal to zero.

$$LST(K) - EST(K) = LFT(K) - EFT(K) \quad \text{...(8)}$$

Equation (8), represents, a latitude. This latitude (or slack) is zero if and only if $K$ is a critical activity.

So, we can identify critical activities by checking the value of *LST*'s and *EST*'s. If value of *LST*'s and *EST*'s are equal for any activities, that activities are called as critical activities.

**Example:** Consider two tables, Tables 15.5(*a*) and (*b*).

**Table 15.5 (a):** Table of Activity, Duration, *EST*, *EST*, *LFT* and *EFT*.

| *Activity (K)* | *Duration* | *LST* | *EST* | *Difference (LST-EST)* | *LFT* | *EFT (LFT-EFT)* | *Difference* |
|---|---|---|---|---|---|---|---|
| 1(*A*) | 5.6 | 0 | 0 | 0 | 5.6 | 5.6 | 0 |
| 2(*B*) | 7.7 | 2.9 | 0 | 2.9 | 10.6 | 7.7 | 2.9 |
| 3(*C*) | 6.5 | 17.4 | 5.6 | 11.8 | 23.9 | 12.1 | 11.8 |
| 4(*D*) | 5.0 | 5.6 | 5.6 | 0 | 10.6 | 10.6 | 0 |
| 5(*E*) | 16.3 | 10.6 | 10.6 | 0 | 26.9 | 26.9 | 0 |
| 6(*G*) | 0 | 23.9 | 10.6 | 13.3 | 23.9 | 10.6 | 13.3 |
| 7(*F*) | 3.0 | 23.9 | 12.1 | 11.8 | 26.9 | 15.1 | 11.8 |

In above Table 15.5(*a*). If the difference between LST and *EST*, and *LFT* and EFT is equal to zero; then that activity no. is called as critical activity. So in above table activity no. 1, 4, 5 or *A*, *D*, *E* is critical activity.

**Note:** Difference between (*LST* and *EST*) and (*LFT* and *EFT*) is called as latitude.

**Table 15.5:** (*b*) Table for Node No., *LNT*, *ENT* and Slack.

| *Node No. (i)* | *LNT* | *ENT* | *Slack (LNT-ENT)* |
|---|---|---|---|
| 1. | 0 | 0 | 0 |
| 2. | 5.6 | 5.6 | 0 |
| 3. | 10.6 | 10.6 | 0 |
| 4. | 23.9 | 12.1 | 11.8 |
| 5. | 26.9 | 26.9 | 0 |

In above Table 15.5(*b*), the difference between *LNT* and *ENT*, represents as slack at node (*i*). Clearly value at node 1, 2, 3 and 5 is equal to zero. It means, this node will be fall on critical activity.

While, at node 4, slack value is 11.8, it means, by this amount of time (11.8), the node 4 can slip and still the project can be completed within time.

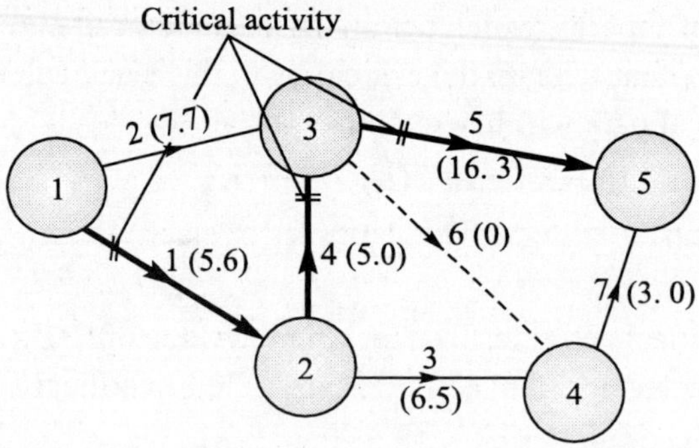


**Fig. 15.7.**

- Activity 1, 4 and 5 are critical activity.
- Node 1, 2, 3 and 5 are critical node and 1, 4 and 5 activity fall on that node.

## 15.4 UNCERTAINTIES IN DURATION OF ACTIVITIES

We have seen in case of *CPM*, that each activity takes a fixed amount of time to complete and this time is already known precisely in advance. But we cannot fix duration of each activity in advance for all the project. These projects are related to research and development type activities or we can say that, in this type of projects, there may be good deal of uncertainties about the duration of various activities.

So each activity duration is considered a random sample from a probability distribution associated with that particular activity.

If the durations of activities are random variables, then critical path and its total length will depend on particular sample that is drawn to show the different activity durations.

Critical path is different for different set of sample for these project, where we are facing uncertainties in activity duration. So one should attempts to obtain the best possible estimate of each duration distribution.

### 15.4.1 PERT

PERT (which was developed by the US Navy in 1957 to plan and schedule the development of the Polaris missiles), it was developed to take account of uncertainty surrounding estimates of task durations.

The method is very similar to the CPM technique but, instead of using a single estimate for the duration of each activity, PERT needs three estimates.

- Most likely time — Duration of activity under normal circumstances we are representing this by letter (mt).
- Optimistic time — The shortest time or duration in which we could expect to complete the activity, we are representing this by letter (ot).
- Pessimistic time — The worst possible time or duration in which we could expect to complete the activity, we are representing this letter (pt).

PERT then combines these estimates to form a single expected duration, $T_e$, using the formula.

$$T_e = \frac{ot + 4mt + pt}{6} \qquad \ldots(9)$$

**Standard Deviation**

A quantitative measure of the degree of uncertainty of an activity duration estimate can be obtained by calculating the standard deviation $S$ of an activity time; by using formula.

$$S = \frac{pt - ot}{6}$$

* S (standard deviation) is proportional to the difference between optimistic and pessimistic estimates.

### 15.4.2 Variance

$$\text{var}(k) = \left(\frac{pt(k) - ot(k)}{6}\right)^2 \quad \text{for } k = 1, 2, \ldots, N \quad \ldots(10)$$

The square of the standard deviation is known as the variance.

By using equation (9) and equation (10) the project completion time and the critical path for each activity are calculated. In a typical large network, a critical path consist of a large number of branches and so we can make following calculation:

- The project completion time will be normally distributed.
- Whose expected value would be the sum of average duration of the activities in the critical path.
- Whose variance is the sum of the variances of all activities on the critical path.

This is very popular technique and give an optimistic results. Because network may have many paths which are not critical but slightly shorter than critical, on the basis of their average durations.

Due to the randomness of durations, these paths, under some combination of activity durations, could become longer than the average longest path. Such path would be ignored if we calculate only one critical path based on the $T_e$.

**Example:**

| | ***ot*** | ***mt*** | ***pt*** |
|---|---|---|---|
| Estimated by *ABC* project | 4 | 7 | 9 |
| Estimated by *XYZ* project | 5 | 8 | 11 |

Calculate the variance we have

(*i*) Variance of *ABC* project $= \left(\frac{9-4}{6}\right)^2 \Rightarrow \left(\frac{5}{6}\right)^2$

$$= \frac{25}{36} = 0.6944$$

Variance for $ABC = 0.6944$

(*ii*) Variance for *XYZ* project $= \left(\frac{11-5}{6}\right)^2 \Rightarrow \left(\frac{6}{6}\right)^2 \Rightarrow 1$

*XYZ* is more uncertain about his estimation compare to ABC.

### 15.4.3 Beta Distribution

It has been seen that Beta distribution curve give satisfactory results with PERT method. This curve is different from normal curve and has represented as follows.

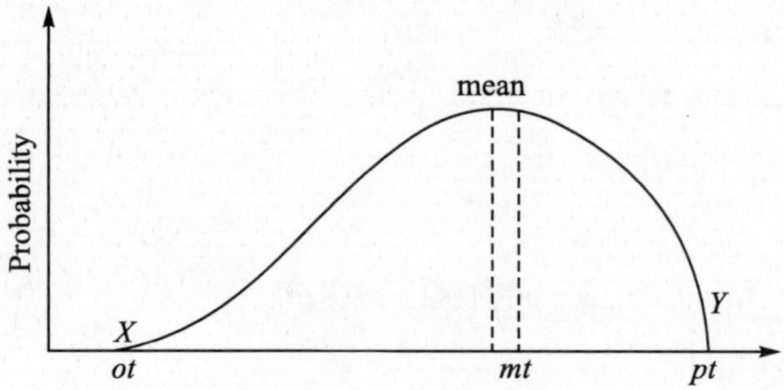


**Fig. 15.8.** Beta distribution.

We have assume a theoretical curve (Fig. 15.8), point $X$ to coincide with the optimistic time $ot$ and point $Y$, with pessimistic time $pt$, *i.e.*, the range is equal to $(pt - ot)$

For a distribution of this type the standard deviation is approximately one-sixth of the range.

$$\text{Standard deviation} \quad (SD) = \frac{pt - ot}{6}$$

$$\text{Variance} \quad = \left(\frac{pt - ot}{6}\right)^2$$

Now we can say that:

- Variance depends on the range $(pt - ot)$.
- If the variance is larger that means there will be greater uncertainty in project.

**Example 1:** *Consider following network (Fig. 15.9).*

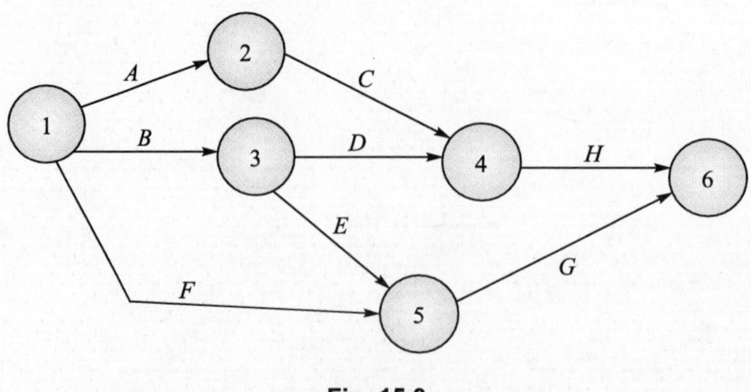


**Fig. 15.9**

**Table 15.6**

| *Activity* | *Activity durations* | | |
|---|---|---|---|
| | *Optimistic (ot)* | *Most likely (mt)* | *Pessimistic (pt)* |
| *A* | *6* | *7* | *9* |
| *B* | *4* | *5* | *6* |
| *C* | *3* | *4* | *4* |
| *D* | *4* | *5* | *6* |
| *E* | *2* | *4* | *5* |
| *F* | *9* | *11* | *16* |
| *G* | *3* | *4* | *5* |
| *H* | *3* | *3* | *3.5* |

*Calculate:*

*(a) Expected time*

*(b) Standard Deviation of each event.*

**Solution:** By using PERT procedure we are having following Table 15.7.

**Table 15.7**

| **Activity** | **Activity duration** | | | **Expected** | **Standard** |
|---|---|---|---|---|---|
| | **Optimistic (*ot*)** | **Most likely (*mt*)** | **Pessimistic ($p_t$)** | **time ($T_e$)** | **deviation(*s*)** |
| *A* | 6 | 7 | 9 | 7.16 | 0.50 |
| *B* | 4 | 5 | 6 | 5.00 | 0.33 |
| *C* | 3 | 4 | 4 | 3.83 | 0.17 |
| *D* | 4 | 5 | 6 | 5.00 | 0.33 |
| *E* | 2 | 4 | 5 | 3.83 | 0.50 |
| *F* | 9 | 11 | 16 | 11.50 | 1.17 |
| *G* | 3 | 4 | 5 | 4.00 | 0.33 |
| *H* | 3 | 3 | 4 | 3.16 | 0.16 |

Expected time $T_e$ for activity *A* is calculated as follows:

$$T_e = \frac{ot + 4(mt) + pt}{6} \Rightarrow \frac{6 + 4 \times 7 + 9}{6} \Rightarrow \frac{6 + 28 + 9}{6}$$

$$T_e = \frac{43}{6} \Rightarrow \mathbf{7.16} \qquad \text{(from table)}$$

Standard deviation (for activity)

$$A = \frac{pt - ot}{6} = \frac{9 - 6}{6} = \frac{3}{6} \Rightarrow 0.50$$

and so on calculate for all other activity that are calculated in above table.

(*ii*) **Standard Deviation for Each Node or Event**

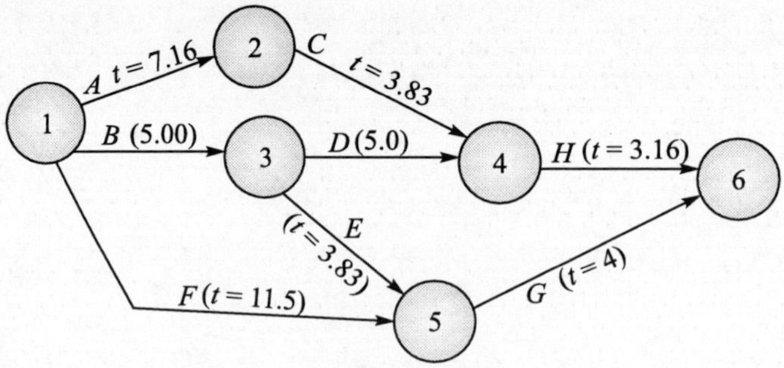


**Fig. 15.10**

1. **For Node 1**

   Standard deviation (*SD*) of node or event 1 is 0.

2. **For Node 2**

   Standard deviation of node 2 – solely depends on activity *A*.

   *SD* of activity *A* = 0.50 = *SD* of event 2.

**Note:** No other activity is reaching at node 2 from node 1 except activity *A*.

3. **For Node 3**

   Similarly *SD* of node 3 is = *SD* of activity *B*.

   = 0.33

4. **For Node 4**

   There are two path to reach node 4 from node 1.

   (*i*) *AC* (*ii*) *BD*

$$SD \text{ of } AC = \sqrt{(SD \text{ of } A)^2 + (SD \text{ of } C)^2}$$

$$= \sqrt{(0.50)^2 + (0.17)^2} \Rightarrow 0.53$$

$$SD \text{ of } BD = \sqrt{(SD \text{ of } B)^2 + (SD \text{ of } D)^2}$$

$$= \sqrt{(0.33)^2 + (0.33)^2}$$

$$= 0.33$$

   *SD* of node or event 4 is = Max *SD* (*AC*, *BD*)

   = *AC* ⇒ 0.53

5. **For Node 5**

   There are also two path to reach node 5 from node

   (*i*) *BE* (*ii*) *F*

   Total *SD* for path $BE = \sqrt{(0.33)^2 + (0.50)^2} \Rightarrow 0.6$

   *SD* for path *F* = 1.17

   *SD* of node 5 is = max (0.6, 1.17) ⇒ 1.17

6. **For Node 6**

There are four path from node 1 to node 6

(i) *ACH* (*ii*) *BDH*

(*iii*) *BEG* (*iv*) *FG*

Calculate total *SD* across above path and set maximum as *SD* of node 6.

**Advantages of PERT**

- PERT focuses attention on the uncertainty of forecasting.
- Technique to calculate the standard deviation for each task and use this to rank them according to their degree of risk.

## 15.5 SIMULATION OF AN ACTIVITY NETWORK

Let we have following three things:

- Arranged list of activities in a topological order, 1, 2, …, *N*.
- Select the start node *SN*(*k*) and finish node *FN*(*k*) for each activity *k,* for $k = 1, 2, \ldots N$.
- The probability distribution with appropriate parameters for the duration of each activity.

We generate one sample for the duration of each activity randomly from the corresponding probability distribution to simulate this stochastic situation consider following point:

- Let we denote these time samples by *TS*(*k*) for $k = 1, 2 \ldots, N$.
- Based on these durations we compute the project completion time $T_{min}$ and identify the critical activities as usual.
- Next, take another sample from the duration distribution of each activity is randomly generated and, by using this new value, the critical activities and project completion time $T_{min}$ are calculated again.
- This procedure is repeated thousand times and maintained the record of the project completion time $T_{min}$ for each sample.
- Then a frequency plot of the project completion times will usually turn out to be normally distributed because of the central limit theorem (CLT).

**Note:** Central limit theorem in statistics tells us that the sample mean $\bar{x}$ is itself a random variable approximately normally distributed with mean $\mu$ and standard deviation $\sigma/\sqrt{n}$, so long as $n$ is sufficiently large.

- Its average and standard deviation are easily calculated. This estimate of the probabilities of completing the project at many times is more reliable compare to usual PERT method. Because we are taking into account the effects of near critical paths also, which may be critical if the activities forming them take more time than their average durations.

## 15.6 CRITICALITY INDEX

As we discussed, that every time we have to calculate critical activities from each sample, this provide a very useful information *i.e.,* criticality index. How many times an activity become critical by keeping this record we can determine percentage of simulation runs in which an activity become critical. This is called Criticality Index of that particular activity.

**Example:** *If an activity k is on a critical path 4,000 times out of 20,000 simulation runs then calculate criticality index of an activity, k.*

**Solution:** Total simulation run = 20,000

No. of time when $k$ became critical is = 4,000

$$\text{Criticality index} = \frac{\text{No. of time when } k \text{ activity is critical} \times 100}{\text{Total simulation run}}$$

$$= \frac{4,0\not{0}\not{0}}{2\not{0}\not{0}\not{0}\not{0}} \times 1\not{0}\not{0} = 20\%$$

- In usual PERT calculations based on average duration, this activity $k$ might not have appeared at all on the critical path. But by using simulation, $k$ is 20% time critical. Simulation procedure helps:

  (*a*) Extend the notion of uncertainties to the critical path itself.

  (*b*) Instead of identifying a single critical path, it identifies activities with criticality index.

  (*c*) It reduces the chance for the near critical activities of being ignored by project manager.

**Consider following procedure for simulating an activity network is given below:**

1. Read network, duration distribution, and number of runs, *i.e.*, no_runs.
2. Order the activity in topological order if necessary.
3. Initialize–

   (*a*) critical activity

   (*b*) frequency – count – array $FREQ\_ARR = 0$

   (*c*) run = 1
4. Generate time samples $TS(k)$ for all activities $k = 1$ to $N$.
5. More forward pass to compute project completion time $T_{\text{MIN}}$.
6. Make backward pass and determine critical activities.
7. Increment appropriate element, *i.e.*,
8. $FREQ\text{-}ARR = FREQ - ARR + 1$

   Store $T_{\text{MIN}}$ for this run. Increment in run by one.

   $$\text{run} = \text{run} + 1$$
9. If run > no-runs

   (*a*) No – go to step 4th

   (*b*) Yes – go to step 10th.
10. Print all $T_{\text{MIN}}$'s and criticality-index.

    But in practice for 20,000 simulation run, printing different values of $T_{\text{MIN}}$ is not useful.

## 15.7 RESOURCE ALLOCATION AND COST CONSIDERATION

So for we have seen in PERT network that only one factor we are considering; TIME. Because we have assumed that there was no limitation on other resources, needed for any number of activities simultaneously resources like:

(*i*) Manpower (*ii*) Work space (*iii*) Equipment, etc.,

Practically, resources are often limited means only fixed amount of resource of each type is available for the project. On the basis of the fixed amount of resources required by each of the activities, we can determine the sequence of performing these activities so that the project is completed in the shortest time.

As there is a limitation on only one resource: **time** (other resources are unlimited), we can use **classical resource allocation approach** as follows:

1. First network is analyzed to determine the critical activities, assuming no resource limitations.
2. Secondly, we tabulated the resource requirement along the critical path.
3. These resources are made available to critical activities otherwise project can be delayed.
4. The rest of the resource is assigned to non-critical activities on the basis of slack or criticalities.
5. This is good deal of flexibility to assign rest resources to non-critical activities.
6. It is try to keep resource requirement as uniform as possible throughout the project period at the time of resource allocation.
7. No. of computer algorithms are proposed and implemented for this type of **resource levelling,** but none of resource levelling program provide an exact optimal solution.
8. Each uses certain heuristics to obtain a near-optimal allocation this problem is further complicated, if no. of resources are to be considered simultaneously, *e.g.,* in one rule shortest duration activity is taken first and in some other rule, longest duration activity is taken first and so on.

   There is no deterministic tool available to determine which rule is best or give optimal solution. It has therefore been suggested that before selecting any specific rule, the network must be simulated by using different rules and select the one which completes the project in the shortest time.

⇒ Remember, this simulation would be deterministic, and no random numbers would be required.

### Cash-flow

We know that not only time duration and resource requirement is necessary for each activity but also it has some cost associated with it. So it is often important to know the amount of cash-flow at different times while the project is in progress. If cost of each activity has been determined then we can easily get the curve of time versus the total money spent for various points in the project.

### Expediting

Most activities can be expedited if more money is spent on them. In general, one can determine a cost versus completion time curve for each activity. Generally it is straight line based on two points:

(*i*) A normal time–normal cost point (*ii*) A crash time–crash cost point

The total project can be expedited by shortening the times of the critical and near-critical activities. One can determine the completion time versus the cost curve for the entire project by repeated calculation. On this curve one may then pick a satisfactory point for which cost and time requirements are both met.

## SUMMARY

- *CPM* is used for projects in which we need duration of each activities is deterministic.
- PERT is used when there is a good deal of uncertainty regarding the time taken by various activities in the project.
- The longest path is called the critical path in the activity network.
- The activities along a critical path are called critical activities.
- Activity network is traversed in forward and backward direction.

- In PERT procedure, the average value of the duration ($T_{av}$) was given by

$$T_{av}(k) = \frac{ot(k) + 4mt(k) + pt(k)}{6} \text{ for } k = 1, 2, \ldots, N \text{ and}$$

Various variance *VAR* are calculated as $VAR(k) = \left(\frac{pt(k) - ot(k)}{6}\right)^2$ for $k = 1, 2, \ldots, N$ and PERT procedure computes only a single critical path based on $T_{av}(k)$ and $VAR(k)$. Simulation plays an extremely valuable role in planning large construction projects.

## SOLVED QUESTIONS

**Q. 1. Define the terms:** [UPTU 2005-2006]

**1. Activity network**

**2. Length of a path**

**3. Critical path in network.**

**Ans.** **1. Activity Network:** Generally any non-trivial project can be thought of as consisting of number of separate activities. There will be precedence relationships between activities. Each activity will require certain time to complete called its duration. The activities the precedence relationships and durations can be conveniently represented by means of an arrow diagram called an activity network of any project is shown in Fig. 15.11.

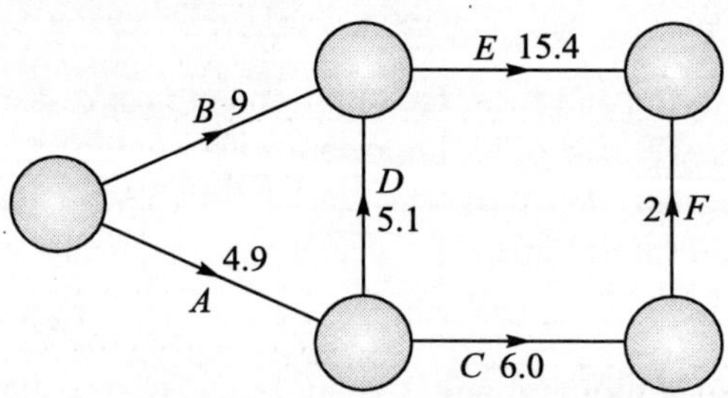


**Fig. 15.11.** Activity network.

**Q. 2. Length of a Path:** The length of path is the sum of the durations of all the activities along that path

**Ans.** In Fig. 15.11.

Length of $ACF = 4.9 + 6.0 + 2 = 12.9$ days

Length of $BE = 9 + 15.4 = 24.4$

Length of $BDCF = 9 + 5.1 + 6 + 2 = 22.1$

**3. Critical Path in Network:** The longest path is called the critical path in the network. The activities along a critical path are called critical activities.

Activities *B* and *E* are critical activities and path *BE* is critical path in Fig. 15.11.

**Q. 2. What is program evaluation and Review (PERT) technique?**

**Ans.** PERT is based on three time estimates for each activity: (*i*) most optimistic time, (*ii*) most pessimistic time, (*iii*) most likely time, using three time estimates often leads to overly optimistic results. PERT is used when there is a good deal of uncertainty regarding the time taken by various activities in project.

**Q. 3. What is topological order?**

**Ans.** A list of activities from a project is said to be in topological order if no activity appears in the list before all of its immediate predecessors have appeared. Topological ordering is possible if and only if there is no cycle in the network.

**Q.4. What are the uses of PERT/CPM for management.**

**Ans.** There are following use of PERT/CPM:

(*a*) PERT/CPM techniques help the management in properly planning the complicated projects, controlling the working plan and also keeping the plan up-to-date.

(*b*) These techniques help the management in reaching the goal at minimum time and minimum cost.

(*c*) Application of network techniques has resulted in better managerial control, better utilisation of resources, improved communication and progress reporting, and better decision-making.

(*d*) Application of PERT/CPM techniques have resulted in saving of time which directly results in the saving of cost.

**Q. 5. Discuss some application areas of PERT/CPM techniques.**

**Ans.** PERT/CPM are widely used in the following areas:

(*i*) **Manufacturing:** The design, and development and testing of new machines, installing and plants layouts are a few examples of PERT/CPM applications to the manufacturing function of a firm.

(*ii*) **Inventory Planning:** Installation of production and inventory control, acquisition of spare parts, and others, have been greatly helped by network techniques.

(*vi*) **Building Construction:** It is one of the largest areas in which the network techniques of project management have found applications.

(*iv*) **Administration:** Networks are used by the administration for streamlining paper work, for making major administrative changes in the system, for long-range planning and for developing staffing plans and so on.

(*v*) **Marketing:** Networks are also used for advertising programmes for development and launching of new products and for planning their distribution.

**Q. 6. A statue is to be erected in a park square on a stone bare which is to be built on a cement concrete foundation. The statue is to be made at another place moved to the base and erected. The various activity of the entire project and the three time estimates for each activity are given in the following table:**

| *Activity Time* | *Time taken for completion in days* | | |
|---|---|---|---|
| | *ot* | *mt* | *pt* |
| *A* : Make statue | 30 | 44 | 54 |
| *B* : Lift statue into place | 1 | 2 | 3 |
| *C* : Construct concrete foundation | 8 | 10 | 12 |
| *D* : Compact and level the site | 2 | 3 | 5 |
| *E* : Move statue to park square | 8 | 12 | 16 |
| *F* : Construct stone base | 14 | 22 | 25 |

(*a*) Draw PERT diagram

(*b*) Compute and identify critical path

(*c*) Variance of critical path

**Ans.** From the given operations it is clear that the operations *A* and *E* are independent of the operations *D*, *C* and *F*, *A*, *E* and the three operations *D*. *C* and *F* can be conducted simultaneously. Hence, we can draw a network in following topological order completion of all operation in two rows.

We construct a table which gives the three time estimates, expected values and variance for each operation.

| **Operation** | ***ot*** | ***mt*** | ***pt*** | $t = (ot + 4mt + pt)/6$ | $V(t) = \left(\frac{pt - ot}{6}\right)^2$ |
|---|---|---|---|---|---|
| *A* : Statue made | 30 | 44 | 54 | 43.3 | 16.0000 |
| *E* : Statue moved | 8 | 12 | 16 | 12.0 | 1.7777 |
| *B* : Statue lifted | 1 | 2 | 3 | 2.0 | 0.1111 |
| *D* : Site completed | 2 | 3 | 5 | 3.2 | 0.2500 |
| *C* : Foundation constructed | 8 | 10 | 12 | 10.0 | 0.4444 |
| *F* : Base constructed | 14 | 22 | 25 | 21.2 | 3.3611 |

Here we construct the PERT network diagram in which:

(*i*) Nodes are shown by squares (or rectangles).

(*ii*) The three time estimates are written at the top of the arrows connecting the operations.

(*iii*) The expected value *t* is written at the bottom of each arrow.

We have following PERT network diagram.

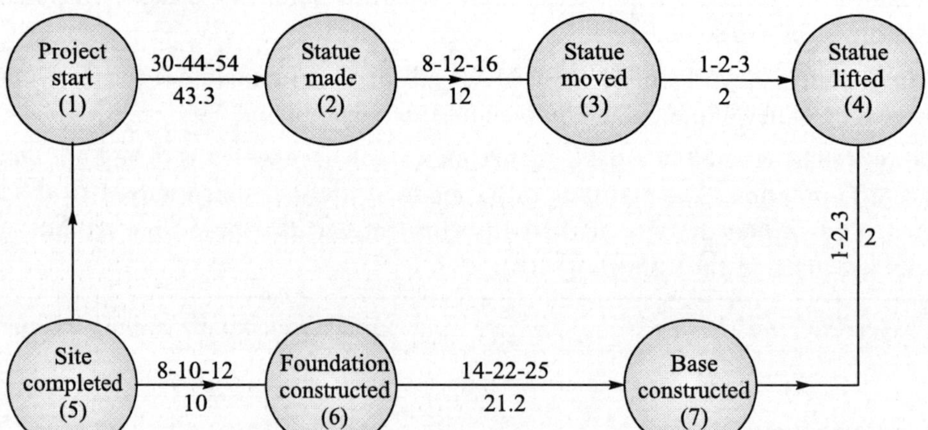


**Fig. 15.12.** PERT network diagram.

Activities are numbered as follows:

| Activity No. | Activity Name |
|---|---|
| 1 | Project start |
| 2 | Statue made (*A*) |
| 3 | Statue moved (*E*) |
| 4 | Statue lifted (*B*) |
| 5 | Site completed (*D*) |
| 6 | Foundation constructed (*C*) |
| 7 | Base constructed (*F*) |

It can be easily computed the critical path is

$1 \to 2 \to 3 \to 4$(see in Fig. 15.12)

(as longest path)

The total estimated time 43.3 + 12 + 2 = 57.3 days and the variance of the critical path

= Sum of the variances of the estimated times

= 16.0000 + 1.7777 + 0.1111 = 17.8888

## EXERCISES

**1.** Compute critical path of following network.

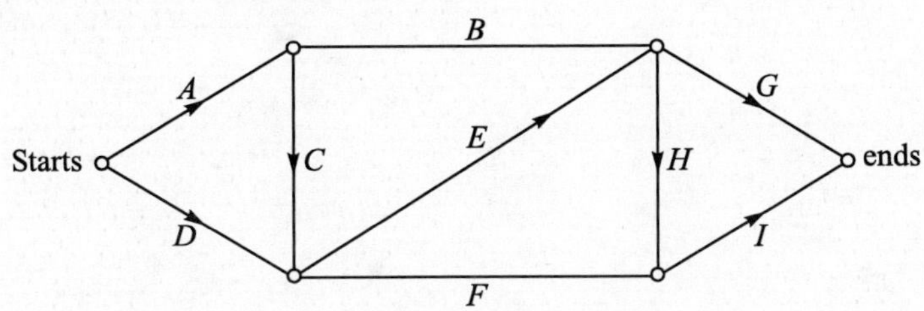


**Fig. 15.13.**

The values for the durations of activities *A*, *B*, *C*, *D*, *E*, *F*, *G*, *H*, *I* are 7, 8, 2, 8, 5, 7, 13, 2, 12 respectively.

**2.** Define following term:

1. Criticality index
2. Cash flow
3. Expediting

3. Show that the activities in a project network can be ordered topologically if and only if the network contains no cycles.
4. Discuss resource allocation and cost consideration.
5. Explain in brief simulation of an activity network.
6. What do you mean by uncertainties in activity durations?
7. What is Beta distribution?
8. How to implement forward pass and backward pass explain with example? Also draw flowcharts of both the method.
9. How you will analyses an activity network.
10. Define following term:

    (*a*) Critical path

    (*b*) Dummy activity

    (*c*) Duration

    (*d*) Activity graph

■■

# 16

# Real World Application of Simulation

**INSIDE THIS CHAPTER**



## 16.1 JUST-IN-TIME MODEL

***Just-in-time (JIT)*** is defined as "a philosophy of manufacturing based on planned elimination of all waste and on continuous improvement of productivity". It also has been described as an approach with the objective of producing the right part in the right place at the right time (in other words, "just-in-time"). Waste results from any activity that adds cost without adding value, such as the unnecessary moving of materials, the accumulation of excess inventory, or the use of faulty production methods that create products requiring subsequent rework. JIT (also known as *lean production* or *stockless production*) should improve profits and return on investment by reducing inventory levels (increasing the inventory turnover rate), reducing variability, improving product quality, reducing production and delivery lead times, and reducing other costs (such as those associated with machine setup and equipment breakdown). In a JIT system, underutilized (excess) capacity is used instead of buffer inventories to hedge against problems that may arise.

JIT applies primarily to *repetitive manufacturing* processes in which the same products and components are produced over and over again. The general idea is to establish flow processes (even when the facility uses a jobbing or batch process layout) by linking work centers so that there is an even, balanced flow of materials throughout the entire production process, similar to that found in an assembly line. To accomplish this, an attempt is made to reach the goals of driving all inventory buffers toward zero and achieving the ideal lot size of one unit.

The basic elements of JIT were developed by Toyota in the 1950's, and became known as the Toyota Production System (TPS). JIT was well-established in many Japanese factories by the early 1970's. A very simple shop floor control system that was developed by Toyota in the 1970s specifically for their Just-In-Time assembly plant, has received considerable attention in the Western World, and is known as the 'kanban' system (literally translated as 'card' system). Japan's success has prompted many scholars and practitioners, to turn their attention to Japanese management practices.

### 16.1.1 Elements of JIT

- Stabilize and level the MPS with uniform plant loading.
- Reduce or eliminate setup times.

- Reduce lot sizes (manufacturing and purchase).
- Reduce lead times (production and delivery).
- Preventive maintenance.
- Flexible workforce.
- Require supplier quality assurance and implement a zero defects quality program.
- Small lot (single unit) conveyance.

Modern manufacturing systems consist of many discrete operations that occur randomly and non-linearly for mathematical models or other methods to be practical. Simulation modeling in the form of discrete event simulation has evolved to become one of the most popular and cost-effective means of analysing complex systems . The technique of discrete-event simulation consists of developing a model of a system whose state changes at discrete intervals of time.

Most manufacturing systems, exhibit discrete production changes and therefore discrete-event simulation is an ideal methodology to apply to these systems. The decision to use simulation for manufacturing systems studies requires that a structured simulation study approach is adhered to, so that the designers can identify within the model, elements of the design which are critical to success and as Rucker proposes, this can then ensure that the simulation will greatly reduce the risks involved when planning automated manufacturing systems. A number of authors have investigated the merits of using simulation to assist with plant layout and factory prototyping and discusses these issues by proposing a simulation-based planning tool to help with decisions concerning the amount of capacity needed in a plant, the buffer size requirement and viewing the effects on throughput time as the plant design changes.

Before one begins to apply simulation techniques to manufacturing systems it is important to be aware of the problems that have to be overcome when attempting to provide solutions to real world situations. The task of simulating a system is broken down into many key items, each of which are diffcult to accomplish satisfactorily. These problems which need to be addressed comprise such tasks as problem and objective definitions through to model building, validation and data analysis.

Over the years there has emerged many simulation-based tools which can be used to help novice simulationists to conduct a simulation study with confidence. These tools have to some extent enabled a reduction to be made in the overall model development and analysis time scales, but many equally important issues can be seriously overlooked.

### 16.1.2 Modeling Issues for JIT Production

- Design of manufacturing cells–cell formation.
- Leveling production.
- Balancing for mixed-model lines.
- Sequencing for mixed-model lines.
- Modeling of pull systems with kanban.
- Determination of container size.
- Determination of number of kanban cards.
- Economic justification of setup reduction.
- Economic frequency of changeovers.
- Improvement curve analysis of changeovers.
- Lot-sizing models.
- Vendor evaluation.

## 16.2 COMPUTER NETWORK MODEL

### 16.2.1 Introduction

The Open Systems Interconnection (OSI) model is a reference tool for understanding data communications between any two networked systems. It divides the communications processes into seven layers. Each layer both performs specific functions to support the layers above it and offers services to the layers below it. The three lowest layers focus on passing traffic through the network to an end system. The top four layers come into play in the end system to complete the process. The TCP/IP has been used for network communications before the adoption of the OSI model, it supports the same functions and features in a differently layered arrangement.

### 16.2.2 OSI Model

The Open System Interconnection(OSI) model is a seven-layer structure that specifies the requirements for communications between two computers. The ISO (International Organization for Standardization) standard 7498-1 defined this model. This model allows all network elements to operate together, no matter who created the protocols and what computer vendor supports them.

- Physical layer
- Data Link layer
- Network layer
- Transport layer
- Session layer
- Presentation layer
- Application layer

#### Layer-1: Physical Layer

The **physical layer** of the OSI model defines connector and interface specifications, as well as the medium (cable) requirements. Electrical, mechanical, functional, and procedural specifications are provided for sending a bit stream on a computer network.

Components of the physical layer include:

- Cabling system components
- Adapters that connect media to physical interfaces
- Connector design and pin assignments
- Hub, repeater, and patch panel specifications
- Wireless system components
- Parallel SCSI (Small computer system interface)
- Network interface card (NIC)

#### Layer-2: Data Link Layer

Layer 2 of the OSI model provides the following functions:

- Allows a device to access the network to send and receive messages
- Offers a physical address so a device's data can be sent on the network
- Works with a device's networking software when sending and receiving messages

- Provides error-detection capability

Common networking components that function at layer 2 include:

- Network interface cards
- Ethernet and token ring switches
- Bridges.

## Layer 3: The Network Layer

Layer 3, the network layer of the OSI model, provides an end-to-end logical addressing system so that a packet of data can be routed across several layer 2 networks (Ethernet, Token Ring, Frame Relay, *etc*.). Note that network layer addresses can also be referred to as logical addresses.

To make it easier to manage the network and control the flow of packets, many organizations separate their network layer addressing into smaller parts known as subnets. Routers use the network or subnet portion of the IP addressing to route traffic between different networks. Each router must be configured specifically for the networks or subnets that will be connected to its interfaces. Routers communicate with one another using routing protocols, such as routing information protocol (RIP) and open version of shortest path first (OSPF), to learn of other networks that are present and to calculate the best way to reach each network based on a variety of criteria (such as the path with the fewest routers). Routers and other networked systems make these routing decisions at the network layer.

## Layer 4: The Transport Layer

Layer 4, the transport layer of the OSI model, offers end-to-end communication between end devices through a network. Depending on the application, the transport layer either offers reliable, connection-oriented or connectionless, best-effort communications. Some of the functions offered by the transport layer include:

- Application identification
- Client-side entity identification
- Confirmation that the entire message arrived intact
- Segmentation of data for network transport
- Control of data flow to prevent memory overruns
- Establishment and maintenance of both ends of virtual circuits
- Transmission-error detection
- Realignment of segmented data in the correct order on the receiving side
- Multiplexing or sharing of multiple sessions over a single physical link.

The most common transport layer protocols are the connection-oriented TCP transmission control protocol and the connectionless UDP (user datagram protocol).

## Layer 5: The Session Layer

Layer 5, the session layer, provides various services, including tracking the number of bytes that each end of the session has acknowledged receiving from the other end of the session. This session layer allows applications functioning on devices to establish, manage, and terminate a dialog through a network. Session layer functionality includes:

- Virtual connection between application entities
- Synchronization of data flow
- Creation of dialog units

- Connection parameter negotiations
- Partitioning of services into functional groups
- Acknowledgements of data received during a session
- Retransmission of data if it is not received by a device.

### Layer 6: The Presentation Layer

Layer 6, the presentation layer, is responsible for how an application formats the data to be sent out onto the network. The presentation layer basically allows an application to read (or understand) the message. Examples of presentation layer functionality include:

- Encryption and decryption of a message for security
- Compression and expansion of a message so that it travels efficiently
- Graphics formatting
- Content translation
- System-specific translation.

### Layer 7: The Application Layer

Layer 7, the application layer, provides an interface for the end user operating a device connected to a network. This layer is what the user sees, in terms of loading an application (such as web browser or e-mail); that is, this application layer is the data the user views while using these applications.

Examples of application layer functionality include:

- Support for file transfers
- Ability to print on a network
- Electronic mail
- Electronic messaging
- Browsing the World Wide Web

## 16.2.3 TCP/IP Model Overview

The OSI model describes computer networking in seven layers. While there have been implementations of networking protocol that use those seven layers, most networks today use TCP/IP. But, networking professionals continue to describe networking functions in relation to the OSI layer that performs those tasks.

**Layers of TCP/IP Model**

- Process layer
- Host to host layer
- Internet layer
- Network access layer

The TCP/IP model uses four layers to perform the functions of the seven-layer OSI model. The network access layer is functionally equal to a combination of OSI physical and data link layers (1 and 2). The Internet layer performs the same functions as the OSI network layer (3). Things get a bit more complicated at the host-to-host layer of the TCP/IP model. If the host-to-host protocol is TCP, the matching functionality is found in the OSI transport and session layers (4 and 5). Using UDP equates to the functions of only the transport layer of the OSI model. The TCP/IP process layer, when used with

TCP, provides the functions of the OSI model's presentation and application layers (6 and 7). When the TCP/IP transport layer protocol is UDP, the process layer's functions are equivalent to OSI session, presentation, and application layers (5, 6, and 7).

Now-a-days, networks is providing ubiquitous communication services to a large number of users. The objective is to suggest /develop new approach or model in reduced cost through better QoS. Simulation is an important techniques to design such type of network. It will focus principally on four research areas:

- understanding network structure
- understanding network dynamics
- predictive modeling and simulation for complex networked systems, and
- design, situational awareness and control of complex networks.

**Study approaches:**

1. **Experiment:** Put all routers together and let people use them.
2. **Mathematic model:** Model routers using a graph theory
3. **Simulation:** Use programming (*e.g.*, C++, or NS2) to represent routers.

Simulation overview in various layer of network model:

**Physical Layer:**

- Simulate channel model BER, PER etc.
- Matlab, Labview, ADS.

**Network Layer:**

- **Free:** NS2, GloMoSim
- **Commercial:** Opnet, QualNet

## 16.3 RELIABILITY ESTIMATION

Reliability and other reliability characteristics of a system can be expressed as some functions of the parameters of the distributions of the random variables (*r.v.'s*), strength and stress associated with the functioning of the system. Estimate these parameters and substitute these values in the expressions for reliability and other characteristics to get their estimates. The estimates of parameters used here are **maximum likelihood estimators** and as such from the invariance property of MLE's, the corresponding estimators of reliability are also MLE's. In absence of hard data the numerical values of the estimators can be obtained from simulation. There exists extensive literature for estimation of reliability analytically for single component systems. But the reliability expressions for multi-component systems are not easy to facilitate analytical estimation of reliability and its other characteristics. Also due to lack of hard data, one way out is simulation, in particular Monte Carlo simulation.

With simulation technique it is possible to estimate reliability (or probability of failure) or other reliability characteristics without going into the analytical techniques. The availability of personal computer and software makes the process comparatively simple. In fact, to evaluate the accuracy of the sophisticated analytical techniques or to verify a new technique, simulation is routinely used to independently evaluate the underlying probability distributions.

### 16.3.1 Monte Carlo Simulation

The Monte Carlo simulation method is an artificial sampling method which may be used for solving complicated problems in analytic formulation and for simulating purely statistical problems. In the simplest form of simulation, each *r.v.* in a problem is sampled several times to that indicates one realization of the problem itself. Solving the problem deterministically for each realization is known as a simulation cycle, a trial or a run. Using many simulation cycles we get the overall probabilistic characteristics of the problem, particularly when the number of cycles N is sufficiently large. Simulation, using a computer, is an inexpensive way (compared to laboratory testing) to study the uncertainty in a problem. The primary components of a Monte Carlo simulation include the followings:

(*i*) *Probability distribution function* (*probability density function*): The physical (or mathematical) system must be described by a set of probability distribution functions.

(*ii*) *Random number generator*: A source of random numbers uniformly distributed on the unit interval must be available.

(*iii*) *Sampling rule*: A prescription of sampling from specified distribution function, assuming the availability of random numbers on the unit interval.

(*iv*) *Scoring* (*tallying*): The outcome must be accumulated into overall tallies or scores for the quantities of input.

## 16.4 JOB SHOPS MODEL

Production scheduling is concerned with the allocation of resources and the sequencing of tasks to produce goods and services. In both theory and practice, much of the work on sequencing problems has been related to manufacturing. Consequently, the job shop has become a favourite theoretical construct to study the various components and interactions of complex sequencing problems.

### 16.4.1 Components of a Job Shop

A job shop model may include the following components:

(*a*) *operations*–elemental tasks

(*b*) *jobs*–one or more related operations that comprise a basic task module

(*c*) *events*–occurrences corresponding to the movement of jobs in the shop

(*d*) *machines*–the facilities which perform the operations

(*e*) *workers*–the resources which operate the machines

(*f*) *queues*–the sets of jobs waiting at machines

(*g*) *routes*–lists of the order and the corresponding machines in which the operations of a job have to be performed

(*h*) *bill of materials*–lists of parts and their quantities that are required for different operations of a job

(*i*) *time*

*arrival time*–the time at which the job is ready for processing at the shop

processing time–the time it takes for a machine to perform an operation for a particular job. Due-date—the time by which the job is supposed to be finished.

(*j*) *dispatching rules*–the methods that specify how machine operators choose which job in their queue to process next.

(*k*) *schedule*–the order in which the jobs are processed by the machines.

(*l*) *performance measures*–the criteria by which the schedule is evaluated.

### 16.4.2 Classification of Job Shops

Sequencing problems have been studied in a variety of job shop settings. Since conditions vary, different solution approaches have been required. The following classification illustrates the diversity of settings and common solution approaches. For a broad classification of various scheduling problems and a review of important theoretical developments of the different classes of problems. Job shops can be defined in the following terms:

(*a*) **The time environment:** The time environment of the shop can be deterministic or stochastic. In a deterministic shop, all times (*i.e.,* arrival, processing, due-date) are known and fixed. In a stochastic shop, any of the times may be random variables with a specified probability distribution.

(*b*) **The job arrival process:** In a static shop, all the jobs arrive simultaneously and thus are ready for processing at the same time. On the other hand, in a dynamic shop, jobs arrive at different times. In the static case, since all jobs are completely known and available, a fixed schedule can be made. Graphical, enumerative or mathematical programming methods are commonly used. However, in the dynamic case, the schedule can change whenever a new job arrives. Thus, heuristic methods are preferred for scheduling a dynamic shop.

(*c*) **The machine configuration:** The shop can have one to several different machines. The number of distinct machines describes an $n$-stage shop, where $n$ is the number of distinct machines. There can also be machines with identical functions and which are grouped into machine centers. This configuration is termed parallel machines. Shop configurations with one or two distinct machines can be studied with algebraic and probabilistic methods. Optimum schedules and dispatching rules have been found for this class of problems for some performance measures.

Optimal solutions have been found for some highly restrictive problems involving three machines. In general, however, for shops with more than two distinct machines, analytical methods have failed to find optimal solutions. This is due to the computational complexity of such problems.

(*d*) **The operation flow process:** If, at one extreme, all jobs follow the same route through the machines, the shop is termed as a pure flow shop. If, at the other extreme, each job has a unique route, the shop is termed as a pure job shop. In the pure flow shop, most solution approaches involve permutation schedules and heuristic procedures. In the pure job shop, branch and bound procedures, enumeration and sampling methods for static and deterministic conditions, and simulation experiments with dispatching rules are used. If each job may have only one of a fixed set of routings (which implies a fixed line of products), the shop is called a closed job shop. On the other hand, if a job may have any arbitrary route, the shop is termed an open job shop. Open shops are concerned with sequencing jobs through the machines and use similar solution approaches as pure job shop problems. Closed shops involve the additional problem of lot-sizing because jobs are generated by inventory replenishment decisions. In addition to the approaches used in open shop problems, closed shop problems also use lot-sizing methods.

### 16.4.3 Dispatching Rules

One of the earliest and most extensive studies on dispatching rules. The dispatching rules are classified on the basis of information requirements into local rules and global rules. Local rules only require information on the jobs waiting at a machine, while global rules require additional information about jobs or machines in other parts of the shop. Another way to classify the rules are as simple rules and composite rules. Simple rules are based on only one kind of information (local or global) while composite rules are combinations of simple rules. Dispatching rules are techniques by which a value is assigned to each waiting job and the job with the minimum value is selected. Heuristics are some "rule of thumb" while a scheduling rule is a combination of one or more priority rules and/or one or more heuristics. Static rules are those in which job priority values do not change with time while dynamic rules are the juxtaposition of static rules.

### 16.4.4 The Simulation Methodology

Dispatching rules in multi-stage open job shops are evaluated using simulation. A brief overview of the simulation methodology explains its appropriateness for this class of sequencing problems. Simulation is a strategy evaluation technique which uses an abstract representation of reality (*i.e.*, a model) and studies its behaviour through time. The behaviour may be influenced by certain or uncertain factors. For models which consider uncertainties, the technique involves the following steps:

(*a*) Describe the system to be studied.

(*b*) Formulate simplifying assumptions about the system.

(*c*) Under the set of assumptions, identify:

- *Parameters:* System attributes which are held constant during the period that the system is being studied.
- *Exogenous variables:* System attributes which are subject to random variations through time. The variations are represented by appropriate probability distributions.
- *Endogenous variables:* System attributes whose values are generated by changes in the exogenous variables.

(*d*) Develop a model which embodies the interrelationships among the parameters, the exogenous and the endogenous variables.

(*e*) Use a random number generator to generate a set of intertemporal events based on the random variation of the exogenous variables.

(*f*) Run the model.

(*g*) Collect statistics on the resulting values of the endogenous variables.

(*h*) Analyze results with descriptive and inferential statistical methods.

(*i*) Draw conclusions and/or propose new sets of model attributes to observe and analyze.

Simulation is an alternative for problems where analytical solutions are practically impossible to compute. Alternatives to the use of simulation are mathematical analysis, experimentation with either the actual system or a prototype of the actual system, or reliance upon experience and intuition. All, including simulation, have limitations. Mathematical analysis of complex systems is very often impossible. Experimentation with actual or pilot systems is costly and time consuming, and relevant variables are not always subject to control. Intuition and experience are often the only alternatives to (computer) simulation available but can be very inadequate. Simulation problems are characterized by being mathematically intractable and having resisted solution by analytical methods.The problems usually involve many variables, many parameters, functions which are not well-behaved mathematically, and random variables. Thus simulation is a technique of last resort. Simulation applies to the design and analysis of system behaviour. For systems design, alternative designs can be generated by using different sets of parameters and exogenous variables, and their results compared to some norm. In systems analysis, the object of study is how real transformations take place. Simulation involves developing a model which embodies the hypothesis for the transformation, supplying the model with real data and comparing the model's results with real outcomes to verify the model's hypothesis.

## 16.4.5 Job Shop Simulation Models: Assumption

**Model Assumption**

A manufacturing system consists of 5 workstations say 1,2... .5 consists of 3, 2, 4, 3 and 1 identical machine, respectively. The system is a network of 5 multiserver queues. Assume that jobs arrive at the system with interarrival times that are IID exponential random variables with mean 0.25 hr. There are 3 types of jobs, and arriving jobs are of type 1, 2, and 3 with respective probabilities 0.3, 0.5, and 0.2. Job type 1, 2, and 3 require 4, 3, and 5 task to be done, respectively, and each task must be done at a specified station and in a prescribed order. The routings for the different job types are:

| *Job types* | *Workstation in routing* |
|---|---|
| 1 | 3, 1, 2, 5 |
| 2 | 4, 1, 3 |
| 3 | 2, 4, 1, 4, 3 |

type 2 jobs, first have a task lone at station 4, then 1 and last at 3. If a jobs arrives at a particular station and finds all machines in that station already busy, the job joins a single FIFO queue at that station. We chose Erlang disribution to represent service time.

**Flow chart**

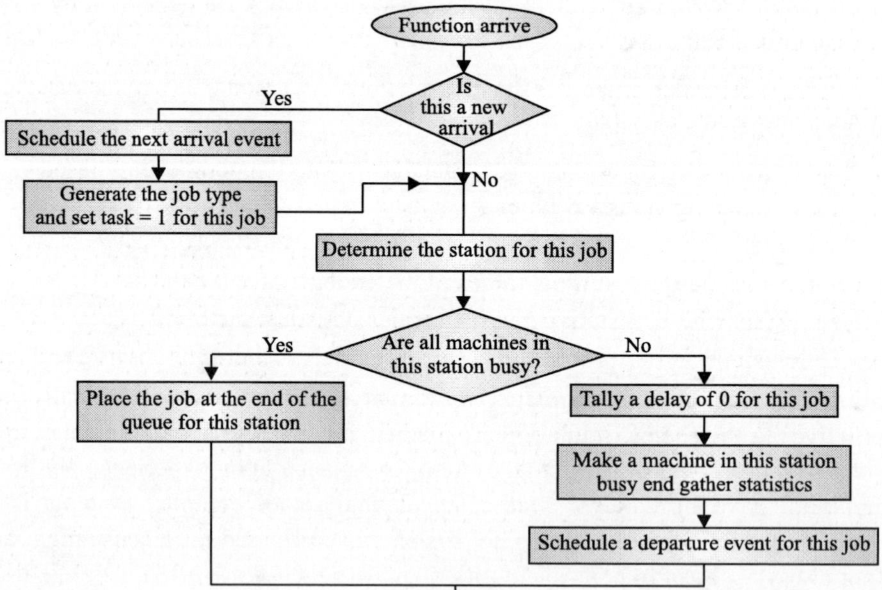


Now convert this flow chart in suitable programming language such as C, SIMLIB, GPSS, etc., for output analysis of simulation run.

We can discuss following point as a simulation output.

- Overall average job total delay
- Average utilization
- Average number in queue
- Average delay in queue

## 16.5 WARRANTY ESTIMATION PROBLEM USING SIMULATION

Almost all products, whether sold directly to the customer or to a producer for assembly into a customer product, now carry a warranty of some kind. the nature and extent of the warranty affect the sales, market share, costs and profits of many businesses.

A warranty can be defined as an assurance from a seller to a buyer that the product sold under waranty is guaranteed to perform satisfactorily up to a certain length of time, the warranty period.

waranties are important to buyers and sellers alike. Buyers need waranties to satisfy their need for assurance that the product will perform satisfactorily. These are a number of the reasons why a buyer is skeptical about satisfactory performance of the product. some reasons are:

- Imperfect standardizations
- Imperfaction due to product innovation
- Inability on the part of the buyer to determine the quality of the product, & so on.

Sellers use warranties mainly for promotional & protectional purposes. A protective warranty is designed to guard sellers from unreasonable claims of purchases.

### Simulation

The use of mathematical functions to determine the life of a product and ultimately the cost of a specific rebate plan enables the user to have a very precise answer. If the life of a product cannot be written in a mathematical formula format, where the probability of occurrence is obtained through integration then the function would be declared to be inappropriate and would be difficult to make a warranty analysis using a mathematical approach. This would not be true in a simulated environment. These simulations were run on a mainframe system, GPSS is available on the PC and would permit the instant analysis of alternate payout plans and alternative length of life distributions. The basic thrust of the simulated approach is the building of a cumulative probability distribution that basically represents the complete length of life product. The simulation is run for a specified number of units, and the individual warranty costs are accumulated for all the units and represent a part of the output at the completion of the simulation.

## 16.6 CAPITAL RECOVERY MODEL

### Definition of 'Capital Recovery'

- The earning back of the initial funds put into an investment. Capital recovery must occur before a company can earn a profit on its investment.
- A euphemism for debt collection. Capital recovery companies obtain overdue payments from individuals and businesses that have not paid their bills. Upon obtaining payment and remitting it to the company to which it is owed, the capital recovery company earns a fee for its services.
- A company's recouping of the money it has invested in machinery and equipment through asset disposition and liquidation.

### Building Recovery Capital

Recovery capital has four components:

**Social capital:** Is about building relationships that include two-way obligations and support.

**Physical capital:** Is about tangible assets that improve recovery options, *e.g.* money and paid work.

**Human capital:** Is about skills, good health, aspirations, hopes and confidence.

**Cultural capital:** Is values, beliefs and attitudes.

### Estimating Capital Recovery Costs

Estimating capital recovery costs requires several steps:

The first is to gather the information on the cost of recovering the capital from an investment. This includes the initial cost of the investment (in private and social terms), an estimate of the useful life of the machinery, its salvage value, and the total number of "horsepower" hours it is expected to provide. The initial cost and useful life of the machine represent the two most important parameters of the capital cost recovery calculations. After its useful life has expired, the machine may still have a salvage value as scrap and a source of parts. The salvage value is received several years in the future and, therefore, must be discounted using data on the private and social interest rates found in the existing Prices tables. It is then deducted from the initial cost to derive today's net cost.

Perhaps the most complicated derivation is the recovery ratio, which involves adjusting the interest rate by the life expectancy of the investment. As defined by Monke and Pearson, this is the share of the net cost that must be recovered each year "to repay the cost of the fixed input at the end of its useful life." Once the recovery ratio is determined, the actual monetary sum is calculated. This figure, the annual recovery cost, is then prorated to an hourly basis.

Once calculated, the private and social costs of recovering investment capital are integrated into the spreadsheet in the same manner as was used to decompose nontradables in the previous chapter. The Input-Output table is expanded to account for the capital input. The newly-derived cost data are similarly incorporated in the Prices tables. Budget tables are likewise expanded. The PAMs adjust accordingly.

## 16.7 INSURANCE-CLAIM MODEL

The insurance industry can be modeled as stochastic modeling which looks at a single policy, an entire portfolio or an entire company. But rather than setting investment returns according to their most likely estimate, for example, the model uses random variations to look at what investment conditions might be like.

"Stochastic" means being or having a random variable. A stochastic model is a tool for estimating probability distributions of potential outcomes by allowing for random variation in one or more inputs over time.

Based on a set of random outcomes, the experience of the policy/portfolio/company is projected, and the outcome is noted. Then this is done again with a new set of random variables. In fact, this process is repeated thousands of times.

A distribution of outcomes is available which shows not only the most likely estimate but what ranges are reasonable too. The most likely estimate is given by the distribution curve's center of mass which is typically also the peak(mode) of the curve, but may be different *e.g.* for asymmetric distributions. This is useful when a policy or fund provides a guarantee, *e.g.* a minimum investment return of 5% per annum. A deterministic simulation, with varying scenarios for future investment return, does not provide a good way of estimating the cost of providing this guarantee. This is because it does not allow for the volatility of investment returns in each future time period or the chance that an extreme event in a particular time period leads to an investment returns less than the guarantee. Stochastic modelling builds volatility and variability (randomness) into the simulation and therefore provides a better representation of real life from more angles.

The claims arising from policies or portfolios that the company has written can also be modelled using stochastic methods. This is especially important in the general insurance sector, where the claim severities can have high uncertainties. Depending on the portfolios under investigation, a model can simulate all or some of the following factors stochastically:

- Number of claims
- Claim severities
- Timing of claims

Claims inflations can be applied, based on the inflation simulations that are consistent with the outputs of the asset model, as are dependencies between the losses of different portfolios.The relative uniqueness of the policy portfolios written by a company in the general insurance sector means that claims models are typically tailor-made. Estimating future claims liabilities might also involve estimating the uncertainty around the estimates of claim reserves.

## 16.8 SIMULATION OF A INVENTORY

[See Chapter 3, Sec. 3.14]

## 16.9 Pi VALUE ESTIMATION

[See Chapter 3, Sec. 3.15]

## SUMMARY

- Just-in-time (JIT) is defined as "a philosophy of manufacturing based on planned elimination of all waste and on continuous improvement of productivity". Modern manufacturing systems consist of many discrete operations that occur randomly and non-linearly for mathematical models or other methods to be practical. Simulation modelling in the form of discrete event simulation has evolved to become one of the most popular and cost-effective means of analysing complex systems.
- The Open System Interconnection model is a seven-layer structure that specifies the requirements for communications between two computers. The ISO (International Organization for Standardization) standard 7498-1 defined this model. This model allows all network elements to operate together, no matter who created the protocols and what computer vendor supports them.
- With simulation technique it is possible to estimate reliability (or probability of failure) or other reliability characteristics without going into the analytical techniques.
- The job shop has become a favorite theoretical construct to study the various components and interactions of complex sequencing problems. Rules in multi-stage open job shops are evaluated using simulation. The simulation methodology explains its appropriateness for this class of sequencing problems.
- Simulation is useful in warranty estimation problem.

## SOLVED QUESTIONS

**Q. 1. Discuss the simulation methodology of job-shops model.**

**Ans.** Dispatching rules in multi-stage open job shops are evaluated using simulation. A brief overview of the simulation methodology explains its appropriateness for this class of sequencing problems. Simulation is a strategy evaluation technique which uses an abstract representation of reality (*i.e.* a model) and studies its behavior through time. The behavior may be influenced by certain or uncertain factors. For models which consider uncertainties, the technique involves the following steps:

(*a*) Describe the system to be studied.

(*b*) Formulate simplifying assumptions about the system.

(*c*) Under the set of assumptions, identify:

(*i*) **Parameters:** System attributes which are held constant during the period that the system is being studied.

(*ii*) **Exogenous variables:** System attributes which are subject to random variations through time. The variations are represented by appropriate probability distributions.

(*iii*) **Endogenous variables:** System attributes whose values are generated by changes in the exogenous variables.

(*d*) Develop a model which embodies the interrelationships among the parameters, the exogenous and the endogenous variables.

(*e*) Use a random number generator to generate a set of intertemporal events based on the random variation of the exogenous variables.

(*f*) Run the model.

(*g*) Collect statistics on the resulting values of the endogenous variables.

(*h*) Analyze results with descriptive and inferential statistical methods.

(*i*) Draw conclusions and/or propose new sets of model attributes to observe and analyze.

**Q. 2. Discuss the primary components of monte carlo simulation.**

**Ans.** The primary components of a Monte Carlo Simulation include the followings:

(*i*) **Probability distribution function (or probability density function):** The physical (or mathematical) system must be described by a set of probability distribution functions.

(*ii*) **Random number generator:** A source of random numbers uniformly distributed on the unit interval must be available.

(*iii*) **Sampling rule:** A prescription of sampling from specified distribution function, assuming the availability of random numbers on the unit interval.

(*iv*) **Scoring (or tallying):** The outcome must be accumulated into overall tallies or scores for the quantities of input.

## EXERCISES

1. Discuss the simulation methodology in job shop model.
2. what is Monte Carlo simulation?
3. Discuss the computer network model. How does simulation suitable for its modeling.
4. What is capital recovery model?
5. Discuss the key elements of JIT model.

# A

# Appendix

## A. USE OF DATABASE IN SIMULATION

We know that simulation program are typically voluminous. It requires heavy data for input, output and for the validation and verification of simulation models and programs. Managing these data using external flat files is difficult. So we need a powerful database. A modern database management system (DBMS) offers powerful and convenient features for the organization, storage, and retrieval of simulation data.

A DBMS provides standard and customizable interfaces for data entry and enforces integrity constraints for data validation. Simulation input may come from existing database; for example, Randell and Bolmsjo used a database describing a factory to simulate the factory system.

Following figure represents the interaction between a SIMSCRIPT II.5 program and a DBMS.

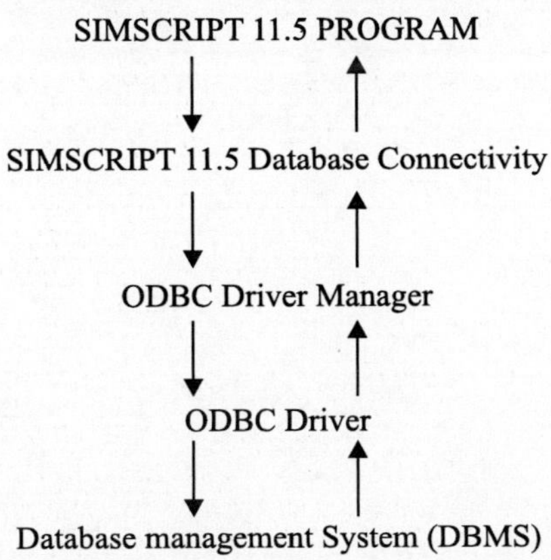


**Interaction between a SIMSCRIPT II.5 program and a DBMS.**

Following advantages are there to use DBMS in simulation:

1. The strength of DBMS query facility can be applied to retrieve and aggregate and portion of the output from one or may simulation runs and models.
2. The DBMS enables concurrent access to simulation data by multiple users and programs.
3. The DBMS prevents unauthorized access to data.
4. The DBMS provides backup and recovery of simulation data.

## A. USE OF AI IN SIMULATION

Simulation and AI can be used to help to invent new technologies and improve the use of the existing ones for example—Evolving Anti-Terrorist Scenarios for wargaming. Simulation and AI are also being used to assist contingency planning for military operations and their logistic support. Simulation serves mostly to explore the feasibility of human-generated plans, and AI serves mostly for assembling rational plans from prestablished component. There would seem to be many other ways in which AI could be used to expand and enhance. The benefits of simulation both in training and in the execution of war against terrorism.

**Note:** AI can be used in team training simulators.

■■

# B

# Appendix

## Deadlock Detection Model

### B. INTRODUCTION TO DEADLOCKS

For many applications, a process needs exclusive access to not one resource, but several. Suppose, for example, two processes each want to record a scanned document on a CD. Process A requests permission to use the scanner and is granted it. Process B is programmed differently and requests the CD recorder first and is also granted it. Now A asks for the CD recorder, but the request is denied until B releases it. Unfortunately, instead of releasing the CD recorder B asks for the scanner. At this point both processes are blocked and will remain so forever. This situation is called a deadlock. Deadlock can be defined formally as follows:

A set of processes is deadlocked if each process in the set is waiting for an event that only another process in the set can cause.

In most cases, the event that each process is waiting for is the release of some resource currently possessed by another member of the set. In other words, each member of the set of deadlocked processes is waiting for a resource that is owned by a deadlocked process. None of the processes can run, none of them can release any resources, and none of them can be awakened. The number of processes and the number and kind of resources possessed and requested are unimportant. This result holds for any kind of resource, including both hardware and software.

Deadlocks can occur when processes have been granted exclusive access to devices, files, and so forth. To make the discussion of deadlocks as general as possible, we will refer to the objects granted as **resources**. A resource can be a hardware device (*e.g.*, a tape drive) or a piece of information (*e.g.*, a locked record in a database). A computer will normally have many different resources that can be acquired. For some resources, several identical instances may be available, such as three tape drives. When several copies of a resource are available, any one of them can be used to satisfy any request for the resource. In short, a resource is anything that can be used by only a single process at any instant of time.

Resources come in two types: preemptable and nonpreemptable. A preemptable resource is one that can be taken away from the process owning it with no ill effects. Memory is an example of preemptable resource.

A **nonpreemptable resource**, in contrast, is one that cannot be taken away from its current owner without causing the computation to fail. If a process has begun to burn a CD-ROM, suddenly taking the CD recorder away from it and giving it to another process will result in a garbled CD, CD recorders are not preemptable at an arbitrary moment.

**Conditions for Deadlock:** There are four conditions must hold to be a deadlock[1]:

1. **Mutual exclusion condition:** Each resource is either currently assigned to exactly one process or is available.
2. **Hold and wait condition:** Processes currently holding resources granted earlier can request new resources.
3. **No preemption condition:** Resources previously granted cannot be forcibly taken away from a process. They must be explicitly released by the process holding them.
4. **Circular wait condition:** There must be a circular chain of two or more processes, each of which is waiting for a resource held by the next member of the chain.

All four of these conditions must be present for a deadlock to occur. If one of them is absent, no deadlock is possible. It is worth noting that each condition relates to a policy that a system can have or not have.

Deadlock occurs in simulation systems due to the dynamic interaction of entities and resources in the course of the simulation. Deuermeyer et al. [5] a graph formalism that represents this ynamic interaction to define deadlock characteristics in discrete simulation systems. The procedure presented detects and resolves deadlocks under the condition that entities are restricted to single-unit resource requests. A distinction is made between self-resolvable and nonself-resolvable deadlocks in simulation systems. Self-resolvable deadlocks are called transient deadlocks, whereas non self-resolvable deadlocks are called permanent deadlocks.

Transient deadlocks have the potential to resolve themselves over time. Such a situation can occur when an entity currently using a multi-unit resource returns the resource unit(s) in its possession. The re-introduction of additional resource units into the simulation can allow other deadlocked entities to use the freed resource and, consequently, relieve the blockage situation in the simulation. Although transient deadlocks have the potential for self-resolution, they cause a loss of processing time and consequent degradation in the simulation generated statistics. In the case of permanent deadlocks, external intervention is required in order for the deadlocked resources to be freed and the entities to continue through the simulation.

In the multiple-unit resource request case, there is a similar distinction between self-resolvable and non self-resolvable deadlocks. However, unlike the single-unit request case, we must also distinguish between two categories of permanent deadlocks. These are called pending deadlock and total deadlock.

A transition diagram of the simulation deadlock states under a multiple-unit request condition is shown in Figure B.1. The vertices in the figure represent possible deadlock states in the natural evolution

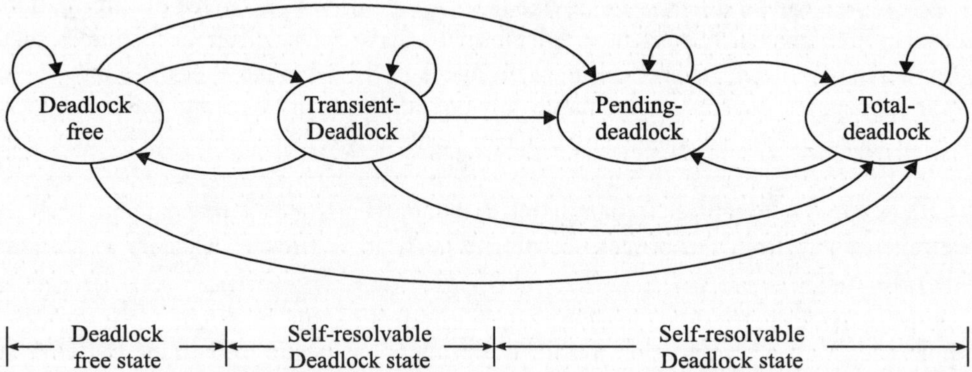


**Fig. B.1** Transition diagram for simulation deadlock states.

of a simulation realization and the edges define feasible state transitions. The simulation starts in a deadlock-free state and can move through the different deadlock states as resource requests and assignments are made during the course of the simulation. Note that once the simulation enters the nonself-resolvable states of either pending or total deadlock, external intervention is required for the deadlocked resources and entities to be freed.

## B.1 Deadlock Modeling

Holt [1] showed how these four conditions can be modeled using directed graphs. The graphs have two kinds of nodes: processes, shown as circles, and resources, shown as squares. An arc from a resource node (square) to a process node (circle) means that the resource has previously been requested by, granted to, and is currently held by that process. In Fig. B.2 (*a*), resource R is currently assigned to process A.

An arc from a process to a resource means that the process is currently blocked waiting for that resource. In Fig. B.2 (*b*), process B is waiting for resource S. In Fig. B.2 (*c*) we see a deadlock: process C is waiting for resource T, which is currently held by process D. Process D is not about to release resource T because it is waiting for resource U, held by C. Both processes will wait forever. A cycle in the graph means that there is a deadlock involving the processes and resources in the cycle (assuming that there is one resource of each kind). In this example, the cycle is C–T–D–U–C.

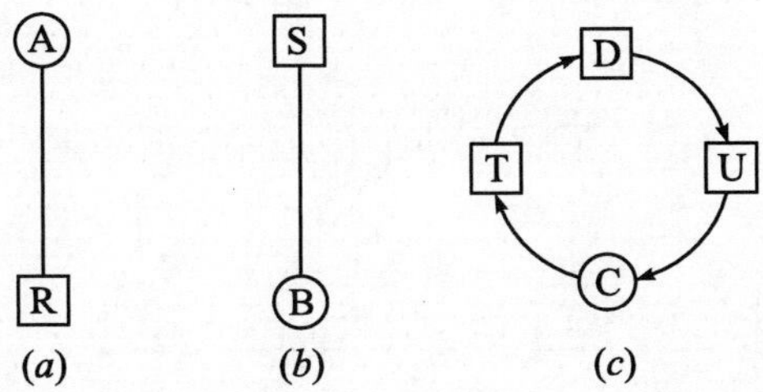


**Fig. B.3.** Resource allocation graphs. (*a*) Holding a resource. (*b*) Requesting a resource. (*c*) Deadlock.

Now let us look at an example of how resource graphs can be used. Imagine that we have three processes, A, B, and C, and three resources R, S, and T. The requests and releases of the three processes are given in Fig. B.2 (*a*).(*c*). The operating system is free to run any unblocked process at any instant, so it could decide to run A until A finished all its work, then run B to completion, and finally run C.

This ordering does not lead to any deadlocks (because there is no competition for resources) but it also has no parallelism at all. In addition to requesting and releasing resources, processes compute and do I/O. When the processes are run sequentially, there is no possibility that while one process is waiting for I/O, another can use the CPU. Thus running the processes strictly sequentially may not be optimal. On the other hand, if none of the processes do any I/O at all, shortest job first is better than round robin, so under some circumstances running all processes sequentially may be the best way.

Let us now suppose that the processes do both I/O and computing, so that round robin is a reasonable scheduling algorithm. The resource requests might occur in the order of Fig. B.3 (*d*). If these six requests are carried out in that order, the six resulting resource graphs are shown in Fig. B.3(*e*)-(*j*), After request 4 has been made, A blocks waiting for S, as shown in Fig. B.3 (*h*). In the next two steps B and C also block, ultimately leading to a cycle and the deadlock of Fig. B.3 (*j*).

However, as we have already mentioned, the operating system is not required to run the processes in any special order. In particular, if granting a particular request might lead to deadlock, the operating system can simply suspend the process without granting the request (*i.e.*, just not schedule the process) until it is safe. In Fig. B.3, if the operating system knew about the impending deadlock, it could suspend B instead of granting it S. By running only A and C, we would get the requests and releases of Fig. B.3 (*k*) instead or Fig. B.3 (*d*). This sequence leads to the resource graphs of Fig. B.3 (*l*).(*q*), which do not lead to deadlock.

After step (*q*), process B can be granted S because A is finished and C has everything it needs. Even if B should eventually block when requesting T no deadlock can occur. B will just wait until C is finished.

| A | A | C |
|---|---|---|
| Request R | Request S | Request T |
| Request S | Request T | Request R |
| Release R | Release S | Release T |
| Release S | Release T | Release R |
| (*a*) | (*b*) | (*c*) |

1. A requests R
2. B requests S
3. C requests T
4. A requests S
5. B requests T
6. C requests R

deadlock
(*d*)

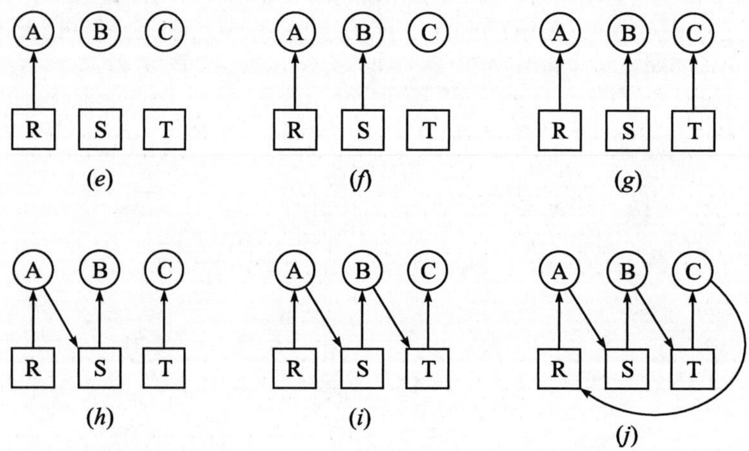


1. A requests R
2. C requests T
3. A requests S
4. C requests R
5. A releases R
6. C releases S

no deadlock
(*k*)

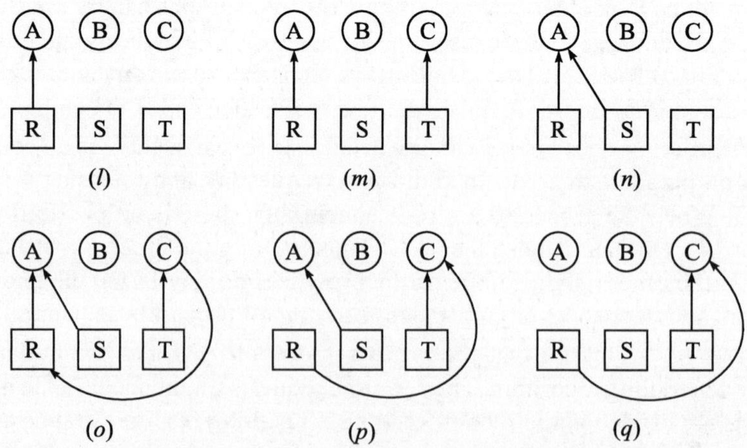


**Fig. B.3** An example of how deadlock occurs and how it can be avoided.

Later in this chapter we will study a detailed algorithm for making allocation decisions that do not lead to deadlock. For the moment, the point to understand is that resource graphs are a tool that let us see if a given request/release sequence leads to deadlock. We just carry out the requests and releases step by step, and after every step check the graph to see if it contains any cycles. If so, we have a deadlock; if not, there is no deadlock. Although our treatment of resource graphs has been for the case of a single resource of each type, resource graphs can also be generalized to handle multiple resources of the same type (Holt, 1972). Researchers have identified three standard approaches to manage deadlock situations:

1. Prevention, by structurally negating one of the four conditions necessary to cause a deadlock.
2. Dynamic avoidance by careful resource allocation.
3. Detection and resolution. Let deadlocks occur, detect them, and take action, prevention, avoidance, and detection and resolution.

While the first two approaches ensure that deadlock will never occur, the last approach allows deadlock to occur and resolves it appropriately. Deadlock prevention averts deadlock by imposing constraints on how the resource requests are made, ensuring that the necessary conditions for deadlock can not occur. Examples of prevention strategies include allocating all of the resource requests before each entity begins execution or a waiting entity returns resources requested by an active entity. Examples of preventive measures in manufacturing systems control include providing a large number of in-process buffers and uni-directional batching of jobs . In the case of simulation, such an approach could potentially restrict the level of realism that can be implemented in the simulation model. For example, in a simulation of a manufacturing situation, this would mandate acquiring both the operator and the machine during a routine setup, even if the machine is not required for the setup.

Deadlock avoidance works by ensuring that the sufficient conditions for deadlock are not met. In deadlock avoidance, every resource request is analyzed prior to execution, and only allowed to occur if it does not cause a deadlock. This analysis requires knowledge of future events of the current system realization. Habermann [8] presents a deadlock avoidance algorithm in which each entity declares the maximum possible number of units of each resource it may need. Given this information, as each resource request is made, an assignment is authorized only if there exists at least one sequence of executions that does not evolve to a deadlock. The stochastic nature of simulation, however, makes it impossible to analyze the effects of future events. Furthermore, the general deadlock avoidance problem has been shown to be NP-Complete even when times are known with certainty [6].

In the deadlock detection and resolution strategy, resource requests are granted without considering the potential for deadlock. At appropriately chosen times, a deadlock detection procedure is invoked. If the procedure identifies a deadlock, the deadlock is resolved. Given that deadlock prevention can be restrictive and deadlock avoidance is infeasible in most simulations, we adopt the detection and resolution strategy for handling deadlocks. The deadlock detection problem has been studied extensively by previous researchers (Cho *et al.* [2], Coffman *et al.* [3], Holt [1], and Singhal [20] present comprehensive surveys of deadlock detection procedures in distributed systems. In the context of the real-time control of manufacturing systems with no alternate part-routings, deadlock detection reduces to the detection of cycles in the graph model (Cho *et al.* [2].

Many algorithms presented to detect/prevent deadlocks reorder the waiting entities into a nondecreasing list of entities by their request size. The algorithms then attempt to find at least one possible sequence of executions that does not result in deadlock (Habermann [8], Coffman *et al.* [3], Holt [10]). In the case of simulation systems, re-ordering of the waiting entities may violate the reallocation rule of the resource. The reallocation rule decides whether the resource queue allows (or disallows) late arrivals in the queue to pass (or not pass) stalled entities. In the case of multiple-unit requests, such a policy can alter the outcome of the deadlock detection by erroneously designating a deadlocked simulation to be free from deadlock. This will be demonstrated by examples in the following section. In this paper, we present a mechanism by which we can model different reallocation policies in a general simulation system. This transformation allows the use of existing procedures for deadlock detection and, further, allows for developing a generic deadlock detection procedure ignoring nuances due to different resource allocation rules.

Resolution is the second half of the detection/resolution strategy. The problem of optimally resolving deadlocks is known to be NP-hard (Gray [7], Leung and Lai [14]). In the case of real-time control of manufacturing systems, resolution consists of removing a deadlocked entity from the resource (machine) it holds, placing it in a temporary buffer, and reallocating the released resource to a waiting entity (Cho *et al.*[2]. The resolution procedure in computer applications typically chooses a set of entities to be aborted and restarted or partially rolled-back in order to break the deadlock. For example, an operating system would resubmit a job to start all over again or a database system would abort a set of transactions to be restarted.

Due to the varied nature of deadlocks in a general simulation system, such removal or roll-back of entities are not directly applicable in the context of modeling manufacturing systems. Deuermeyer *et al.* [5] present two situation-based deadlock resolution schemes that enable the simulation to automatically recover from deadlock situations without any additional burden on the simulation modeler. Later in this paper we present a resolution scheme for multiple-unit requests.

**B.2 Assumptions and Deadlock Examples**

We are primarily concerned with entity and resource relationships that are modeled using the time-dependent graph described by Deuermeyer *et al.* [5]. Attention is restricted to the seize and release statements that generate these relationships in a general simulation model. Other aspects of the simulation model are not germane to this analysis and are therefore not considered. In this section we will present three examples of deadlocks in simulation. The first example demonstrates the evolution of a simulation that leads to a total deadlock with entities restricted to single-unit resource requests. The details of the detection and resolution procedures under these conditions are described in Deuermeyer *et al.* [5]. The next two examples demonstrate the effects of multiple unit resource requests and the reallocation rule used by the resource to refill unassigned units on the classification of the deadlock.

The MOR/DS simulation paradigm (Curry *et al.* [4]) that we employ is similar to the approaches taken in both SIMAN (Pegden *et al.* [16]) and GPSS (Schriber [18, 19]). Under this paradigm, entities acquire necessary units of a resource through a simulation structure called the seize statement, and return the acquired resource units through a simulation structure called the release statement. Entities can request one resource type at a time in the simulation and, thus, entities requiring more than one type of resource simultaneously must acquire them sequentially. Resources can have a capacity greater than one and there is a unique queue associated with each resource. Furthermore, we assume that the total number of units simultaneously held and/or requested by a single entity never exceeds the total capacity of the resource.

The seize statement maintains the number of available units and the prioritized queue of entities waiting for a resource. If a sufficient number of resource units are available when an entity makes a request, the seize statement assigns the appropriate number of resource units to the requesting entity and updates the available capacity. If the requested number of units is not available, the requesting entity is inserted in the queue according to the user-specified queue discipline (*e.g.*, first-in, first-out (FIFO), last-in, first-out (LIFO), priority number, *etc.*). The release statement manages the number of available resource units and the resource queue as entities relinquish control of the resource. As resource units are returned by an entity, the release statement reallocates the resource to waiting entities using a resource fill rule.

The fill rule, designated as either pass or no-pass, determines whether entities waiting in a resource queue can pass other entities in the same queue. Passing of entities can be a significant factor in the multiple-unit seize case where the first entity in the queue requests more units than are currently unassigned. Obviously, under these conditions the first entity will be forced to wait in the queue. However, in some cases, requests from entities further back in the queue could be satisfied if those entities were allowed to "pass" the stalled entity. The SIMAN language uses a no-pass fill rule, while GPSS uses the pass fill rule. In the MOR/DS simulation language, the fill rule for each queue is specified by the analyst when developing the model.

In the illustrations that follow, we use rectangles to represent resources and circles to represent entities. Notationally we represent the $i^{th}$ entity by $e_i$ and the $j^{th}$ resource by $r_j$. A resource assignment of $k$ units to an entity is depicted with a directed edge labeled $k$ from the resource to the entity [Figure B.4($a$)], while a resource request of $k$ units is depicted with a directed edge labeled $k$ from the entity to the resource [Figure B.4($b$)].

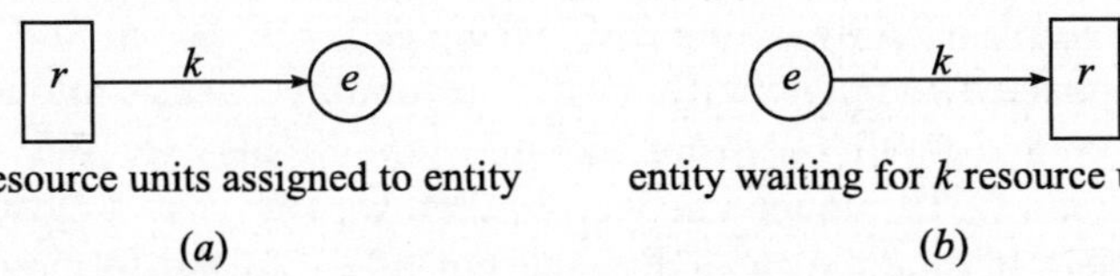


**Fig. B.4** Entity-resource relationship scheme. (a) assignment edge, (b) request or queue edge.

Note that we drop the labeling on the edges with unit requests or unit assignments without ambiguity. We assume that a request for a resource that can be satisfied changes immediately to an assignment edge. Hence, there is no need to represent a request from an entity to a resource that can be satisfied. In text, we use the notation $(e, r, k)$ to represent a request of $k$ units of resource $r$ by entity $e$, and $(r, e, k)$ to represent an assignment of $k$ units of resource $r$ to entity $e$.

**Example 1:** Consider the situation as shown in Figure B.5($a$) but with the associated entity-resource relationship graph as shown in Figure B.5($a$). Here, entities $e_1$ and $e_2$ are currently requesting one unit of resource $r_2$ and $r_1$, respectively, and resources $r_1$ and $r_2$ allow passing of entities. Also, entities $e_5$ and $e_6$ are waiting for two units of resources $r_1$ and $r_2$, respectively. Now, because of the pass filling rule, all of the entities in the different resource queues will eventually gain control of their requested resources. Consequently, the situation depicted in Figure B.5($c$) is in transient deadlock. It should be noted that if neither of the resources allows passing of entities, the simulation system would be in pending deadlock.

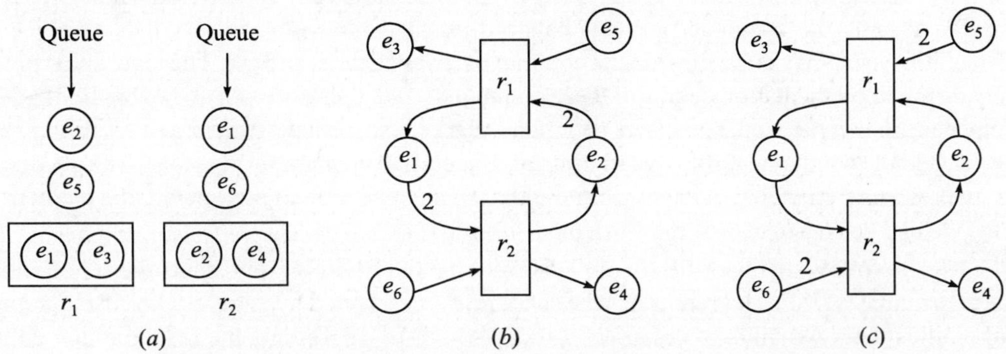


**Fig. B.5(*a*), 5(*b*) and 5(*c*):** Entity and Resource Relationship Graph.

In the above examples deadlocks occurred as a result of the requesting entities being unable to relinquish the resources currently in their possession. This feature is the primary reason for the occurrence of deadlocks in simulation systems. The examples clearly demonstrate the complications inherent in detecting deadlocks in simulation systems. These complications are due to the dynamic nature of the simulation, the multiplicity of request sizes by entities, and the filling rules of the resources.

**B.3 Entity and Resource Relationship Graph**

A directed graph G = (*V*, *E*) has a vertex set *V* and an edge set *E*, where (*u,v,k*)*E* implies there is an edge directed from *uV* to vV with a label *k*N, where *N* is the set of positive integers. A vertex *vV* is called a root vertex (source) of V, if there is no *uV* for which (*u,v,k*)*E* (*i.e.*, the in-degree of *v* is zero). A vertex *vV* is called a leaf vertex (sink) of *V*, if there is no *uV* such that (*v,u,k*)E (*i.e.*, the out-degree of *v* is zero). A vertex is called an interior vertex if both the in-degree and out-degree are non-zero. A vertex is called an isolated vertex if both the in-degree and out-degree are zero.

A directed graph *G′* is said to be a subgraph of directed graph *G*, if *G′* = (*V′*, *E′*), where *V′ V* and *E′ E*. Let *V′ V*, and *G′* be constructed by letting *E′ E*, where (*u,v, k*)*E′* for all *u,vV′* and (*u,v,k*)*E*. The set *V′* is called a strongly connected component (SCC) of *V* if for all vertices u and *v V′*, there is a directed path in *G′* connecting u to *v*, and another path connecting *v* to *u*. In this case, the subgraph *G′* is induced by *V′*. Let *V′* be an SCC and *G′* its induced subgraph; then we say *V′* is a closed strongly connected component (CSCC) if there is no edge in *G* directed out of *V′*. In the formalism developed below, strongly connected components represent the circular wait relationships indicative of deadlock and are therefore critical to succeeding developments.

We represent the entity-resource (ER) relationship by the directed graph *Gt* = (*Vt*,*E*t), for *t*0. Here, *Gt* denotes the ER graph at instant *t*, with vertex set *Vt* and edge set *Et*. The vertex set contains all the resources of the system as well as those entities who own at least one resource unit and/or are waiting for a resource at instant *t*. Let *Gt*, *t*0, denote the ER graph over time, beginning with the configuration *G*0, where *V*0 consists of the set of all resources and *E*0 is empty. Since set *G*t changes over time due to the execution of a series of seize or release statements, only those entities that are waiting for a resource or possess a resource at time *t* will be included in the graph *G*t. Idle resources are represented as isolated vertices. Hence the graph can change configuration only upon execution of a seize or a release statement and we can completely describe the dynamic behaviour of the ER graph by appropriately updating *Gt* after the execution of each seize or release statement. We take the convention that this update process is right continuous (*i.e.*, *G*t denotes the ER graph immediately after the execution of a seize or a release statement).

In the algorithms to follow, we use terminology and pseudo-code notation resembling common programming languages. Bold-face type is used to indicate reserved words within this pseudo-code language. We assume the language provides capabilities to work with arbitrary lists of objects, and provides facilities to modify these lists via the commands insert, delete, and get. The algorithms presented do not rely on the implementation details of these commands and, therefore, we do not explicitly describe an implementation mechanism. The insert command places a specified element at a specified position in the list. The delete command removes and destroys a specified element from the list. The get command removes and returns a specified element from the list, so as to be able to manipulate the element in the algorithm. We use get(*l*,*i*) to access the $i^{th}$ element from list l.

We use the following quantities in our descriptions: QUEUE*t* [*r*] represents the set of entities in the queue at resource *r*, USES*t* [*r*] represents the set of resources owned by entity *e*, OWNS*t* [*e*] represents the set of entities that own units of resource *r*, and WAITS*t* [*e*] represents the resource that entity *e* is waiting for. In addition, CAP[*r*] denotes the capacity of resource *r*, REM*t*[*r*] denotes the unassigned capacity (remaining capacity) of resource *r*, and REQ*t*[*e*,*r*] denotes the number of requested units of resource *r* by entity *e*. We drop the square brackets to represent a vector and the index (*t*) wherever appropriate.

Figure B.6 illustrates the seize statement operation in the notation described above. Lines 1.2 add entity *e* to the vertex set of the graph if entity *e* currently owns no resource. Lines 3.8 check to see if there is available capacity in resource *r* to accommodate the request by entity *e*; if so, entity *e* is added to the ER graph with edge (*r*,*e*,k), else entity *e* is added with edge (*e*,*r*,*k*).

| **seize(*e*,*r*,k) - entity *e* requests *k* units of resource *r*** |
|---|
| 1. if *e* *V* then |
| 2. insert(*V*,*e*); |
| 3. if REQ[*e*,*r*] REM[*r*] then |
| 4. insert(*E*,(*r*,*e*,*k*)); |
| 5. REM[*r*] = REM[*r*] - REQ[*e*,*r*]; |
| 6. else |
| 7. insert(*E*,(*e*,*r*,*k*)); |
| 8. endif |

**Fig. B.6** Definition of the seize statement.

| **release(*e*,*r*,*k*) - entity *e* releases *k* units of resource** |
|---|
| 1. delete(*E*,(*r*,*e*,*k*)); |
| 2. delete(USES[*r*], *e*); |
| 3. REM[*r*] = REM[*r*] + *k*; |
| 4. if OWNS[*e*] = Ø then |
| 5. delete(*V*,*e*); |
| 6. if QUEUE[*r*] Ø then " resource fill-rule " |
| 7. *e′*= firstfit(r); |
| 8. dowhile(*e′* Ø) |
| 9. delete(*E*,(*e′*,*r*,REQ[*e′*,*r*])); |
| 10. insert(*E*,(*r*,*e′*,REQ[*e′*,*r*])); |
| 11. REM[*r*] = REM[*r*] - REQ[*e′*,*r*]; |
| 12. *e′*= firstfit(*r*); |
| 13. endwhile |
| 14. endif |

**Fig. B.7** Definition of the release statement.

Figure B.7 defines the operation of the release statement. Lines 1.3 remove the assignment edge (*r*,*e*,*k*) from the graph and update the set USES[*r*] as entity *e* returns the units of the resource *r* it is holding. Line 3 updates the available units of resource *r*. Line 5 deletes entity *e* from the graph, if it no longer owns any other resource. Lines 6.14 perform the filling function to reallocate the available resource units to waiting entities. The function firstfit returns an entity that satisfies the capacity constraints of resource $r$, or returns a null (d8) if no such entity exists. Details of these functions are provided in Venkatesh *et al.* [22].

1. Conditions for Deadlock and Deadlock Detection
2. Preliminaries.

The notion of an SCC (or CSCC) is important in the ER graph *Gt*, since the existence of an SCC indicates the occurrence of a deadlock in the simulation. It was shown by Deuermeyer *et al.* [5] that with the simulation restricted to single-unit resource requests, total deadlock corresponds to the presence of at least one CSCC in the ER graph. In the multiple-unit request case, we define total deadlock similarly. However, for the multiple-unit request case, some resource units of the deadlocked resource can be idle. The presence of a total deadlock in the simulation indicates the need for external intervention for the deadlocked entities and resources to be freed. Further, the presence of an SCC in the multiple-unit resource requests case can indicate pending deadlock, which also requires external intervention for the deadlock to be resolved. Here, the user can choose not to intervene at the instant of discovery to avoid penalizing non-deadlocked entities during resolution, since some units of the deadlocked resources continue to process entities.

Beginning with *G*0, which is free from deadlock, we track the evolution of the simulation after each execution of seize or release statements to identify the occurrence of deadlocks. The existence of an SCC at instant *t* in the ER graph implies that a deadlock is present. If the SCC is closed, this deadlock can only be resolved by external intervention (total deadlock). If the SCC is not closed then some entities will be delayed and some resource capacities not operational. This situation may be temporary however, and it may be acceptable to let the simulation proceed for a time. If it is temporary and will eventually resolve itself, the deadlock is a transient deadlock. Otherwise, it is a pending deadlock.

Suppose at instant *t* an SCC that is not closed has been found in the simulation ER graph. At this point, entities in the SCC are blocked because every entity in the SCC is waiting for resources held by other entities in the SCC, and some of the entities in the SCC are waiting for resources held by entities not in the SCC. However, it may be possible that some entity not in the SCC might return a resource that could initiate a sequence of events that would cause the SCC to collapse. An example would be the return to operation of a machine under repair or off-shift. This sequence of events might be realized in the near future, might be realized over the long term, or might never be realized, depending on the specific sequence of events within the simulation run.

We use the concept of reduction (similar to that given in Coffman *et al.* [3]) as a method to characterize the fortuitous events that may clear an SCC in the ER graph. Assume for current discussion purposes that the ER graph contains a single SCC. The reduction process is designed to distinguish between pending deadlock and transient deadlock. Basically, reduction attempts to return resource units from entities not waiting for any resource in the ER graph (*i.e.*, sink vertices in *Gt*). As these entities are reduced, additional resource units will be freed, that in turn may allow the reduction of more waiting entities, and so on. The reduction process will eventually reduce entities that are not waiting for any resource in the SCC but holding resource units contained in the SCC. The reduction of such entities return units of resources contained in the SCC, thereby either causing the SCC to clear (indicating transient deadlock), or become closed (indicating pending deadlock).

As an example, consider the situation depicted in Figure 5(*b*). The reduction process would begin with the set of entities ($e_1$, $e_2$, $e_3$, and $e_4$). Initially, entity $e_3$ will return the unit of resource $r_1$ in its possession. Next, entity $e_4$ will return the unit of resource $r_2$ in its possession. Now, since neither entity $e_1$ nor entity $e_2$ can obtain their requested resource units, and, further, since neither entity will relinquish the resource that it holds until it obtains the requested resource, the reduction process will terminate.

The reduction will terminate with both entities, $e_1$ and $e_2$ unable to satisfy their request, indicating a pending deadlock.

Figure B.8 presents the details of reducing entity *e*. Here, HOLDS[*e*,*r*] denotes an entity attribute that indicates the number of units of resource *r* held by entity *e*. The EFF_CAP[*r*] indicates the effective available capacity of resource *r*, which is the sum of the returned units of resource r due to reduction and the unassigned capacity of resource *r*, REM[*r*]. Note EFF_CAP denotes a vector of size *R*, where *R* is the number of resources in the simulation system, and maintains the available resource units during the reduction process.

reduce(*e*)1. for each *r* such that (*r*,*e*,HOLDS[*e*,r]) *E* do 2. EFF_CAP[*r*] = EFF_ CAP [*r*] + HOLDS[*e*,*r*];

Figure B.8. Definition of function reduce.

The reduction process has two possible concluding states: either the SCC in the ER graph has been cleared or there remains a CSCC. In the former case the SCC is said to be reducible, and the simulation is in transient deadlock (Figure B.1). In the latter case, the original SCC in the ER graph can never unravel by natural simulation events, and the SCC is said to be irreducible, so the deadlock is pending (Figure B.1). This leads to the following definitions.

Definition 1 (Total Deadlock): A simulation is said to be in total deadlock at any instant *t* if the simulation ER graph contains a CSCC.

Definition 2 (Pending Deadlock): Given a non-closed SCC at instant *t*, if the reduction of this graph is a CSCC then the simulation is said to be in pending deadlock.

Definition 3 (Transient Deadlock): Given a non-closed SCC at instant *t*, if the reduction of this graph is not a CSCC then the simulation is said to be in transient deadlock.

Definition 4 (Deadlock free): A simulation is said to be deadlock-free if there does not exist an SCC in the ER graph.

It should be noted that if there exists a pending deadlock in the simulation, then the SCC that could not be cleared in the reduction process will eventually evolve into a CSCC (total deadlock). Likewise, an unresolved CSCC (total deadlock) can revert to an SCC (pending deadlock) if an entity request of a later arrival can be satisfied by the idle resource units of a deadlocked resource. Consequently, except when a deadlocked resource allows passing of entities, total deadlock is an absorbing state, and pending deadlock is a transient state in the simulation state transition diagram.

Initially, we check for the existence of a CSCC in the ER graph using the search function. The function search is a depth-first search based algorithm to detect SCCs and CSCCs in a graph and has been adapted from the work of Deuermeyer *et al.* [5]. A key property of the algorithm is that it runs in linear time of the maximum of the number of edges and vertices in the graph $O(\max\{|V|, |E|\})$. If such a closed component is discovered in the ER graph, we know that the simulation is in total deadlock. If we find an SCC that is not closed in the ER graph, we reduce the graph (using the function reduction) to determine if the simulation is in pending deadlock. Of course, if the outcome of the reduction function indicates an absence of pending deadlock then the simulation is in transient deadlock. The function reduction is adapted from the work of Coffman *et al.* [3].

### B.4 Management of Pending and Total Deadlock

In light of the above discussion there are two deadlock cases (total and pending) that require external intervention for the deadlock to be resolved. However, we may choose not to intervene in the case of a pending deadlock at the instant of its discovery in the ER graph. This leads to two possible deadlock handling principles: either we maintain an ER graph that is free from pending and total deadlocks at any instant, or we maintain an ER graph that is free from total deadlocks. In the following discussion we will develop the theory required to handle these two cases.

Assuming that pending and total deadlocks in the ER graph are detected and resolved continuously, we know that the ER graph is free from these deadlocks before the execution of a seize or release statement (although transient deadlocks may exist). In order to retain a simulation free from pending or total deadlocks, it is sufficient to invoke the deadlock detection scheme (and resolve any pending or total deadlocks) after every unsuccessful resource request. A successful resource request at a seize statement necessarily creates a new sink vertex in the ER graph, and hence does not create a new SCC. Further, the execution of a release statement also cannot create a pending or total deadlock, since, if it could, the ER graph would be in pending or total deadlock before the release statement. It should be realized that the ER graph may contain a number of SCCs, however, by the assumption of continuous detection and resolution of pending or total deadlocks, we are assured that none of these SCCs are in pending or total deadlock.

[1] Holt, R.C."Some Deadlock Properties of Computer Systems," ACM Computing Surveys, 4, 3, 179.196 (1972).

[2] Cho, H., T.K. Kumaran, and R.A. Wysk, "Graph-Theoretic Deadlock Detection and Resolution for Flexible Manufacturing Systems," IEEE Transactions on Robotics and Automation, 11, 3, 413-421 (1995).

[3] Coffman, E.G., M.J. Elphick, and A. Shoshani, "System Deadlocks," ACM Computing Surveys, 3, 2, 67.78 (1971).

[4] Curry, G.L., B.L. Deuermeyer, and R.M. Feldman, Discrete Simulation: Fundamentals and Microcomputer Support, Holden-Day, Inc., Oakland, California (1989).

[5] Deuermeyer, B. L., G. L. Curry, A.D. Duchowski, and S. Venkatesh, "An Automatic Approach to Deadlock.

Detection and Resolution in Discrete Simulation Systems," ORSA Journal of Computing (to appear, 1995).

[16] Pegden, C.D., R.P. Sadowski, and R.E. Shannon, Introduction to Simulation Using SIMAN, Systems Modeling Corporation, Sewickley, PA (1990).

[17] Pierce, N.G. and M.J. Drevna, "Development of Generic Simulation Models to Evaluate Wafer Fabrication Cluster Tools," Proceedings of the 1992 Winter Simulation Conference, 874.878, Arlington, VA (December 1992).

[18] Schriber, T.J., Simulation Using GPSS, John Wiley, NY (1974).

[19] Schriber, T.J., An Introduction to Simulation Using GPSS/H, John Wiley, NY (1990).

[20] Singhal, M. "Deadlock Detection in Distributed Systems," IEEE Computer, 22, 12, 37.48 (1989).

**Table A: Normal curve tail probabilities. Standard normal probability in right-hand tail (for negative values of *z*, probabilities are found by symmetry**

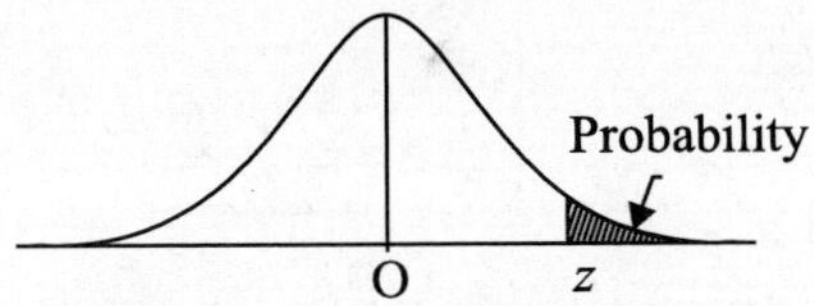


| *z* | .00 | .01 | .02 | .03 | .04 | .05 | .06 | .07 | .08 | .09 |
|---|---|---|---|---|---|---|---|---|---|---|
| | **Second Decimal Place of *z*** | | | | | | | | | |
| 0.0 | .5000 | .4960 | .4920 | .4880 | .4840 | .4801 | .4761 | .4721 | .4681 | .4641 |
| 0.1 | .4602 | .4562 | .4522 | .4483 | .4643 | .4404 | .4364 | .4325 | .4286 | .4247 |
| 0.2 | .4207 | .4168 | .4129 | .4090 | .4051 | .4013 | .3074 | .1936 | .3897 | .3859 |
| 0.3 | .3821 | .3783 | .3745 | .3707 | .3669 | .3632 | .3394 | .1557 | .3520 | .3483 |
| 0.4 | .3446 | .3409 | .3372 | .3336 | .3300 | .3264 | .3128 | .1192 | .3156 | .3121 |
| 0.5 | .3085 | .3050 | .3015 | .2981 | .2946 | .2912 | .2477 | .2843 | .2810 | .2776 |
| 0.6 | .2743 | .2709 | .2676 | .2643 | .2611 | .2518 | .2546 | .2514 | .2483 | .2451 |
| 0.7 | .2420 | .2389 | .2358 | .2327 | .2296 | .2266 | .2136 | .1206 | .2177 | .2148 |
| 0.8 | .2119 | .2090 | .2061 | .2033 | .2005 | .1917 | .1949 | .1922 | .1894 | .1867 |
| 0.9 | .1841 | .1814 | .1788 | .1762 | .1736 | .1711 | .1685 | .1660 | .1635 | .1611 |
| 1.0 | .1587 | .1562 | .1539 | .1515 | .1491 | .1469 | .1446 | .1423 | .1401 | .1379 |
| 1.1 | .1357 | .1335 | .1314 | .1292 | .1271 | .1251 | .1230 | .1210 | .1190 | .1170 |
| 1.2 | .1151 | .1131 | .1112 | .1093 | .1075 | .1056 | .1038 | .1020 | .1003 | .0985 |
| 1.3 | .0968 | .0951 | .0934 | .0918 | .0901 | .0815 | .0169 | .0853 | .0838 | .0823 |
| 1.4 | .0808 | .0793 | .0778 | .0764 | .0749 | .0715 | .0122 | .0708 | .0694 | .0681 |
| 1.5 | .0668 | .0655 | .0643 | .0630 | .0611 | .0606 | .0694 | .0582 | .0571 | .0559 |
| 1.6 | .0548 | .0537 | .0526 | .0516 | .0506 | .0495 | .0485 | .0475 | .0465 | .0455 |
| 1.7 | .0446 | .0436 | .0427 | .0418 | .0409 | .0401 | .0392 | .0384 | .0375 | .0367 |
| 1.8 | .0359 | .0352 | .0344 | .0336 | .0329 | .0332 | .0114 | .0307 | .0301 | .0294 |
| 1.9 | .0287 | .0281 | .0274 | .0268 | .0262 | .0256 | .0250 | .0244 | .0239 | .0233 |
| 2.0 | .0228 | .0222 | .0217 | .0212 | .0207 | .0202 | .0197 | .0192 | .0188 | .0183 |
| 2.1 | .0179 | .0174 | .0170 | .0166 | .0161 | .0158 | .0154 | .0150 | .0146 | .0143 |
| 2.2 | .0139 | .0136 | .0132 | .0129 | .0123 | .0122 | .0119 | .0116 | .0113 | .0110 |
| 2.3 | .0107 | .0104 | .0102 | .0099 | .0096 | .0044 | .0691 | .6089 | .0087 | .0084 |
| 2.4 | .0082 | .0080 | .0078 | .0075 | .0011 | .0011 | .0669 | .0068 | .0066 | .0064 |
| 2.5 | .0062 | .0060 | .0059 | .0057 | .0051 | .0054 | .0052 | .0051 | .0049 | .0048 |
| 2.6 | .0047 | .0045 | .0044 | .0043 | .0041 | .0040 | .0039 | .0038 | .0037 | .0036 |
| 2.7 | .0035 | .0034 | .0033 | .0032 | .0031 | .0010 | .0029 | .0028 | .0027 | .0026 |
| 2.8 | .0026 | .0025 | .0024 | .0023 | .0021 | .0002 | .0021 | .0021 | .0020 | .0019 |
| 2.9 | .0019 | .0018 | .0017 | .0017 | .0016 | .0016 | .0015 | .0015 | .0014 | .0014 |
| 3.0 | .00135 | | | | | | | | | |
| 3.5 | .000233 | | | | | | | | | |
| 4.0 | .0000317 | | | | | | | | | |
| 4.5 | .00000340 | | | | | | | | | |
| 5.0 | .000000287 | | | | | | | | | |

***Scurce:*** F.E. Waipole, *Introduction to Statuaries* (New Your: Macmillan, 1965)

**Table B: t distribution, values corresponding to various right hand tail probabilities**

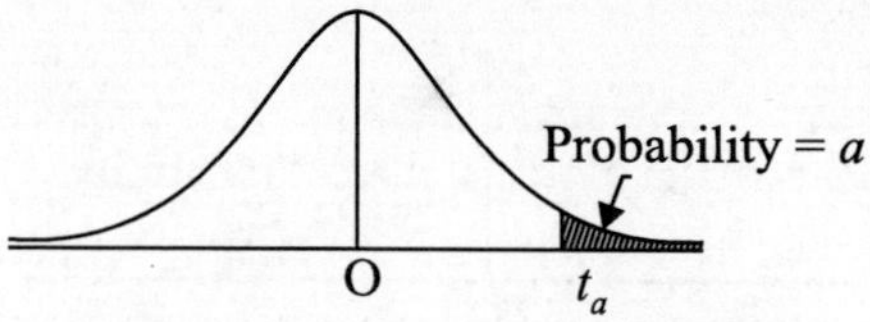


| df | t.100 | t.050 | t.025 | t.010 | t.005 |
|---|---|---|---|---|---|
| 1 | 3.078 | 6.314 | 12.7063 | 31.821 | 63.657 |
| 2 | 1.886 | 2.920 | 4.303 | 6.965 | 9.925 |
| 3 | 1.638 | 2.353 | 3.182 | 4.541 | 5.841 |
| 4 | 1.533 | 2.132 | 2.776 | 3.747 | 4.604 |
| 5 | 1.476 | 2.015 | 2.571 | 3.365 | 4.032 |
| 6 | 1.440 | 1.943 | 2.447 | 3.143 | 3.707 |
| 7 | 1.415 | 1.895 | 2.365 | 2.998 | 3.499 |
| 8 | 1.397 | 1.860 | 2.306 | 2.896 | 3.355 |
| 9 | 1.383 | 1.833 | 2.262 | 2.821 | 3.250 |
| 10 | 1.372 | 1.812 | 2.228 | 2.764 | 3.169 |
| 11 | 1.363 | 1.796 | 2.201 | 2.718 | 3.106 |
| 12 | 1.356 | 1.782 | 2.179 | 2.681 | 3.055 |
| 13 | 1.350 | 1.771 | 2.160 | 2.650 | 3.012 |
| 14 | 1.345 | 1.761 | 2.145 | 2.624 | 2.977 |
| 15 | 1.341 | 1.753 | 2.131 | 2.602 | 2.947 |
| 16 | 1.337 | 1.746 | 2.120 | 2.583 | 2.921 |
| 17 | 1.333 | 1.740 | 2.110 | 2.567 | 2.898 |
| 18 | 1.330 | 1.734 | 2.101 | 2.552 | 2.878 |
| 19 | 1.328 | 1.729 | 2.093 | 2.539 | 2.861 |
| 20 | 1.325 | 1.725 | 2.086 | 2.528 | 2.845 |
| 21 | 1.323 | 1.721 | 2.080 | 2.518 | 2.831 |
| 22 | 1.321 | 1.717 | 2.074 | 2.508 | 2.819 |
| 23 | 1.319 | 1.714 | 2.069 | 2.500 | 2.807 |
| 24 | 1.318 | 1.711 | 2.064 | 2.492 | 2.797 |
| 25 | 1.316 | 1.708 | 2.060 | 2.485 | 2.787 |
| 26 | 1.315 | 1.706 | 2.056 | 2.479 | 2.779 |
| 27 | 1.314 | 1.703 | 2.052 | 2.473 | 2.771 |
| 28 | 1.313 | 1.701 | 2.048 | 2.467 | 2.763 |
| 29 | 1.311 | 1.699 | 2.045 | 2.462 | 2.756 |
| ∞ | 1.282 | 1.645 | 1.960 | 2.326 | 2.576 |

***Source:*** "Table of Percentage Points of the *t*-Distribution."

# Model Test Papers

## Model Test Paper–I

We know that simulation is use in the training of civilian and military personnel. This usually occurs when it is prohibitively expensive or simply too dangerous to allow trainees to use the real equipment the real world. AI is also a useful techniques in military activities, so AI can play a vital role in training of simulators.

**Note:** *Attempt all questions.*

1. Attempt any two parts of the following:
   (*a*) Name four principal entities, attributes and activities to be considered for the simulation of the following systems:
      (*i*) A gasoline filling station
      (*ii*) A cafeteria
      (*iii*) A banking system
      (*iv*) A hospital system
   (*b*) Differentiate between static physical models and dynamic physical mode. Take suitable example to illustrate the use of these models.
   (*c*) Discuss various types of models.
2. Attempt any two parts of the following:
   (*a*) Find the growth in national consumption for five years using the distributed lag model. Assume the initial income $Y_{-1}$ is 80 and take the government expenditure in the 5 year to be as follows:

| Year | G |
|---|---|
| 1 | 20 |
| 2 | 25 |
| 4 | 35 |
| 5 | 40 |

   (*b*) Explain COBWEB model with an example. Demonstrate its use also.
   (*c*) Discuss in detail how an pure pursuit problem is simulated.
3. Attempt any two parts of the following:
   (*a*) Write a program to generate and print 5,000 uniformly distributed numbers between 0.000 and 0.999 (3 significant digits). Test the numbers of uniformly by dividing them into 10 range and using chi-square test.

(*b*) (*i*) Differentiate between analog simulation and digital simulation.

(*ii*) Discuss Montes Carlo computation versus stochastic simulation.

(*c*) How to simulate an Autopilot system? Where we use autopilot system?

**4.** Attempt any two parts of the following:

(*a*) The birth rate of a country is adding 100,000 people a year to an initial population of 5,000,000. The average life expectancy is 65 year. It is estimated that of coal is consumed pre one ton annum for each individual. Draw a diagram showing how the country's resources of 500,000,000 tone are being depleted.

(*b*) How logistic function is combination of exponential and modified function? Explain with example.

(*c*) (*i*) How feedback in socio-economic system occurs?

(*ii*) Discuss multi-segment models.

**5.** Attempt any two of the following:

(*a*) Explain Discrete-system simulation language also discuss CSMP-II and SIMSCRIPT-II. 5.

(*b*) (*i*) Show that the activities in a project network can be ordered topologically if and only if the network contains no cycles.

(*ii*) How to resolve uncertainties in activity duration.

(*c*) Define following term:

(*i*) Expression based language.

(*ii*) Criticality index.

■■

# Model Test Paper–II

**Note:** *Attempt all questions.*

1. Attempt any two parts of the following:

   (*a*) What do you mean by system simulation? Why we need simulation?

   (*b*) Differentiate between static mathematical and dynamic mathematical mode. Take suitable examples to illustrate the use of these models.

   (*c*) Discuss the various segment of corporate model. When we need corporate model?

2. Attempt any two parts of the following:

   (*a*) Define the linear congruential methods (LCG), and use the LCG to generate a sequence of random numbers with $x_0 = 37$, $a = 7$, $c = 29$ and $m = 100$.

   (*b*) Define the terms:

   (*i*) Rejection method

   (*ii*) Poker test

   (*c*) Explain the following:

   (*i*) Simulation of water reservoir system

   (*ii*) Fixed time-step Vs event-to-event model.

3. Attempt any two parts of the following:

   (*a*) Compute following parameters for a single server queuing system when inter-arrival time and service times are given by the table below:

| Customer | Inter-arrival time | Service time |
|---|---|---|
| 1 | – | 2 |
| 2 | 4 | 3 |
| 3 | 1 | 1 |
| 4 | 4 | 4 |
| 5 | 2 | 2 |
| 6 | 4 | 3 |

   (*i*) Average waiting time of customer (minutes).

   (*ii*) Probability that a customer has to wait in the queue.

   (*iii*) Probability of idle time of the server.

   (*iv*) Average service time of customer in minutes.

   (*v*) Average time between the arrival.

   (*vi*) Average time customer spends in the system.

   (*vii*) Probability of server being busy.

(*b*) Draw the cobweb model for the following market

$P_0 = 5.0$ $c = -2.4$

$a = 10.0$ $d = 1.2$

$b = 0.9$ Assume the market is always cleared.

(*c*) Explain Monte-Carlo simulation and find the value of π, by using Monte Carlo method of simulation.

**4.** Attempt any two parts:

(*a*) Two competing companies invest funds in capital equipment to improve their positions. The rate of which each invests funds decreases linearly as their own investment increases but increases linearly as their competitor's investment increases.

Draw a diagram from which to simulate the competition and determine under what conditions the investment will stabilize.

(*b*) Explain exponential growth models? Also describe that there is inverse relationship between *K* the growth rate coefficient and *T*, the time constant.

(*c*) Explain system dynamics model that is studied under the auspices of the club of Rome.

**5.** Attempt any two of the following:

(*a*) Simulation of PERT networks.

(*b*) Define the following terms:

(*i*) Cash flow (*ii*) Critical path in network

(*iii*) MODSIM-III

(*c*) Compare between continuous simulation language and discrete simulation language.

■■

# Model Test Paper–III

**Note:** *Attempt all questions.*

**1.** Attempt All: (2 × 10 = 20)

(*a*) What do you mean by system simulation?

(*b*) Name several entities, attributes, activities, events, and state variables for the fast-food restaurant and cafeteria systems:

(*c*) Compare between Simulation and analytical method.

(*d*) How do you select input model without data?

(*e*) what is data modeling?

(*f*) what is probability mass function?

(*g*) Differentiate between chi $^2$ test and kolmogorav-smirnov test.

(*h*) Define poisson distribution.

(*i*) How do you measure performance of single queuing system?

(*j*) What is capital recovery model?

**2.** Attempt any five parts of the following: (6 × 5 = 30)

(*a*) What are the principles in creating system models? Discuss advantages of simulation.

(*b*) Draw the cobweb model for the following market

$P_0 = 5.0$ $c = -2.4$

$a = 10.0$ $d = 1.2$

$b = 0.9$ Assume the market is always cleared.

(*c*) Estimate value of pi, by using Monte Carlo method of simulation.

(*d*) Discuss verification, calibration, and validation of models for simple systems.

(*e*) The percentage rates of return on 10 items are 18.8, 27.9, 21.0, 6.1, 37.4, 5, 22.9, 1.0, 3.1 and 8.3. Estimate the parameters of lognormal model of this data.

(*f*) Discuss job-shop model.

**3.** Attempt All. (10 × 5 = 50)

(*a*) What do you mean by confidence interval and confidence level? Suppose that a 90% confidence interval states that the population mean is greater than 100 and less than 200. How would you interpret this statement?

*OR*

The sequence of numbers 0.54, 0.73,0.98, 0.11,and 0.68 has been generated. Use the kolmogorov –Smirnow test with $\alpha = 0.05$ to determine if the hypothesis that the numbers are uniformly distributed on the interval [0, 1] can be rejected.

(*b*) Discuss single server queuing system. Compute following parameters for a single server queuing system when inter-arrival time and service times are given by the table below:

| Customer | Inter-arrival Time | Service Time |
|---|---|---|
| 1 | — | 2 |
| 2 | 4 | 3 |
| 3 | 1 | 1 |
| 4 | 4 | 4 |
| 5 | 2 | 2 |
| 6 | 4 | 3 |

(*i*) Average waiting time of customers (minutes).
(*ii*) Probability that a customer has to wait in the queue.
(*iii*) Probability of idle time of the server.
(*iv*) Average service time of customer in minutes.
(*v*) Average time between the arrival.
(*vi*) Average time customer spends in the system.
(*vii*) Probability of server being busy.

*OR*

Why do we need parameter estimation? Records pertaining to the monthly number of job-related injuries at an underground coal mine were being studied by a federal agency. The values for the past 100 months were as follows:

| Injuries per month | Frequency of occurrence |
|---|---|
| 0 | 35 |
| 1 | 40 |
| 2 | 13 |
| 3 | 6 |
| 4 | 4 |
| 5 | 1 |
| 6 | 1 |

Apply the chi-square test to these data to test the hypothesis that the underlying distribution is Poisson. Use a level of significance of $\alpha = 0.05$.

(*d*) Differentiate between discrete and continuous random variable. The number of items bought in a store for a sample of 250 customers are given as follows:

| Number of Items ($x_i$) | Number of customers ($n_i$) |
|---|---|
| 1 | 25 |
| 2 | 128 |
| 3 | 47 |
| 4 | 38 |
| 5 | 12 |

Calculate:
(*i*) Probability distribution for each item.
(*ii*) Cumulative distribution for each item.

*OR*

How do you generate arrival patterns? Explain briefly.

(*e*) Discuss the computer network model. How does simulation suitable for its modeling.

*OR*

Discuss types of system models.

# EXAMINATION PAPERS

## Fourth Semester Examination–2005–06
## (M.C.A.)

*Time:* 3 Hours | **Total Marks:** 100

**Note:** *Attempt all questions.*

**1.** Attempt any two parts of the following:

(*a*) What do you understand by system simulation?

(*b*) Define the terms:

(*i*) Components of System

(*ii*) System Environment

(*iii*) State of System

(*c*) Discuss full corporate model.

**2.** Attempt any parts of the following: (10 × 2 = 20)

(*a*) What do you understand by system analysis, system design and system postulation?

(*b*) Discuss in detail how an inventory system in simulated.

(*c*) Discuss Monte-Carlo computation versus stochastic simulation.

**3.** Attempt any two parts of the following: (10 × 2 = 20)

(*a*) Compute following parameters for a single server queuing system when interarrival time and service times are given by the table below:

| Customer (min) | Interarrival time (max) | Service time |
|---|---|---|
| 1 | 0 | 20 |
| 2 | 10 | 15 |
| 3 | 15 | 10 |
| 4 | 35 | 05 |
| 5 | 30 | 15 |
| 6 | 10 | 15 |
| 7 | 05 | 10 |
| 8 | 05 | 10 |

(*i*) Average waiting time of customers (minutes).

(*ii*) Probability that a customer ha to wait in the queue.
(*iii*) Probability of idle time of the server.
(*iv*) Average service time of customer in minutes.
(*v*) Average time between the arrival.
(*vi*) Average time customer spends in the system.
(*vii*) Probability of server being busy.

(*b*) Compare the analytical method and simulation method of study.

(*c*) Discuss the application areas of simulation.

**4.** Attempt any four parts of the following: $(10 \times 4 = 40)$

(*a*) Discuss Cobweb model. Prove or disprove Cobweb model is a special case of distributed lag model.

(*b*) Two competing companies invest funds in capital equipment to improve their positions. The rate at which each invests funds decreases linearly as their own investment increases but increases linearly as their competitor investment increases. Draw a diagram from which to simulate the competition and determine under what conditions the investment will stabilize.

(*c*) Write a program to generate and print 5,000 uniformly distributed numbers between 0.000 and 0.999 (3 significant digits). Test the numbers for uniformity by dividing them into 10 ranges and using chi-square test.

(*d*) Define the terms:

(*i*) Network model of a project (*ii*) Activity network
(*iii*) Length of a path (*iv*) Critical path in network

(*e*) Compare between continuous simulation language and discrete simulation language.

(*f*) Discuss the time advance mechanism in discrete event system simulation.

■■

# Fourth Semester Examination –2006-07

## (M.C.A)

*Time: 3 Hours* | ***Total Marks:*** *100*

**Note:** Attempt all questions.

1. Attempt any **two** parts of the following:
   (*a*) Differentiate between static mathematical and dynamic mathematical model. Take suitable examples to illustrate the use of these models. 10
   (*b*) Name four principal entities, attributes and activities to be considered for the simulation of the following systems:
   (*i*) University registration system
   (*ii*) Examination system
   (*iii*) Railway reservation system.
   (*c*) Differentiate between continuous and discrete systems with suitable examples. 10
2. Attempt any two parts of the following:
   (*a*) Draw of Cobweb model for the follwoing market:

$$D = \frac{17.9}{p^{1/2}} - 4.6, \quad 9S = 0.5(P_{-1} - 1)$$

   Assume the market is always cleared.
   (*b*) Explain distributed lag model with an example. Demonstrate its use.
   (*c*) What are type of system simulation? Explain each with example.
3. Attempt any two of the following:
   (*a*) Differentiate between Monte-Carlo computation and stochastic simulation. 10
   (*b*) Explain the following:
   (*i*) Simulation of continuous systems 5
   (*ii*) Simulation of water reservoir system. 5
   (*c*) What is the method of testing random number generation of non-uniformly distributed random numbers? 10
4. Attempt any two of the following:
   (*a*) Two competing companies invest found in capital equipment to improve their positions. The rate at which each invest funds decreases linearly as their own investment increases but increases linearly as their competitor's investment increases. Draw a diagram from which to simulate the competition and determine under what conditions the investments will stabilize. 10
   (*b*) What type of a model is the world model? Explain it in detail. 10
   (*c*) Explain exponential decay models. Also describe logistic curves. 10
5. Attempt any **two** of the following:
   (*a*) CSMP-III and MODSIM-III 10
   (*b*) Continuous and discrete simulation language and expression based language. 10
   (*c*) Simulation of PERT networks.

# Fourth Semester Examination –2007-08
# (M.C.A)

*Time: 3 Hours* | *Total Marks: 100*

**Note:** *Attempt all questions.*

1. Attempt any **two** parts of the following: 10

   (*a*) Discuss the basic nature of simulation. Give reasons to simulate. 10

   (*b*) Explain the various steps used in the process of simulation and depict it using flow chart. 10

   (*c*) Name three of the principal entities attributes and activities to be considered if the operation of following to the carried out: 10

   (*i*) Computer Centre (*ii*) Bank

   (*iii*) Reservation Counter (*iv*) Hospital emergency room.

2. Attempt any **two** parts of the following:

   (*a*) How different segments of a corporate model are formulated? 10

   (*b*) Discuss the concept and methodology of Monte Carlo simulation. 10

   (*c*) What do you understand by: 10

   (*i*) Real Time Simulation (*ii*) Hybrid Simulation.

3. Attempt any **two** of the following:

   (*a*) What type of models are called distributed lag models and why? 10

   (*b*) Explain the simulation of auto piolet. 10

   (*c*) Test the following sequence of numbers for uniformity and independence:
   0.594, 0.928, 0.515, 0.055, 0.507, 0.351, 0.262, 0.797, 0.442, 0.097, 0.798, 0.227, 0.127 0.474,0.825, 0.007, 0.185, 0.929, 0.852
   Discuss the procedures used.

4. Attempt any **two** of the following:

   (*a*) Simulate the game between two players as: The two players simultaneously flip a coin each. If both sides are the same (both heads or both tails) player *A* wins a rupee from *B* otherwise *B* wins a rupee from *A*. Each player starts with 10 rupees by conducting the experiment 500 times, find out how long the game will last on the average before one of the player goes broke. 10

   (*b*) State the properties of random numbers. What are different techniques for generating random numbers? 10

   (*c*) Discuss exponential growth model with an example. 10

5. Attempt any **two** of the following:

   (*a*) Show that the activities in a project network can be ordered topologically if and only if the network contains no cycles. 10

   (*b*) Discuss the various factors taken into consideration while selecting discrete system language. 10

   (*c*) How is feedback in socio-economic system carried out? 10

■■

# Third Semester Examination–2013-14
# (M.C.A)

***Time:*** *3 Hours* — ***Total Marks:*** *100*

**Note:** *Attempt questions from each Section as indicated.*

## SECTION-A

**1.** Attempt **all** parts: (10 × 2 = 20)

(*a*) Define the term simulation and experimental frame.

(*b*) Write principles for building models.

(*c*) Differentiate between physical and mathematical model.

(*d*) Give any three situations when simulation is the appropriate tool and not an appropriate tool.

(*e*) Define Model. What are the different types of models? Name them.

(*f*) What is Continuous Simulation?

(*g*) Differentiate between analog and digital simulation.

(*h*) What do you mean by System Dynamics?

(*i*) What is PERT and CPM?

(*j*) What is object-oriented simulation?

## SECTION-B

**2.** Attempt any **three** parts:

(*a*) Name four principal entities, attributes and activities for the following systems:

(*i*) A fast food restaurant

(*ii*) A hospital emergency room

(*iii*) A railway reservation system

(*iv*) University examination system.

(*b*) Explain various techniques of simulation.

(*c*) Explain simulation of water reservoir system.

(*d*) Explain exponential growth model.

(*e*) Explain CSMP-III simulation language.

## SECTION-C

**3.** Attempt any **five** questions: (5 × 10 = 50)

(*a*) What do you mean by a Corporate Model? Discuss full corporate model with the help of a suitable diagram.

(*b*) Describe different steps in simulation study.

(*c*) Define Discrete Simulation. Explain fixed-time step and event-to-event discrete simulation model.

(*d*) What are Pseudo-Random Numbers? Explain methods of testing random number generation of uniformly distributed random numbers.

(*e*) Explain modified exponential growth model. What is generalization of growth models?

(*f*) Explain MODSIM-III simulation language. What are its features?

(*g*) Explain forward pass method to compute the critical path of an activity network.

# Index

## OTHER DETAILS

(i) Who recommended you this book ? Please tick (✓) in the box near the option relevant to you.

☐ Teacher ☐ Friends ☐ Bookseller

..........................................................................................................................................

..........................................................................................................................................

..........................................................................................................................................

..........................................................................................................................................

(ii) Name of the recommending teacher, his designation and address:

..........................................................................................................................................

..........................................................................................................................................

..........................................................................................................................................

..........................................................................................................................................

(iii) Name and address of the bookseller you purchased the book from :

..........................................................................................................................................

..........................................................................................................................................

..........................................................................................................................................

..........................................................................................................................................

(iv) Name and address of your institution (Please mention the University or Board as the case may be)

..........................................................................................................................................

..........................................................................................................................................

..........................................................................................................................................

..........................................................................................................................................

(v) Your name and complete address :

..........................................................................................................................................

..........................................................................................................................................

..........................................................................................................................................

..........................................................................................................................................

(vi) Course you are enrolled in :

..........................................................................................................................................

..........................................................................................................................................

..........................................................................................................................................

..........................................................................................................................................

**The best assessment will be awarded each month : The award will be in the form of our publications as decided by the Editorial Board.**

***Please mail the filled up coupon at your earliest to :***

**S.K. KATARIA & SONS®**
4885/109, Prakash Mahal, Dr. Subhash Bhargav Lane,
Opposite Delhi Medical Association, Daryaganj, New Delhi-110002
☎ +91-11-43551243, Mobile: +91-9871775858, 7678425209
e-mail: katariabook@gmail.com, katariabooks@yahoo.com
Website: www.skkatariaandsons.com